Principles and Appli
Evidence-Based Pub

Practice

Principles and Application of Evidence-Based Public Health Practice

Edited by

Soundappan Kathirvel
Department of Community Medicine and School of Public Health,
Postgraduate Institute of Medical Education and Research, Chandigarh,
Punjab, India

Amarjeet Singh
Department of Community Medicine and School of Public Health,
Postgraduate Institute of Medical Education and Research, Chandigarh,
Punjab, India

Arun Chockalingam
Professor of Medicine & Global Health, University of Toronto, Toronto,
Canada
Professor of Health Sciences, York University, Toronto, ON, Canada
Founding Director, Office of Global Health, National Heart,
Lung and Blood Institute at the National Institute of Health,
Bethesda, Maryland, United States

ELSEVIER

ACADEMIC PRESS
An imprint of Elsevier

Academic Press is an imprint of Elsevier
125 London Wall, London EC2Y 5AS, United Kingdom
525 B Street, Suite 1650, San Diego, CA 92101, United States
50 Hampshire Street, 5th Floor, Cambridge, MA 02139, United States
The Boulevard, Langford Lane, Kidlington, Oxford OX5 1GB, United Kingdom

Notices
Knowledge and best practice in this field are constantly changing. As new research and experience broaden our understanding, changes in research methods, professional practices, or medical treatment may become necessary.

Practitioners and researchers must always rely on their own experience and knowledge in evaluating and using any information, methods, compounds, or experiments described herein. In using such information or methods they should be mindful of their own safety and the safety of others, including parties for whom they have a professional responsibility.

To the fullest extent of the law, neither the Publisher nor the authors, contributors, or editors, assume any liability for any injury and/or damage to persons or property as a matter of products liability, negligence or otherwise, or from any use or operation of any methods, products, instructions, or ideas contained in the material herein.

ISBN: 978-0-323-95356-6

For information on all Academic Press publications visit our website at
https://www.elsevier.com/books-and-journals

Publisher: Stacy Masucci
Acquisitions Editor: Elizabeth A. Brown
Editorial Project Manager: Pat Gonzalez
Production Project Manager: Swapna Srinivasan
Cover Designer: Christian J. Bilbow

Typeset by TNQ Technologies

Working together
to grow libraries in
developing countries

www.elsevier.com • www.bookaid.org

Contents

Section I
Introduction to public health practice

1. Principles and approaches in public health practice
Rajavel Saranya and Soundappan Kathirvel

2. Principles of public health research and writing
Periyasamy Gandhi Aravind and Soundappan Kathirvel

3. Evidence-based public health practice
Rizwan Suliankatchi Abdulkader

4. **Applying evidence-based strategies for public health preparedness and emergency management**
    Vinayagamoorthy Kalaiselvi and Jaya Prasad Tripathy

Section II
Case studies of public health practice

5. ## Public health practice and clinical practice
    Kathiresan Jeyashree

6. ## Effect of public health research on policy and practice
    Punam Bandokar and Madhur Verma

7. Public health policy for social action to ensure better population health
Kalaiselvi Selvaraj

8. Democratization of health care in hospital settings— An integral part of public health practice
Ishwarpreet Kaur and Amarjeet Singh

9. Community participation for improving the coverage and quality of evidence-based public health practice
Hemant Deepak Shewade, Deepak H.J. Murthy and Seetharam Mysore

10. Intersectoral coordination for concerted efforts to improve the population health using evidence-based public health practice

Seetharam Mysore, Deepak H.J. Murthy and Hemant Deepak Shewade

11. Public health approaches to address substance use: An urgent need of multisectoral engagement

Cristina Rabadán-Diehl

12. Healthy aging and quality of life of the elderly

Arun Chockalingam, Amarjeet Singh and Soundappan Kathirvel

Section III
Future of public health practice

13. **The need for a paradigm shift to ensure adequate skilled human resources for effective public health practice**

    Myo Minn Oo

14. Effective use of information technology for the quality of public health practice

Palanivel Chinnakali and Swetha S. Kumar

15. Sustaining population benefit using evidence-based public health

Rama Shankar Rath and Ayush Lohiya

16. Precision medicine and public health practice

Gomathi Ramaswamy

17. Public health practice—A futuristic perspective
Patricio V. Marquez and Arun Chockalingam

Contributors

Rizwan Suliankatchi Abdulkader, Indian Council of Medical Research-National Institute of Epidemiology (ICMR-NIE), Chennai, Tamil Nadu, India

Periyasamy Gandhi Aravind, Department of Community Medicine, ESIC Hospital, Hyderabad, Telangana, India

Punam Bandokar, Indian Council of Medical Research-National Institute of Epidemiology (ICMR-NIE), Chennai, Tamil Nadu, India

Palanivel Chinnakali, Department of Preventive and Social Medicine, Jawaharlal Institute of Postgraduate Medical Education & Research (JIPMER), Puducherry, India

Arun Chockalingam, Medicine and Global Health, Faculty of Medicine, University of Toronto, Toronto, ON, Canada; Health Sciences, York University, Toronto, ON, Canada; Global Health, National Heart, Lung and Blood Institute at the National Institute of Health, Bethesda, MD, United States

Kathiresan Jeyashree, Indian Council of Medical Research-National Institute of Epidemiology (ICMR-NIE), Chennai, Tamil Nadu, India

Vinayagamoorthy Kalaiselvi, Centre for Public Health, Punjab University, Chandigarh, Punjab, India

Soundappan Kathirvel, Department of Community Medicine and School of Public Health, Postgraduate Institute of Medical Education and Research, Chandigarh, Punjab, India

Ishwarpreet Kaur, Department of Community Medicine and School of Public Health, Postgraduate Institute of Medical Education and Research, Chandigarh, Punjab, India

Swetha S. Kumar, Department of Preventive and Social Medicine, Jawaharlal Institute of Postgraduate Medical Education & Research (JIPMER), Puducherry, India; Department of Research, SingHealth Polyclinics, Singapore

Ayush Lohiya, Department of Public Health, Kalyan Singh Super Speciality Cancer Institute, Lucknow, Uttar Pradesh, India

Patricio V. Marquez, World Bank Group, Washington, DC, United States; Johns Hopkins University Bloomberg School of Public Health, Baltimore, MD, United States

Deepak H.J. Murthy, Swami Vivekananda Youth Movement (SVYM), Saragur, Karnataka, India

Seetharam Mysore, Swami Vivekananda Youth Movement (SVYM), Saragur, Karnataka, India

Myo Minn Oo, Department of Medical Microbiology and Infectious Diseases, University of Manitoba, Winnipeg, MB, Canada

Cristina Rabadán-Diehl, Westat, Rockville, MD, United States

Gomathi Ramaswamy, All India Institute of Medical Sciences (AIIMS), Bibinagar, Telangana, India

Rama Shankar Rath, Department of Community Medicine and Family Medicine, All India Institute of Medical Sciences (AIIMS), Gorakhpur, Uttar Pradesh, India

Rajavel Saranya, Department of Community Medicine and School of Public Health, Postgraduate Institute of Medical Education and Research, Chandigarh, Punjab, India

Kalaiselvi Selvaraj, Department of Community and Family Medicine, All India Institute of Medical Sciences (AIIMS), Madurai, Tamil Nadu, India

Hemant Deepak Shewade, Division of Health Systems Research, Indian Council of Medical Research-National Institute of Epidemiology (ICMR-NIE), Chennai, Tamil Nadu, India

Amarjeet Singh, Department of Community Medicine and School of Public Health, Postgraduate Institute of Medical Education and Research, Chandigarh, Punjab, India

Jaya Prasad Tripathy, Department of Community Medicine, All India Institute of Medical Sciences (AIIMS), Nagpur, Maharashtra, India

Madhur Verma, Department of Community and Family Medicine, All India Institute of Medical Sciences (AIIMS), Bathinda, Punjab, India

Foreword

contemplare meliora

Reading this book was a pleasure. It fills an important gap in the literature and will be of significant help to anyone concerned with enhancing public health policy, practice, and research. It will serve as an important resource and practice tool around the globe. Let me briefly explain why.

During and after completing my Ph.D. in population health, I worked as a public health practitioner in Canada for 7 years. I understood that using evidence to inform policy and practice offered the best way to have the largest impact with the modest resources available. However, to my surprise, there was a paucity of high-quality evidence in many important areas. Areas with longer histories, such as the diagnosis, testing, and treatment of communicable and enteric diseases, were relatively advanced. By comparison, we were slow to recognize that noncommunicable diseases have emerged as the largest threat to public health and that they are a function of the complex interactive set of determinants such as genetics, early childhood development, nutrition, physical activity, substance use, to environmental exposures, housing, income and income distribution, social networks, and connection, to name a few. As various chapters in this book note, it wasn't until the mid-1980s that epidemiologist Geoffrey Rose was among the first to emphasize the need to combine clinical interventions with population-level interventions whose aim was to shift entire population distributions of risk factors and conditions. But there was still a paucity of evidence on how to do this. It wasn't until the large community-based heart health trials of the late 20th century such as the North Karelia, and Framingham projects that we began to realize it would take a complex combination of dynamic surveillance, policies, and practices to make and sustain improvements in population health. It has only been more recently that we've understood the importance of health equity on population health and that Rose's approach may be most useful when we understand the need to alter risk distributions of multiple subpopulations within a general population, each with different needs, levels of readiness, risk, resources, values, traditions, and so forth.

In the early 1990s, I asked a research network for help to identify evidence on the most efficient ways to reduce tobacco-related disease across a population. I received feedback telling me what I already knew, which was that smoking is a significant risk to health and that helping people to quit was highly cost-effective. They referenced the effectiveness of various individual types of smoking cessation aids compared to doing nothing. What I needed was evidence on how to maximize resource efficiency (get the biggest population bang for my buck). I asked for evidence to help design an effective and cost-efficient group smoking cessation program. I was told that compared to controls, group cessation was effective, but no one could tell me how large the group should be, who was most effective at leading the group, what topics were responsible for the positive impact, or how many meetings were best. In Canada, as in other nations, cigarette packages had text messages warning of the dangers of smoking. So, I asked if there was evidence to show this was effective. By now, you probably know the answer. Reliable evidence was scarce. With more than 50 years of research at the time to draw upon, how could we know so little about how to address one of the leading causes of death in the entire world? It was then that I changed career paths from trying to be an evidence-informed public health practitioner to becoming a practice-informed public health researcher.

I am pleased to say that many others have joined the quest and today a wider array of evidence is available to inform public health practice, policy, and research. Indeed, there has been an explosion of work developing and perfecting new approaches for improving public health in an affordable and equitable way. The questions others and I asked 30 years ago can now be informed by high-quality evidence. There is a huge body of work aiding the facilitation and translation of evidence into practice. The field has grown so rapidly that it is difficult for busy practitioners and policymakers to keep track of the latest and most important advancements. This is precisely what this book does for its readers.

The editors and authors are world leaders in public health research and practice with experience working in both high- and low-resource settings. Like me, many see themselves as both practice-informed public health researchers and evidence-informed practitioners. These perspectives have enabled them to identify and consolidate the latest high-quality evidence to inform public health practice in multiple settings.

I have known two of the editors for some time. I can't remember precisely when I met Dr. Arun Chockalingam because he is one of those people that you feel you have known most of your life. He is a highly respected public health resource who generously provides stellar support and advice whenever called upon. His career has included time as an academic and as a senior leader in Canadian and American health agencies. For more than 5 decades, often working quietly behind the scenes facilitating and enabling the success of others, Arun has been a major influence in reducing the burden of non-communicable diseases, especially heart disease and stroke. I met Dr. Kathirvel while speaking at an education seminar in Chandigarh, India. I was immediately struck by this young man's intellectual brilliance and desire to constantly become better at his craft. These two people share some important traits. Both are humble, authentic individuals with enormous personal integrity who have dedicated themselves to making the world a better place. It came as no surprise to me that this book and the authorship team they assembled reflect their personal values and objectives: effective, thorough, exceedingly helpful, practical, and unpretentious.

In an era where media and search algorithms direct our attention to what seems like a never-ending stream of bad news, it would be easy to conclude that the challenges we face in public health are insurmountable. However, we must not fall victim to this type of myopia. The fact is that public health has made extraordinary progress in the last 50 years. Life expectancy has improved on every continent, and people enjoy more years in good health. People are less likely to die as infants or children. Just as they have for hundreds of years, the continuing gains in population health are largely due to improvements in public health policies and practices with a focus on altering the factors and conditions which determine health and increase the risk of disease or injury. This is the legacy we have a responsibility to build upon.

When he went on to become the Governor General of Canada, my former colleague at the University of Waterloo the Right Honorable David Johnston chose a coat of arms that included the Latin phrase *contemplare meliora* which roughly translates as "envision better things." To achieve better things requires not only imagination, but the generation, consolidation, and widespread utilization of new and existing knowledge, new perspectives, novel methods, and practices. This is the essence and goal of this book. It is an excellent consolidation of current principles, evidence, and their application to improve public health. I trust you will enjoy and benefit from it, just as I have.

Paul McDonald, Ph.D., FRSPH, FCAHS
Professor and Former Dean, Faculty of Health
and
Senior Fellow, Dahdaleh Institute for Global Health Research
York University, Toronto, Canada

Preface

Public health is a multidisciplinary field and is practiced in a complex environment. It includes the prevention of disease, promotion of health, and prolonging people's life (including curation and rehabilitation) through organized community effort. Various cadres like epidemiologists, policymakers, implementors, program managers, economists, social scientists, nutritionists, clinicians, administrators/executives, data or information managers, and others practice public health. However, the understanding and practice of public health vary across the public health human resource spectrum which is further aggravated by the COVID-19 pandemic. Notably, public health is a neglected area in overall health and hence low sanction of funds. Hence, the limited available resources must be used efficiently. Evidence-based public health practice (EBPHP) is one of the important contemporary solutions for efficiently using the available resources. Its principles and practice revolve around the identification, customization, implementation, evaluation, and sustainability of effective public health interventions. However, no comprehensive content or standard course is available to teach and practice EBPHP especially catering to public health practitioners who are not formally trained in public health and scholars of public health bachelors similar to 'Good Clinical Practice' and 'Good Laboratory Practice'.

In the above background, we have written this book selecting a group of authors from various countries trained in core public health working in academic institutions of repute, national and international developmental organizations, and clinicians based on their vast experience especially the EBPHP that potentially provides a multidimensional perspective.

This book primarily targets administrators/executives, social scientists, economists, nutritionists, clinicians, and others who don't have formal training in public health. Further, the book will help the students, especially those enrolled in undergraduate or postgraduate public health courses, in the conceptual understanding of the public health practice beyond their academic examination. It might also be a useful resource for new employees in any public health agency.

This book will provide a basic understanding of various (psycho-social, environmental, and political) determinants of health and public health approaches (from policy level to local practice level) applicable to EBPHP. The core areas of public health namely epidemiology, biostatistics, social sciences/ health promotion, health management, health economics, environmental and occupational health, nutrition, and public health research are covered in this book differently. In a way, this book is designed to inform public health practitioners about day-to-day practice.

This book is divided into three sections. The first section deals with the principles and basic concepts of evidence-based public health practice followed by case studies on the above principles and basic concepts in the second section. The third section concerns the "Future of Public Health Practice" which discusses the current situation and the need for future reforms of various domains of public health practice. Importantly, the book covers the application and future of emerging technology-based approaches and the practice of precision public health.

Though we attempted to be more wholesome, public health practice is a vast area and hence we focused on the core principles and approaches of EBPHP that may further stimulate the readers to advance their learning in the future.

Soundappan Kathirvel
Amarjeet Singh
Arun Chockalingam

Acknowledgment

First, we thank the almighty who always poured positive energy and provided an enabling environment to complete this important task. Secondly, we must thank all the teachers since our early life for inspiring and continued motivation in shaping this monograph. We are indebted to all authors for sharing their experiences through the included chapters. We sincerely thank Ms. Elizabeth Brown, Senior Acquisitions Editor, and Ms. Patricia Gonzalez, Senior Editorial Project Manager, and their team from Elsevier for their kind support, facilitation, and accommodation of all our requests. We don't have enough words to thank Prof. Paul W. McDonald, Professor and Former Dean, Faculty of Health, York University, Canada, who has written a comprehensive foreword to our book after painstaking reading of all chapters of this book. A special thanks to Mr. Kalyan Sundaram, Senior Fellow, Canada India Foundation, for his expert support on the language editing of the book chapter. Last but not least, thanks to all our family members for their patience, continuous moral, and material support in completing this book.

Section I

Introduction to public health practice

Chapter 1

Principles and approaches in public health practice

Rajavel Saranya and Soundappan Kathirvel
Department of Community Medicine and School of Public Health, Postgraduate Institute of Medical Education and Research, Chandigarh, Punjab, India

Public health and its practice

Public health is "the science and art of preventing disease, prolonging life, and promoting health and efficiency through organized community effort and informed choices of society, organizations, public and private communities, and individuals," as defined by CEA Winslow in 1887. It included environmental sanitation, control of communicable diseases, health education to improve personal hygiene, organizing necessary medical care, and social support to ensure a standard of living for everyone to maintain health which ultimately enables any individual to achieve health and longevity as their birthright [1,2]. Several public health associations worldwide attempted to revise this definition since the scope of public health expanded over a period and is no longer confined to hygiene and sanitation. These revisions predominantly included the expanded disease spectrum, community participation, and social determinants of health. These revisions were in line with the World Health Organization's definition of health: "a state of complete physical, mental, and social well-being and not merely the absence of disease or infirmity" [3]. Further, it emphasizes the attainment of the highest standard of health by all, for which the primary health care (PHC) strategy was introduced and scaled up globally. Currently, a subset of PHC strategy, i.e., universal health coverage (UHC), is emphasized as part of the Sustainable Development Goals 2030. Public health is expected to deliver certain essential services or functions as part of the routine practice defined by various agencies. Understanding the basic principles and approaches in public health practice is essential to develop, implement, evaluate, and scale up evidence-based public health interventions that deliver the essential services to improve and maintain the health and well-being of the individual and population.

Principles and Application of Evidence-Based Public Health Practice
https://doi.org/10.1016/B978-0-323-95356-6.00005-7

Principles of public health practice

Public health practice varies within and across countries. However, the fundamental principles of public health practice should stay the same. Due to its dynamic nature and overlap with public health research and, to a certain extent, routine clinical practice, some of the principles of practice are common among them. The basic principles are equity, fairness and inclusiveness, empowerment, effectiveness, evidence-based practice, and public health ethics. Indirectly, the above principles include the availability, accessibility, and affordability components and cultural sensitivity. Other additional principles, like the precautionary principle and solidarity, are not usually considered fundamental principles of public health practice. However, these are applicable in the different contexts of public health practice.

Equity

Health equity is the "absence of unfair, avoidable, or remediable differences between individuals and groups, which is defined based on demographics, economics, geography, social groups, or other dimensions of inequalities like sex, gender, sexual orientation, ethnicity, language, disability, religion, and others, in achieving the highest levels of health" [4]. Equity and equality are interchangeably used. However, they are different. Equality recommends that everyone in the community is provided the same standard of services. However, not all people in the community are the same and will have the same health needs. Hence, the different needs of the people (combined as various subgroups) in the community need to be addressed to achieve the same level of health and well-being, through which equality in health can be achieved among all community members. This health equity and equality is linked with fair distribution and inclusiveness.

Fairness and inclusiveness

Fairness is "the state, condition, or quality of being fair, or free from bias or injustice" with anyone or the group in the community [5]. To attain this fairness, public health practitioners must identify the inherent bias or injustice and its nature in the community and make the policies or establish the practices acknowledging the same. Such policies and practices, inclusive of various community subgroups, will address most of the community health needs and hence better population health and well-being. Fairness and inclusiveness are basically from the provider side, whether a healthcare provider or policymaker, irrespective of the level of community participation. It will ensure social justice and equity. Further, community participation and empowerment are needed to exploit fairness and inclusiveness to achieve health equity.

Empowerment

Empowerment enables individuals and communities to take control of their health. Empowerment can be achieved through increasing knowledge, i.e., general and specific health literacy among community members. It further needs effective communication models across different community groups. Fairness and inclusiveness also play a role in using various communication strategies and improving health literacy. Improvement of health literacy is not linearly linked with empowerment, rather community participation is the key to achieving empowerment. The avenues for community participation must be considered and created in routine public health practice and the depth of community participation decides the empowerment. Just providing the information at once or multiple times and taking feedback cannot be called community participation and will not lead to empowerment. It must go beyond providing information and consultation, i.e., involvement and collaboration with the community members, because the health and well-being of a community cannot be defined by the policymakers or practitioners but rather by the community itself.

Empowerment facilitates the social relationship between individuals and the community and the communitization of healthcare services, which is the road to ensuring a better quality of care [6]. It facilitates health activism if the policymakers or practitioners do not practice fairness and inclusiveness.

Effectiveness

Effectiveness is delivering services or interventions and achieving the population health in practical and real-time situations. Of course, it will be lower than the efficacy tested in a controlled or ideal situation. A controlled or ideal situation is mostly possible in clinical practice but not in routine public health practice settings. Hence, considering its nature, the effectiveness is more applicable to public health practice areas. Further, public health practitioners must look for the efficiency of the services or interventions, i.e., how to achieve maximum health using the existing minimal resources. Repeated operational research (implicit or explicit) is needed to achieve such efficiency. Efficiency is considered while implementing or scaling up of public health services or interventions. Effectiveness is usually assessed through public health research that is directly linked with the "evidence-based practice" principle.

Evidence-based practice

Evidence-based public health practice is collecting, compiling, analyzing, and using the available scientific evidence to make population health decisions. The scientific evidence should not confine to only peer-reviewed publications and include the analysis of routinely collected data since public health practice generates a huge quantity of data. However, the evidence generation in public health practice is low which is due to the complexity of the practice area, funding and other resource availability, research capacity, and others. The available evidence needs to be assessed based on the type or level (systematic review of randomized control trial to case studies or expert opinion), GRADE (Grading of Recommendations, Assessment, Development, and Evaluations) of recommendation and its interpretation level [7]. In the absence of evidence in the concerned field, the available evidence in similar areas can be adopted after necessary modification for the contextual factors and pilot testing. However, the decision-making in evidence-based practice is not just dependent on the availability of evidence. It depends on other contexts like population needs, values and preferences, availability of resources and expertise, and organization and other enabling environments [8]. Further, public health practice must also generate new evidence for future use during the process of the use of evidence.

Ethics in public health practice

Though fundamental ethical principles followed in public health practice are like clinical practice, any breach will affect the whole community. Hence, it must adhere to ethical principles strictly. The fundamental ethical principles are autonomy, beneficence, nonmaleficence, and justice [9]. The autonomy of an individual is relative in public health practice, rather it applies to the autonomy of the community, including the rights and dignity of the community. Beneficence focuses on community benefit against individual benefit. For example, vaccination of a person with a low risk for an infection or disease may not benefit him directly, rather it will benefit the family and community in controlling the transmission.

Nonmaleficence or 'do-no-harm' recommends minimal or no harm to any community member. However, it is relative to the context of the practice area. For example, the harm principle, a related principle of nonmaleficence, advocates restricting the liberty of a person or group to prevent harm to others, e.g., banning smoking in public places to prevent exposure among nonsmokers. Public health professionals must assess and quantify the expected and unexpected harms caused to individuals and the community while implementing public health interventions. Such harm must be reported along with the effectiveness of the intervention for informed scale-up of the interventions.

The proportionality principle used in ethics checks the balance between selecting the services or interventions and the burden in the community [10]. Further, there are other ethical principles like social justice (equitable distribution), reciprocity (other benefits), solidarity (for collective welfare), accountability and transparency, and efficiency might be considered during public health practice.

Other overlapping principles

Availability, accessibility, acceptability, and affordability of services or interventions are the overlapping principles in public health practice. Availability and accessibility are predominantly linked to equity, fairness, inclusiveness, and effectiveness principles. Similarly, the acceptability or cultural sensitivity of a service can be linked with inclusiveness and empowerment and hence with equity.

Domains and functions of public health

The domains and scope of public health are wide. Almost everything related to improving and maintaining health and well-being is directly or indirectly linked to public health. The public health practice includes three broad domains: health improvement, protection, and healthcare service quality improvement. All three domains are interdependent and work in cohesion. Health improvement focuses on the wider social determinants to reduce the inequities and psychological aspects of health. Health protection focuses on specific disease control and prevention from environmental, occupational, chemical, radiation, nuclear, and other threats. In a way, health improvement and protection can be linked with primordial and primary prevention in public health practice. Healthcare service quality improvement deals with providing quality and cost-effective clinical care, an efficient healthcare system and the practice of evidence-based care. These domains inform the list of essential functions or services to be delivered by public health (Fig. 1.1) [11].

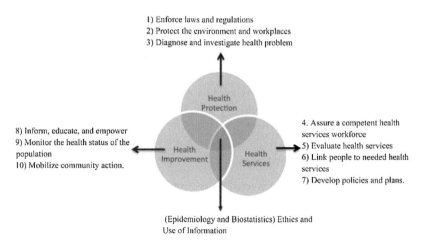

FIGURE 1.1 The domains and functions of public health practice [11].

The essential functions of public health are classified under three core areas: assessment, policy development, and assurance. Public health needs to perform a wide range of functions to maintain individual and population health; hence, listing the functions of public health may be difficult. However, certain essential functions are identified globally and in different country contexts [12]. These are:

1. Surveillance of disease burden and pattern to identify any potential outbreaks and health risks
2. Diagnose and confirm the potential health threats and investigate the burden and determinants to identify effective interventions
3. Educate and empower people to adopt preventive and promotive practices through informed choices
4. Enhancing community participation and collaborations in identifying, understanding, planning, and managing local health issues and emergencies
5. Health policy development, implementation, and evaluation to improve the health of the individual, community, and regional level, and assist in emergency preparedness and response.
6. Advocate for evidence-based public health practices, policies, laws, and amendments in the existing practices, policies or laws to address the social determinants of health.
7. Ensure the availability and acceptability of healthcare services and enable the population to access healthcare and allied services, and identify the potential barriers and strategies to overcome barriers.
8. Ensure the availability of a trained and dedicated workforce with the highest ethical standards and technical quality to undertake the public health practice, leadership, and management of health programs
9. Monitoring, supervision, and evaluation of the ongoing interventions and providing actionable recommendations at multiple levels of healthcare systems
10. Collaboration with research academia for introducing cutting-edge and novel research methods and for translating the research findings into practice.

The list of essential functions might vary according to the agency which proposes it and based on the role of public health discipline in managing the population health in their setting.

Public health interventions

The functions or services delivered as part of the public health practice are termed 'public health interventions,' and both can be easily correlated. The broad areas of 17 population-based public health interventions (as wheel) identified by the Minnesota Department of Health cover all spectrums of public health practice. Further, the practice area can be at individual, family, community, health system, or beyond health systems level (Fig. 1.2) [13]. The public health intervention wheel has five wedges or groups. The first group includes surveillance and investigation of health events, outreach, and screening. The second group includes medical care of cases, referral and follow-up, and delegated functions. The third group is about consultation, counseling, and health teaching of the individual and population. The fourth wedge is on engagement and collaboration with the community and other stakeholders. The fifth group includes advocacy, policy development, and implementation and social marketing. These interventions can be social, economic, political, developmental, environmental, behavioral, and health systems interventions or simply preventive, therapeutic, and other interventions.

Quantifying and prioritizing the public health problem is the first step in identifying potentially effective interventions. The effective interventions must

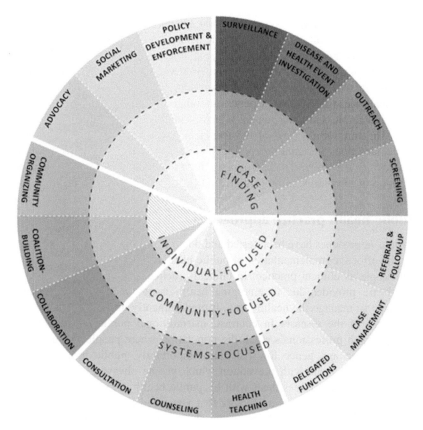

FIGURE 1.2   Population-based public health interventions [13].

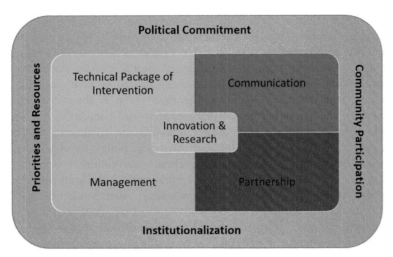

FIGURE 1.3   Essential components for effective implementation of public health intervention.

be identified through a systematic assessment of available evidence and modified to the local context. The public health intervention, i.e., the technical package must be simple, culturally acceptable, and cost-effective. Complex interventions are difficult to implement. However, the implementation and the effectiveness of a public health intervention are not just dependent on the technical package. It also depends on other components like partnership, communication, and management which promote the uptake of the intervention. The innovations beyond the technical package will facilitate effective implementation. In addition, political commitment and availability of resources are the important determinants for the effective implementation and scale-up of any intervention (Fig. 1.3) [14].

Public health practice approaches

Various approaches have been tested and documented in public health practice for the effective implementation of public health interventions. However, it is not simple as clinical practice. In clinical practice, patients reach out to the healthcare provider due to emerging health issues and listen well to the clinician to manage the health issue. It predominantly involves diagnosis and treatment of the disease. In contrast, in traditional public health practice, public health professionals reach the population where people may not have the felt need and hence low listening. Importantly, public health practice predominantly deals with prevention which people always give low importance. Further, the issues dealt with by public health are complex and there could be multiple attributing factors that also need to be tackled by public health professionals. In addition to the above issues, people or communities have never been the same across geography and over a period in the same geography. A public health intervention that is acceptable and effective in one community may not be necessarily acceptable and effective in another community in contrast to the traditional clinical practice, where the same medicine will work for a particular disease across patients. Notably, public health practice is limited with resources or funding; hence, it must identify strategies for achieving efficiency using available resources. As mentioned before, public health is practiced at various levels from the individual, community and beyond the health sector. Hence, the public health practice approaches vary depending on the abovementioned context.

Primary health care and universal health coverage

PHC is defined as a "whole-of-society approach to health that aims at ensuring the highest possible level of health and well-being and their equitable distribution by focusing on people's needs and as early as possible along the continuum from health promotion and disease prevention to treatment, rehabilitation and palliative care, and as close as feasible to people's everyday environment" [15]. Equitable distribution of services, use of appropriate and cost-effective technology, involvement of all stakeholders and engagement, and empowerment of the community are the key principles of PHC. UHC aims to provide equitable access to all types of healthcare services with ensured quality and without financial hardship to people [16]. UHC focuses only on the availability, acceptability, accessibility, and affordability of the spectrum of healthcare services (preventive, promotive, curative, and rehabilitative) and does not cover the intersectoral coordination, community participation, and other social determinants of health similar to PHC. Hence, UHC must be considered a substrategy to achieve the PHC goals.

Recently, addressing the social determinants of health through the involvement of all relevant sectors beyond health has been given importance. The broad structural social determinants are macroeconomic policies, social policies like housing, public policies like education, health and insurance, and governance (Fig. 1.4). These determine the socio-economic position of an individual and population, which influences the intermediary social determinants like behavior, psychosocial factors, and standard of living. The health system is just one factor in the list of intermediate social determinants; hence, active involvement of sectors beyond health is essential to improve and maintain individual and population health [17]. Health in All Policies (HiAP) is one such approach focusing on social determinants of health. It emphasizes the health considerations in developing and implementing comprehensive policies across sectors to address health determinants. HiAP is an integrated governance model with the common objective of 'population health,' potentially an efficient and resource-saving strategy [18].

The 'Planetary Health' and 'One Health' approach are the upgraded version of PHC concept, including the sustainable environment. Planetary health is defined as the "achievement of the highest attainable standard of

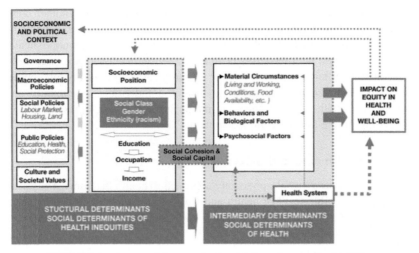

FIGURE 1.4   Conceptual framework of social determinants of health [17].

health, well-being, and equity worldwide through judicious attention to the human systems—political, economic, and social—that shape the future of humanity and the Earth's natural systems that define the safe environmental limits within which humanity can flourish" [19]. Similarly, One Health is "the collaborative effort of multiple health science professions, together with their related disciplines, and institutions—working locally, nationally, and globally—to attain optimal health for people, domestic animals, wildlife, plants, and our environment" [20].

"Analyze, measure and intervene" approach

Public health practice involves observation, collection, compilation, quantification, and comparison of information, a core component of epidemiology and an essential function-surveillance. Such information will facilitate the measurement of health or disease and identify the potential interventions for the same. For example, the routine surveillance indicates an impending outbreak of a communicable disease; public health practitioners might investigate in detail to identify suitable interventions and implement them which is a routine day-to-day practice in public health. Information technology eased the task of measuring and analyzing and health and diseases and at times; it also suggests potentially effective interventions.

Population versus high-risk approach

In a population-based or universal approach, all the members of the population are provided interventions or services irrespective of the risk profile. Though it is a holistic approach, it is resource-heavy and uptake of the services might be low due to lower felt need by the population. However, it will make people even at low risk aware of the risk factors and hence prevent them from risk factor exposure if implemented effectively. The high-risk or targeted approach focuses on only a selected population group based on their high-risk profile. Risk assessment and classification is a prerequisite to practicing the high-risk approach. Here, the chances of uptake of the interventions or services might be higher and more cost-effective than with the population-based approach. However, it has its limitations. Using it as a long-term strategy may not be cost-effective and will not prevent low-risk people from future exposure to risk factors. It will focus only on one small proportion of the population at high risk of developing the disease. However, most cases could be from the so-called 'low-risk' population which forms the major proportion of the population, the 'Prevention Paradox.' Further, the high-risk approach potentially will not prevent more people from entering high-risk groups since the low-risk group is not provided interventions. Notably, behavior change is one of the difficult outcomes to achieve, which may impact the overall effectiveness of the intervention if only a high-risk approach is practiced. Since the risk classification informs who could develop the disease early, it must be used as a "short-term" strategy and a population-based approach as a "long-term" strategy.

Community versus facility-based approach

Public health practice is not just confined to the community setting, like clinical practice is confined to the physician consultation room and the health facility. Health protection and improvement domain-related functions are predominantly delivered through a community approach and health service quality improvement domain-related functions are mostly delivered through a facility-based approach. The community-based approach focuses on the whole population and the facility-based approach focus on the individuals visiting the health facilities. However, both are interlinked and difficult to practice alone. Health facilities work as an important surveillance point to quickly assess the effectiveness of any intervention implemented at the community level. Similarly, surveillance at a facility might indicate the presence of emerging diseases that directs to implementing a preventive intervention at the community level. It is important to understand that public health delivers both preventive and curative services in an integrated manner, i.e., the community approach is more preventative and the facility-based approach is more curative.

The integrated (community and facility-based) public health practice model is the most common model practiced worldwide. Community health workers or volunteers deliver basic and predominantly preventive healthcare services in this model. They assess the health status (risk assessment, screening, and presumptive disease) of the people periodically and link the high-risk or screen-positive population with the health facilities. Health facilities usually follow a three-tier system namely primary, secondary, and tertiary care in most countries globally. Based on the type, severity, and progression of the disease, there will be forward and back-referrals between health facilities of various levels and between health facilities and the community.

Regulatory versus non-regulatory approach

The majority of public health services are delivered through a non-regulatory approach. Regulatory approaches protect the population from important risks that endanger their health. It is used when (a) the behavior of an individual affects the members of the community, like smoking in public places; (b) immediate and mass-level behavior change is warranted like the use of helmets or seat belts to prevent injuries or accidents; (c) the vested interest of business affects the population like food manufacturing and related services, tobacco industry, and others; and (d) in case of epidemics or pandemics to prevent and control further disease transmission. Otherwise, the regulatory approach is used when there is a threat to the country's economy directly or indirectly. The majority of the regulatory approach needs multisectoral involvement beyond public health from the development of the regulation to effective enforcement.

Public health—A multidisciplinary field

Public health is a young and multidisciplinary field. It is interchangeably called as "Community Medicine," "Preventive Medicine," "Preventive and Social Medicine," "Public Health Medicine," and "Community Health" across the world even now. Though public health evolved from the above terminologies/fields, it provides equal importance to medical, social, environmental, statistical, and other fields in contrary to them. Identifying domains and segregating the functions are linked with the academic and practice subspeciality area of public health (Fig. 1.5) [21]. Public health includes epidemiology, biostatistics, preventive medicine, environmental and occupational health, social and behavioral sciences (psychology, anthropology, economics, political science), health policy, planning, program and management, nutrition, global or international health, and public health informatics areas. All the subspeciality areas of public health are interdependent and practiced together.

Epidemiology and biostatistics

Epidemiology (epi-upon; demos-people; logos-study) is defined as the "study of the distribution and determinants of health-related states or events in a specified population, and the application of this study to the control of health

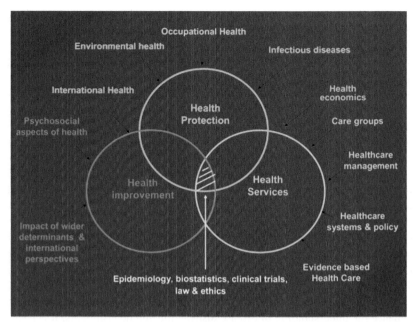

FIGURE 1.5 The domain-wise academic and practice subspeciality areas of public health [21].

problems" [22]. It is one of the core areas of public health primarily based on unbiased observation and analysis of health-related events among the population and action to control them. The study of the distribution of the disease can be termed descriptive epidemiology. The distribution of disease/s for time ("When"), place ("Where"), and person ("Whom") is essential to understand public health problems and to broadly identify the potential interventions. The ongoing pattern of the disease occurrence among people and time period plotted across geographical areas will guide public health professionals in triaging the risk population, resource allocation and implementing the potential intervention.

Epidemiology looks beyond the distribution of any health-related events. It wants to assess the determinants of the disease, i.e., cause or risk factors—"Why" and "How" components. Traditionally, the epidemiological triad of agent, host, and environment concept is used to ascertain the cause of disease. Though it is still used in communicable disease epidemiology, the multifactorial causation of disease is observed not only in noncommunicable diseases but also in communicable diseases. Hence, assessing these multiple determinants is essential to understand the disease better and identifying effective interventions. Such assessment of determinants and comparison to delineate the cause or risk factor is called analytical epidemiology. Ultimately, epidemiology is linked with action, i.e., applying the findings to control the disease. The action could be developing, testing, and scaling up effective interventions, advocacy, policy making and implementation, and others. This is called applied epidemiology [23].

Since the last two decades, the molecular and genetic factors have been assessed invariably with all diseases, bringing the molecular and genetic epidemiology field into practice. Molecular epidemiology assesses the changes at the molecular level due to genetic and environmental risk factors that will improve the understanding of disease causation and distribution. Though overlapping, genetic epidemiology assesses the role of genetic factors in the causation and distribution of the disease.

There are other subspeciality areas or terminologies within epidemiology like social epidemiology, circular epidemiology, legal epidemiology, and managerial epidemiology. Social epidemiology assesses the social determinants of health, linking the occurrence of disease and its distribution [24]. Circular epidemiology is repeating a similar kind of epidemiological assessment to understand the changing pattern of risk factor associations over time [25]. Legal epidemiology studies the association of law with the occurrence and distribution of diseases [26].

Biostatistics is an integrated component of epidemiology that helps to enhance the understanding of disease distribution and determinants. It is a vast field similar to epidemiology, ranging from simple descriptive statistics to advanced analytical statistics. It guides all the types of assessment done as part of epidemiology. Though information technology has eased the use of biostatistics, conceptualizing the various biostatistics methods applicable to epidemiology is mandatory for better public health practice.

Preventive/community/family medicine

Case management and quality of care are part of the public health practice. Public health focuses on PHC, including preventive, promotive, curative, and rehabilitative care. 'Community physician' or 'family physician' is expected to provide the same in primary or secondary healthcare settings. In some places, preventive specialists provide a wide spectrum of preventive services. Though public health professionals with medical backgrounds primarily provide it, preventive interventions can be provided by nonmedical experts like nutritional interventions. Case management is linked with surveillance, assessment of disease burden, understanding the natural history of endemic and emerging diseases, and preventing further disease transmission. The wide spectrum of healthcare services provided as part of public health practice can be associated with various levels of prevention. Primordial prevention, a holistic concept, prevents people from exposure to risk factors. Primary prevention deals with the prevention of the occurrence of disease through health promotion and specific protection like vaccines. Secondary prevention includes early diagnosis (including screening) and treatment; tertiary prevention includes disability limitation and rehabilitation. Quaternary prevention prevents people from overmedicalization and related ill health [27]. Public health professionals must have expertise in dealing with all these levels of prevention.

Health promotion, social, and behavioral sciences

Health promotion is enabling and empowering people to control over and improve their health [28]. Its practice area covers physical, mental, social, and spiritual areas and is practiced in all levels of disease prevention (primordial to quaternary prevention). Further, its focus ranges from developing and implementing public policies to changing behavior at the individual level. The same is reflected in the health promotion emblem released by Ottawa Charter for health promotion (Fig. 1.6) [28]. Health promotion at the individual and population level is linked with various social science subspecialities like psychology, anthropology, economics, and politics. At times, health promotion is considered one of the subspeciality areas under social and behavioral science. Understanding the social structures, institutions, culture, customs, and practices plays an important role in addressing the social determinants of

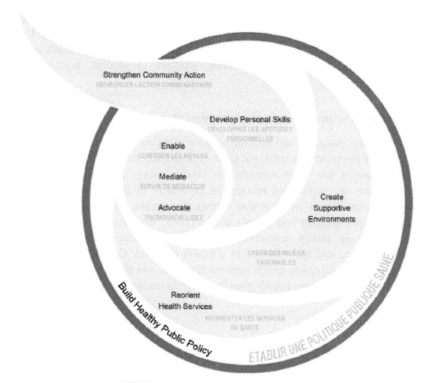

FIGURE 1.6 Health promotion emblem [28].

health. This field makes public health think beyond the traditional biomedical causation of disease to psycho-social causation that will help to decode the causes of diseases. It makes the public health professionals understand the health or disease from people's perspective (EMIC perspective).

Environmental and occupational health

Environmental and occupational health is a public health practice area linked with the causation of various diseases. The environmental cause of the disease has been discussed for ages, i.e., even before the miasmatic theory and the linkage of water contamination to cholera epidemic in London. Beyond air, water, and soil, other macro- and microenvironments are assessed including weather, climate, use of pesticides, food, living standards, sanitation and hygiene, presence of disease vectors and agents, and others. Various environmental exposures are linked with the molecular changes in the body which is assessed as part of molecular epidemiology. Further, environmental assessment needs specialized skills and sophisticated instruments most of the time.

Occupational health deals with the health of the people indulged in a specific occupation. Workers' health and safety are directly linked with a country's economy. It deals with industrial safety practices like a safe working environment and other enabling factors to maintain the worker's health and workers' behavior in preventing and managing exposures and diseases. This field does the periodic health assessment for understanding and estimating the known and unknown diseases among workers to specific occupational exposure.

Health policy, program, and management

Health policy, program, and management are other core areas of public health. It is one of the most powerful, but least explored subspeciality areas in public health practice due to its complexity. Health policy, health program, and management are sometimes recognized separately in public health practice with different levels of expertise among the public health professionals. However, they overlap and grossly deal with the health systems. It includes health planning, policy development, advocacy, implementation, monitoring and evaluation, and maintenance of a health program. Operational research, implementation research, health systems research, and dissemination research enable the process from health planning to maintenance of health program.

Health economics

Health economics, a social science subject initially identified as part of the 'Health Policy, Program, and Management' area, is now recognized as a separate subspeciality area in public health. The economic principles are applied in healthcare, in disease prevention, causation and cure. It deals with cost-effectiveness, cost—benefit analysis, cost-minimization analysis, and efficiency. Behavior economics is another subarea under health economics that deals with the association of people's behavior and the utilization of healthcare and other services in maintaining their health [29]. Though multidisciplinary, health technology assessment predominantly deals with economic evaluation of the interventions to make decisions in public health [30].

Public health nutrition

Public health nutrition applies the principles of nutrition in public health practice integrated with social and behavioral science. This field assesses the association between nutrition and health or disease at individual and population levels in detail. Further, it assesses the effect of nutritional interventions using social and behavioral sciences on health and disease. It is one of the major public health practice areas recognized worldwide and its application is useful in both maintaining health and preventing diseases.

Public health informatics

Public health informatics is the application of information technology in all areas of public health practice which is not limited to surveillance, prediction, preparedness, and management of diseases. Recent technologies like telemedicine, artificial intelligence, and its components like machine learning or deep learning technologies have proven to be effective for timely prediction of disease transmission and outbreak, better diagnosis, and treatment using algorithms and providing clinical care to people from remote areas. The huge public health data generated across healthcare and other settings including genetic, molecular, and environmental data can be easily integrated using these technologies to practice precision public health.

Global health or international health

Global health is a recently growing field of public health and deals with health-related events impacting countries across borders. It must be considered as an extension of international health which was focusing only on issues impacting the participating countries from non-participating countries [31]. Global health enhances the understanding of public health issues and their causes to find solutions for local situations.

Conclusion

Public health practice is aimed at preventing disease, prolonging life, and promoting the health of the population using various social, behavioral, and health systems interventions. Equity, fairness and inclusiveness, empowerment, effectiveness, evidence-based practice, and public health ethics are the fundamental principles of public health practice. It tested and used various strategies like PHC, UHC, HiAP, and others to deliver public health services or interventions and deal with the complex population health determinants. Public health is a dynamic and multidisciplinary field. The subspeciality area includes epidemiology, biostatistics, preventive medicine, environmental and occupational health, social and behavioral sciences (psychology, anthropology, economics, political science, and others), health policy, planning, program and management, nutrition, public health informatics, and global health. Newer and more advanced technologies have the potential to improve the efficiency of public health service delivery and hence improve the population health.

References

[1]   Singh AJ. What is (there) in a name. Indian J Community Med 2004;29(4):151−4. https://journals.lww.com/ijcm/Fulltext/2004/29040/What_is__There__in_a_Name.1.aspx.

[2]   Ahmed FU. Defining public health. Indian J Public Health October 2011;55(4):241. https://www.ijph.in/article.asp?issn=0019-557X;year=2011;volume=55;issue=4;spage=241;epage=245;aulast=Ahmed.

[3]   World Health Organization. Constitution of the World Health Organization. Available from: https://www.who.int/about/governance/constitution. (Accessed 26 February 2023).

[4]   World Health Organization. Health equity. Available from: https://www.who.int/health-topics/health-equity#tab=tab_1. (Accessed 26 February 2023).

[5]   Sandro G. Fairness and public health | SPH. 2017. Available from: https://www.bu.edu/sph/news/articles/2017/fairness-and-public-health/. [Accessed 26 February 2023].

[6]   Lechyd Cyhoeddus Cymru Public Health Wales. Principles of community engagement for empowerment. 2019. p. 1−18.

[7]   Reed S, Gordon G. What is GRADE? BMJ best practice. Available from: https://bestpractice.bmj.com/info/us/toolkit/learn-ebm/what-is-grade/. (Accessed 26 February 2023).

[8]   Brownson RC, Fielding JE, Maylahn CM. Evidence-based public health: a fundamental concept for public health practice. Annu Rev Public Health 2009;30:175−201. https://pubmed.ncbi.nlm.nih.gov/19296775/.

[9]   Coughlin SS. How many principles for public health ethics? Open Public Health J January 1, 2008;1:8−16.

[10]   Have M ten, de Beaufort ID, Mackenbach JP, van der Heide A. An overview of ethical frameworks in public health: can they be supportive in the evaluation of programs to prevent overweight? BMC Publ Health December 22, 2010;10(1):638.

[11]   Karkee R. Public health education in South Asia: a basis for structuring a master degree course. Front Public Health July 21, 2014;2. pmc/articles/PMC4104799/.

[12]   Centers for Disease Control and Prevention. 10 essential public health services. Available from: https://www.cdc.gov/publichealthgateway/publichealthservices/essentialhealthservices.html. (Accessed 27 February 2023).

[13]   Minnesota Department of Health. Public health interventions, definitions, and practice levels (public health intervention wheel). Available from: www.health.state.mn.us. (Accessed 26 February 2023).

[14]   Frieden TR. Six components necessary for effective public health program implementation. Am J Public Health January 2014;104(1):17−22.

[15] World Health Organization. Primary health care. Available from: https://www.who.int/news-room/fact-sheets/detail/primary-health-care. (Accessed 26 February 2023).

[16] World Health Organization. Universal health coverage (UHC). Available from: www.who.int/news-room/fact-sheets/detail/universal-health-coverage-(uhc). (Accessed 26 February 2023).

[17] World Health Organization. A conceptual framework for action on the social determinants of health. Available from: https://www.who.int/publications/i/item/9789241500852. (Accessed 26 February 2023).

[18] Greer SL, Falkenbach M, Siciliani L, McKee M, Wismar M, Figueras J. From health in all policies to health for all policies. Lancet Public Health August 2022;7(8):e718−20.

[19] Whitmee S, Haines A, Beyrer C, Boltz F, Capon AG, de Souza Dias BF, et al. Safeguarding human health in the Anthropocene epoch: report of the Rockefeller Foundation-Lancet Commission on planetary health. Lancet November 14, 2015;386(10007):1973−2028. http://www.thelancet.com/article/S0140673615609011/fulltext.

[20] Roger F, Caron A, Morand S, Pedrono M, de Garine-Wichatitsky M, Chevalier V, et al. One Health and EcoHealth: the same wine in different bottles? Infect Ecol Epidemiol 2016;6(1). https://pubmed.ncbi.nlm.nih.gov/26899935/.

[21] Thorpe A, Griffiths S, Jewell T, Adshead F. The three domains of public health: an internationally relevant basis for public health education? Publ Health February 2008;122(2):201. pmc/articles/PMC7111666/.

[22] Last JM. Epidemiology and ethics. Law Med Health Care April 29, 1991;19(3−4):166−74.

[23] Brownson RC, Samet JM, Bensyl DM. Applied epidemiology and public health: are we training the future generations appropriately? Ann Epidemiol February 1, 2017;27(2):77. pmc/articles/PMC5578705/.

[24] Honjo K. Social epidemiology: definition, history, and research examples. Environ Health Prev Med September 2004;9(5):193. pmc/articles/PMC2723602/.

[25] Kuller LH. Circular epidemiology. Am J Epidemiol November 1, 1999;150(9):897−903. https://pubmed.ncbi.nlm.nih.gov/10547134/.

[26] Centers for Disease Control and Prevention. Legal epidemiology. Available from: https://www.cdc.gov/dhdsp/policy_resources/legal_epi.htm. (Accessed 26 February 2023).

[27] Kisling LA, Das JM. Prevention strategies. StatPearls Publishing; May 8, 2022. p. 1−4. Available from: https://www.ncbi.nlm.nih.gov/books/NBK537222/. [Accessed 26 February 2023].

[28] World Health Organization. Health promotion. Available from: https://www.who.int/teams/health-promotion/enhanced-wellbeing/first-global-conference/emblem. (Accessed 26 February 2023).

[29] Reed DD, Niileksela CR, Kaplan BA. Behavioral economics. Behav Anal Pract June 1, 2013;6(1):34−54.

[30] Millar R, Morton A, Bufali MV, Engels S, Dabak SV, Isaranuwatchai W, et al. Assessing the performance of health technology assessment (HTA) agencies: developing a multi-country, multi-stakeholder, and multi-dimensional framework to explore mechanisms of impact. Cost Eff Resour Allocation December 2, 2021;19(1):37.

[31] Chen X, Li H, Lucero-Prisno DE, Abdullah AS, Huang J, Laurence C, et al. What is global health? Key concepts and clarification of misperceptions: report of the 2019 GHRP editorial meeting. Glob Health Res Policy December 1, 2020;5(1):1−8. https://ghrp.biomedcentral.com/articles/10.1186/s41256-020-00142-7.

Chapter 2

Principles of public health research and writing

Periyasamy Gandhi Aravind[1] and Soundappan Kathirvel[2]
[1]*Department of Community Medicine, ESIC Hospital, Hyderabad, Telangana, India;* [2]*Department of Community Medicine and School of Public Health, Postgraduate Institute of Medical Education and Research, Chandigarh, Punjab, India*

Introduction

A sound public health practice is based on robust public health research. Research enables answering the challenges posed in routine public health practices. In addition, research identifies newer questions and the unknown aspects in the domain to explore further. The findings from research also influence the policies of various sectors beyond health, e.g., tobacco regulation and universal iodization in India. Hence, public health must generate and collate the evidence necessary for research leading to informed policy decisions and program implementations [1]. Considering its vital role, public health research is acknowledged as one of the 10 listed essential public health services globally [2].

Principles of public health research

Unlike traditional biomedical/clinical research, public health research overlaps more with routine public health practices. Delineating the practice and research is difficult in public health and hence its principles [2,3]. The fundamental difference between public health research and clinical research is the unit of research. Clinical research focuses only on individuals, whereas public health research focuses on communities, blocks, states, and regions beyond individuals. Accordingly, public health research principles vary from those of clinical research. While it is essential to know the principles of public health research, it is also pertinent to understand where they overlap with and where they differ from clinical research.

Principles and Application of Evidence-Based Public Health Practice
https://doi.org/10.1016/B978-0-323-95356-6.00011-2

Principle of respect for autonomy, rights, and dignity

An individual's autonomy, right, and dignity have been kept absolute and paramount in clinical research. Though public health research respects the individual's autonomy, it is not absolute and subservient to the community's autonomy. The community can participate in part or full or not participate at all, allowing certain aspects of intervention but restricting other aspects of public health research. Public trust is paramount for better community participation in public health research [1]. Hence, the community must be involved in all stages of the continuum, such as planning to disseminate or change the policy/practice. Here, communication is key to understanding the community's local cultural practices and autonomy and improving community participation through their earned trust. Community consultation can be achieved through a research-specific ad-hoc committee or existing community participatory mechanisms like village health and sanitation committees, women's health committees, or other self-help groups. Importantly, the inclusion of all sections of the community, especially the socially vulnerable and marginalized community, is crucial. It is essential that the researcher respects the rights and dignity of the community while conducting the research [3].

Individual autonomy and community welfare have been at odds more than a few times worldwide. Though an individual's autonomy and rights need to be equally respected, conflicts can arise between individual autonomy and community welfare. For example, during the COVID-19 pandemic, wearing face masks was compulsory, which was considered an infringement of an individual's choice of whether or not to wear masks. However, community welfare or rights triumphed over individual autonomy. Though public health research does not have the power as public health practice to impose a measure at the community level, the community's decision to exercise its autonomy in participating in the research supersedes any decision of a particular individual in the community.

Principle of beneficence

The concept of beneficence implies a moral commitment to aid other people (for instance, the duties of medical professionals toward their patients) or confer benefits upon others [4]. A public health practitioner owes a moral obligation to do good for the community in which they work and conduct research. Achievement of the public good by maximizing the social benefits is the goal of public health research. Here again, individual benefits take a back seat to the collective good. Exposure to an infectious agent and disease occurrence may not happen to everyone. Irrespective of the risk, all individuals in the community are vaccinated to prevent community transmission and protect the high-risk population. The vaccination provided as part of public health measures may or may not benefit the individual. However, it will facilitate the achievement of effective herd immunity to stop further transmission of the infection in the community. Similarly, universal salt iodization may not benefit persons who consume adequate iodine from other sources. However, it will ensure that people with iodine deficiency are protected against its related disorders.

Principle of nonmaleficence

"Nonmaleficence" is derived from the Latin phrase "primum non nocere" meaning "first, do no harm" and is one of the four pillars of research. It dictates that the public health researcher must make every effort to minimize the harm done to communities and individuals during any study. The harm may include physical and mental harm due to the interventions, social harms such as discrimination, stigmatization, and subsequent isolation of people with certain diseases such as HIV, TB, leprosy, and mental health disorders due to their participation in the study. Physical and mental harm can be prevented and reduced by adhering to the protocols addressing the foreseen harms and responding to any adverse event with an adequate standard of care. Social harm can be prevented by maintaining the individual participant's and community's identification details securely and confidentially.

The harm, least infringement, and proportionality principles are all linked with public health research. As per the harm principle, the liberty of the group or individual within the group can be regulated or restricted only to prevent injury to others or other groups, and strong ethical reasoning is a must for restraining action. Even during such scenarios, the researcher must make all efforts to have the minimum restriction on the curtailment of liberty in accordance with the principle of least infringement [3].

In addition to the intended benefits, interventions may have certain undesirable adverse effects on individuals and groups. The principle of "primum non nocere" has been modified regarding the risk–benefit ratio while planning to test and implement a public health intervention. Thus evolved the principle of proportionality, wherein public health researchers must weigh the risks and benefits of a particular intervention at the individual and community levels. The community versus individual liberty curtailment, breach of autonomy, and denial of privacy need to be balanced against the potential benefits it can have for the community. For example, quarantining the close contacts of COVID-19 patients thereby restricting their free movements had a greater public interest at large, i.e., preventing the spread of the disease and thus reducing the morbidity and mortality among vulnerable populations.

Principle of social justice

Social justice is the bedrock that provides stability to an unequal society. Although a political science concept, social justice is one of the principles of public health practice which is a confluence of multiple domains that influence human health [5]. The United Nations calls social justice the basic principle for "peaceful and prosperous coexistence within and among nations" [6]. In essence, it talks about the distributive justice and recognition of special groups regarding the benefits and burden of public health research among various groups. This distribution must be done equitably between the study groups. Thus, public health researchers should ensure that the risks posed by their research must not worsen the existing vulnerabilities and the social disadvantages already experienced by the group of people included in the study. Social justice principle prohibits unscrupulous exploitation of the vulnerable population such as people with poverty, poor literacy, and others as part of public health research.

Principle of reciprocity

Reciprocity is "a state or relationship in which there is mutual action, influence, giving and taking, correspondence, etc., between two parties or things" [7]. In health, people and communities must be provided additional benefits for participating in research. The benefit can be preferential treatment in subsequent healthcare access, access to food, shelter, and monetary compensation for wages lost and injuries borne. For example, at the end of the vaccine trial, priority and free-of-cost vaccination must be given (provided the vaccine is effective) to the eligible participants who were included in the placebo arm of the vaccine trial.

Principle of solidarity

Solidarity is a community concept wherein any issues—risks and benefits—are seen as a joint and common concern between the community members [8]. Public health researchers can leverage this principle of solidarity to promote research that focuses on the collective welfare and common good, despite risks at the individual level. They must respect and strengthen the existing relationship between individuals and communities where the research is undertaken [3].

Principle of accountability and transparency [3]

Transparent reporting of the purpose of the research, the risks and benefits, and conflicts of interest to the community is needed for making an informed participation decision. Only a completely informed person/community can exercise their autonomy to the fullest extent in the research. Researchers should take responsibility and accountability for the actions and adverse events related to the research. Further, the researcher must inform the community of the potential adverse events and the remedies and alternatives available to the community. The research shall be fair and honest for all intentions and purposes. These measures improve the trust the community under study places with the researchers and research, enhancing participation and completion rates.

Research methodologies in public health research

Adhering to the above-discussed principles of public health research, researchers should conduct the studies using standard methods. The researcher decides the study methodology based on the research question/objective, available time and resources, and the capacity of the research team.

Research question

The "Research Question" is the primary driver of any study. A well-stated research question has the potential to solve half of the problem. Hence, public health researchers should spend adequate time identifying and framing appropriate and relevant research question/s. A systematic literature review on the topic of interest facilitates further refining the research question/s. A research question must contain feasible, interesting, novel, ethical and relevant features within it [9]. Of the components of a research question, various formats such as PICO (T), PECOS, and SPICE have been suggested [10]. PICO (T) is a simple and widely used format to phrase each component of a research question. The study **P**opulation, the **I**ntervention or the exposure of **I**nterest being tested/studied, the **C**omparator for the intervention tested or exposure of interest, and the main **O**utcome of interest. In some study designs, the **T**imeline or duration of the study is included. Not all components of PICO will be part of all types of research questions which primarily depend on the context of the question and linked study design. For example, in a descriptive cross-sectional study, the research question should have at least the study population and the outcome. In case of randomized control trial, it will have all PICO components. Some example research questions are given in Table 2.1.

Aims and objectives

The aim is a broad statement of the outcome/s of the research phrased in a statement format. Objectives are a more specific and structured form of the research question phrased in an actionable and practical format. The acronym 'SMART' describes the essential characteristics of any objective. The objective should be very **S**pecific on the outcome, population, and intervention/exposures. It should be **M**easurable, objectively, to quantify the exposure and or outcomes, **A**chievable—which includes the feasibility of the research in the given research environment, **R**ealistic with the available resources, and within a specified **T**imeline [11].

TABLE 2.1 Example research questions and their PICO components.

Sn	Research question	PICO components	Linked study design
1	What is the prevalence of tuberculosis among under-five children in Chandigarh (India)?	Population: Under-five children from Chandigarh Intervention: Not applicable Comparator: Not applicable Outcome: Prevalence of tuberculosis	Descriptive, cross-sectional study
2	What are the dental caries rates among adolescents (13—19 years) from a rural community in India who use underground water and river water for drinking?	Population: Adolescents from a selected rural community in India Intervention/exposure of interest: Underground water Comparator: River water Outcome: Dental caries rates	Case-control or cohort study
3	What is the effect of digital and conventional (paper-based) self-management educational intervention on the quality of life of patients with newly diagnosed type 2 diabetes mellitus in India?	Population: People with type 2 diabetes mellitus Intervention: Digital self-management educational intervention Comparator: Paper-based self-management educational intervention Outcome: Quality of life	Randomized control trial

A few examples of objectives used in a public health research:

1. To assess the prevalence of tuberculosis among children under-five attending the outpatient departments of selected urban primary health-care centers of Chandigarh.
2. To assess the effect of digital and conventional (paper-based) self-management educational intervention on the quality of life of patients with newly diagnosed type 2 diabetes mellitus from selected urban primary healthcare centers of Chandigarh (India)

Study designs

Public health research uses various epidemiological study designs. The type of study design principally depends on the nature of the research question. It can be classified using different methods, like quantitative versus qualitative;

observational versus experimental; study unit: individual versus communities. The summary of the various study designs is depicted in Fig. 2.1.

Quantitative versus qualitative

Broadly, the research methods can be classified into quantitative and qualitative research. Quantitative research quantifies the health or the disease and tests the hypotheses or assumptions. In contrast, qualitative research tends to explore the process and the reason that underlie the relationship between the various health-related factors [12]. It enables the researchers to understand the social and behavioral issues linked with health status. Hence, quantitative research traditionally answers the 'What' and 'How' part of the quantification of the research question, and qualitative research especially answers the 'Why' part of the research question [13]. Quantitative research objectively measures the exposures or outcomes using validated tools, while qualitative research uses subjective, open-ended questions to capture the phenomenon. For example, the coverage of COVID-19 vaccine in a community is assessed using a quantitative method. However, why a certain proportion of people did not take the vaccine can be explored through qualitative methods. Given these variations, are these methods contradictory to each other? No. These study methods are, either independently or integrated, used in different scenarios based on the research question. Integration (data collection, analysis, or interpretation of the results) of these two main research methods is called a

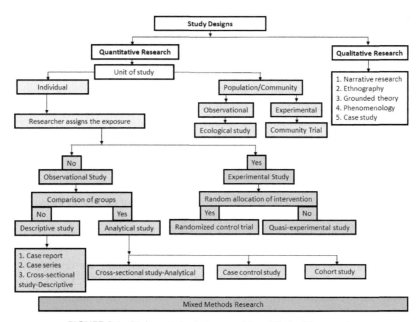

FIGURE 2.1 Various study designs used in public health research.

"Mixed Methods" study design. The mixed methods studies provide information about the research question more comprehensively than either research method alone [14]. In essence, quantitative and qualitative methods are not at odds but two sides of the same coin.

Observational versus experimental study designs

Observational and experimental study designs are the two categories of quantitative research. It is based on natural assignment or researcher's assignment of exposures to participants. In observational studies, the researchers assess the exposures or outcomes among individuals in their natural environment and relationships [15]. Descriptive and analytical study designs are the subtypes of observational studies. Case reports, case series, and descriptive, cross-sectional study (prevalence studies) designs are part of descriptive study design. Similarly, analytical cross-sectional studies, case-control, and cohort studies are analytical studies since they compare one group with the other. The direction of measurement of exposure and outcome decides the analytical study designs. In case-control study, the researcher starts with the outcome or disease, assesses the exposure, and measures the odds ratio (ratio between odds of exposure among people with disease and odds of exposure among people without the disease). However, the researcher moves from exposure to outcome in cohort study and measures the risk of disease among the exposed (risk ratio or relative risk). In an analytical, cross-sectional study, prevalence odds ratio or prevalence ratio is calculated. Two more association measures measured by cohort study are attributable risk (AR) and population-attributable risk (PAR). The risk ratio is pertinent and more of academic interest, whereas AR and PAR imply the policy and practice levels. AR provides information on the reduction in the percentage of acquiring the problem, provided the elimination of the exposure. Similarly, PAR provides information about the number and proportion of saved in the population if exposure is eliminated and is a useful measure for policymakers.

In experimental studies, the researcher assigns the exposure or intervention randomly (randomized control trial) or nonrandomly (quasiexperimental study). People with the disease are provided interventions in traditional experimental studies. However, healthy people are provided intervention to prevent the disease occurrence in field trials (preventive trials). The community trial is a subset of a field trial where the entire community is the unit of study instead of the individual. Notably, all experimental studies must be registered with a trial registry to facilitate evidence-based public health practice.

In addition to identifying the study designs, determining the discipline of the research methods used in public health research is essential, for example, operational research (OR), implementation research (IR), health systems research (HSR), and costing/economic analysis. Operational research is "*any research producing practically useable knowledge (evidence, findings, information, etc.) which can improve program implementation (e.g., effectiveness, efficiency, quality, access, scale up, sustainability) regardless of the type of research (design, methodology, approach) falls within the boundaries of operations research"* [16]. Use of existing program generated data to identify the operational issues and provide solutions for a public health program is the classical trademark of any OR. Though implementation and HSR deal with a broader policy level compared to OR, there is another school of thought that all the three research methods, i.e., OR, IR and HSR, are the same. Further, cost-effectiveness analysis studies are an integral part of public health research, influencing the policy and practice at all levels of healthcare system.

Sample size and sampling technique

Sample size and sampling are crucial parts of the study methodology, which decides the validity of the findings and their generalizability to the population. Based on the primary outcome of the study, the sample size is calculated to detect the measure of interest with adequate power and confidence. Based on the type of primary outcome variable, the sample size calculation broadly falls into three domains, namely, (a) single proportion or mean or (b) compare two (or more) proportions or means, and (c) compare matched groups [17]. For qualitative research, information saturation primarily decides the final sample size.

Probability and nonprobability sampling are the two important methods of sampling. Probability sampling, otherwise known as random sampling, is predominantly used in quantitative research. Simple random sampling, systematic random sampling, stratified random sampling, and cluster random sampling are part of probability sampling. Nonprobability sampling techniques include convenience, purposive, quota, and snow-ball sampling that are predominantly used in qualitative research.

Outcomes and its assessment

Outcome assessment is an essential step in public health research. The outcome should be measurable, quantifiable, and standardized for comparison. The researcher must identify the primary and secondary outcomes wherever applicable and provide operational definitions, including its measurement. Biases during outcome measurements can be avoided by strictly following the data collection/capture, entry, analysis, interpretation, and reporting protocols.

Standards for conducting research using various study designs

Study design—specific checklists are available to ensure the presence of essential elements to conduct research (Table 2.2) [15].

TABLE 2.2 Various study designs employed in public health research and standard checklists.

Study design	Checklists
Observational (cross-sectional, case-control and cohort)	STROBE (STrengthening the Reporting of OBservational studies in Epidemiology)
Randomized control trials	CONSORT (CONsolidated Standards Of Reporting Trials)
Systematic review and metaanalysis	PRISMA (Preferred Reporting Items for Systematic reviews and Meta-Analyses)
Qualitative research	COREQ (COnsolidated criteria for REporting Qualitative research) ENTREQ (ENhancing Transparency in REporting the synthesis of Qualitative research)
Economic evaluations	CHEERS (Consolidated Health Economic Evaluation Reporting Standards)

Summary of the steps in conducting public health research

Summarizing the points discussed above, the essential steps to conduct research in public health are as follows [18]:

- Identifying and formulating a public health problem, i.e., research question
- Review of literature and refinement of the research question
- Formulating the hypothesis and objectives
- Defining the outcomes and methods of assessment
- Adopting appropriate study design
- Sample size and sampling technique
- Planning the analysis
- Pretesting or pilot study
- Implementation and data collection
- Data management: cleaning and analysis
- Interpretation and reporting

Public health ethics

Ethics is one of the pillars of research on human beings. Public health research should be scientifically sound and ethically approved before its implementation among the target population. The principles of public health research discussed in the initial parts of the chapter explain some of the underlying values for ethics in public health research. Key components of ethics in public health research are discussed below.

Risk–benefit analysis

One of the fundamental principles of public health research is to assess and weigh the risks and benefits of a particular intervention. The risks for the community/individual can be physical, mental, or social. The rational determination of the risks of a specific intervention must be done based on the best available evidence and compared with the expected benefits from the intervention/procedure of the research. Further, the researcher must understand that the risk can extend beyond the study period and the community/participants. For example, an adverse physical reaction from an intervention may have direct physical and mental effects on him. It will also have social effects on their family members, owing to the loss of productivity and activity. Thus, the risk profile and its environment must be established and assessed comprehensively. Similarly, the study may not directly benefit the participants. However, it may benefit nonparticipants in the future.

Risks must be assessed context-specific. Risks for the same intervention vary between age groups, social class, race, and community. No or minimal risk for adults can be a risk in the pediatric population, even if benefits are equal. Prediction and quantification of risks and benefits is a challenging task. The researcher must estimate it using the best available evidence and establish monitoring systems to quantify the predicted and identify the new adverse effects during research. Ethical or data safety and monitoring boards must review the interim reports and conduct an independent assessment of risk and benefit [2].

Informed consent

Informed consent forms the bedrock for human involvement in public health research. Considering the unit of research in public health, which in most cases are the community/population, the process of informed consent differs from clinical research. Study designs like cluster randomized trials, community trials, and IR interventions deal with the whole community. In this scenario, a two-level, hierarchical informed consent, i.e., at first level with the community leaders or gatekeepers and at second level with the individuals of the community must be included [3]. In follow-up studies, reconsenting is essential whenever there is a change in protocol in the form of a new intervention or new data are required for the research or new information concerning the safety of the participants emerges. Effective communication and consultative mechanism with the community and members is essential for a meaningful and high recruitment rate of communities/individuals in public health research [2]. Community consultation is essential to achieve four ethical goals, namely enhanced protection, enhanced benefits, legitimacy, and shared responsibility [19].

Written informed consent is the norm. The ethics committee (EC) may allow verbal consent depending on the research context. It may waive the informed consent in case of analysis of secondary data of an existing program. Importantly, it has the duty and prerogative to consider and accord such waiver of consent, or otherwise, for public health research [3].

Protection of vulnerable population

Not all communities and populations have the same characteristics and vulnerabilities. Vulnerabilities to the anticipated risks and adverse outcomes vary according to the education levels, economic status, social status, race, ethnicity, social class, gender, age, sexual orientation, disease/disability states, etc. The vulnerabilities also make them potentially exploitable. Specific protection measures to ensure the rights and welfare of the vulnerable population must be outlined in the research protocol and approved by the EC. Not just inclusion, unjust exclusion of the vulnerable population needs great attention while conducting research. Such exclusion, without strong rationales, such as increased risks and reduced direct benefits to the vulnerable group, violates the principle of justice [2].

Conflict of interest

Conflict of Interest (COI) is "a set of conditions whereby professional judgment concerning a primary interest, such as participant's welfare or the validity of research, either is or perceived to be unduly influenced by a secondary interest." The secondary interests are financial, personal, academic, or political benefits [3]. Funding mechanism for research and the collaborator's role in the research are potential factors that compromise the research integrity. An example would be funding or involvement of the tobacco industry in any health research or soft drinks manufacturers in obesity research [2]. Though their participation per se may not be wrong, the secondary interests could potentially influence the research questions, data collection methods, and other aspects of the research till the dissemination of results. Standard operating procedures must be established to identify, mitigate, and manage the existing and potential conflicts of interest of the researchers. Disclosing all conflicts of interest in the protocol and research publications is one way of addressing the COI [3].

Funding for public health research

Financial support is essential for any research, and more so for public health research. Multiple government and nongovernment organizations are providing funding opportunities for public health research. Internationally, the

National Institutes of Health is the largest funder of health research. Others include the European Commission, the United Kingdom Medical Research Council Bill, Wellcome Trust, and the Bill and Melinda Gates Foundation [20]. In India, the Indian Council of Medical Research, the nodal body for health research, is the principal funding agency. The agencies provide grants mostly on prioritized and focused areas (NCDs, Malaria, HIV, and others) of public health. However, the researcher should understand and confirm the COI of the funding agency before accepting the grant. Further, not all research needs funding. For example, OR based on existing programmatic data may not require funding, and one can execute it using existing resources.

Role of Ethics Committees in public health research

ECs play a decisive role in public health research. Crucial issues of assessing the risk—benefit ratios, deciding the waiver of consents, type of consents, reviewing the COI, reviewing, approving and ensuring the safety protocols of the study participants, and adequacy of a protection mechanism for the vulnerable population are the primary responsibilities of the ECs. Clearance from the local ethical committee is important since the local committee understands the existing psychosocial and biological environment more than a distant ethical committee.

Dissemination of results

Irrespective of the type of study results (positive or negative), public health researchers must disseminate it to individuals and communities in addition to policy makers, program managers, and research academia. Further, the researcher must ensure the confidentiality of individual identifiers and of vulnerable communities during dissemination. In certain studies, the participants will be informed of the results if they are significant. For example, anonymous testing is conducted in HIV sentinel surveillance, wherein the researcher must inform the participants if their sample is positive for HIV. However, it is an ethical obligation in the above scenario.

The choice of dissemination of the results at the individual and community level needs to be 'offered' and not routinely 'informed.' The participants can be offered a choice to be informed or not-informed of their results [2]. Thus, data dissemination requires much deeper contemplation and consultation at the individual level.

Principles of public health writing

Public health writing is the art of effectively communicating public health research findings and practices to various target groups. One of the major challenges in research-practice gap is convincing the policy makers, political, and institute leaders to adopt the research evidence. This can be addressed by means of effective communication of the proven and potential impact of the tested interventions. Public health writing is a powerful tool to present the evidence in a simplified manner with various subgroups.

Writing for policy/advocacy—Policy briefs

A policy brief is a summary document that lists the problem and its magnitude, the rationale of the intervention, and its effectiveness and the actionable point. The primary target audience for the policy briefs is bureaucrats, ministers of government, political leaders, and healthcare philanthropists. Evidence from economic evaluations is of great value in public health intervention policy briefs. The policy briefs empower healthcare advocates and activists and garner the support of the media to disseminate it to gain public support for the program. Policy briefs help secure funding and implement the proposed public health intervention in routine [21]. The policy brief should not exceed one page and should be better presented as infographics. It should identify the problem and the proposed intervention to deal with the problem, available alternatives, how the proposed intervention is better than the alternatives, approximate cost of implementation and expected savings, and a final paragraph on the key points favoring the intervention to address the problem. The policy brief must be free from scientific/research jargon.

Writing for scientific journal or organization—Research publications

Writing in a scientific journal is the most widely known and practiced form of writing undertaken by public health researchers. The research articles in public health can be from any of the following types: Original articles based on the primary data or secondary data analysis, reviews (narrative or systematic review with or without meta-analysis), editorials, viewpoints, letter to the editor, case studies, commentaries, infographics, photographs and others.

For example, the anatomy of an original article and the descriptions are given below:

Abstract: It is a structured/unstructured summary of the problem studied, the objectives of the study, brief methodology, key findings and conclusion, which usually ranges between 150 and 300 words. The end of the abstract is generally linked with five to six relevant keywords (MeSH terms).

Introduction or Background: It should briefly describe the burden of the public health problem studied, along with the latest evidence available on the topic of study and the existing lacunae. Concisely, it is the justification or rationale of any research. The introduction should end with the objective/s of the study, stated in clear terms. The usual length of an introduction is between 400 and 750 words.

Material and Methods: It should have a clear description of (a) study design, (b) study setting, (c) study population-inclusion/exclusion criteria, (d) sample size and sampling technique, (e) description of interventions/controls (if it is an intervention study), exposures (if observational), (f) study procedure-data collection and tools used, (g) the specific outcomes of the study, (h) statistical analysis, and (i) ethics. The methods section should be transparent enough to replicate the study by other researchers in their settings.

Results: This section should enumerate the characteristics of the study participants, report the absolute and relative measures of the outcomes, the strength of association between the exposure/intervention and outcomes as applicable. Tables, graphs, and flow charts can be used in this section to convey the results. However, the result text should not repeat the findings mentioned in the table. Instead, it should provide the key findings. The citation of all illustrations in the text is important.

Discussion: In this part, the author should discuss (a) summarize the key findings; (b) reasons or speculation for key findings and its comparison with relevant literature; (c) the strength of the study; (d) the limitations of the study; and (e) policy and programmatic implications ± conclusion.

References: It is imperative to acknowledge all the referred sources for conducting the research. The references are cited in the text as per the journal style and listed at the end of the article. Electronic reference manager software such as Mendeley, Endnote, Zotero, etc., are available for easing the referencing and citation in public health writing.

The journals' specific guideline decides the place for mentioning the COI, funding source, and ethical/regulatory approval status (beginning or end of the article). The type of documents varies from journal to journal, which the researchers must confirm with the journal guideline before submission. Generally, the editorial board of the journal assesses the submitted article's suitability with the journal's scope and decides the next step. The majority of scientific journals are peer-reviewed (blinded or unblinded).

Writing for the population at large

Public health concerns population health, and it is vital to communicate evidence-based public health practices to the target population. Public health writing aimed at the general and target population for the concerned public health intervention improves their acceptability. Opinion pieces in popular print and social media, blogs, and commentaries are a few important channels to disseminate public health practices to the population. Such public health writing must be simple, concise, interesting, context-specific, and engage the population to read, understand, and inculcate in practice. The writing can start with a case study to highlight the problem at hand, the overall burden of the issue, available solutions, the best available solution, and its pros and cons. The communication material must have anecdotes and analogies to which the target audience can relate and be devoid of scientific jargon. The use of popular media and communication in the local language are important strategies to reach the maximum population. The communication material should include the author's credentials to validate its authenticity and improve trust among the people.

Influence of public health research on public health policy or practice

Public health research must influence the current policy or practice on population health. Some research directly impacts the policy or practice, and others indirectly by observation of novel findings. The other way, public health professionals should review the available evidence while making policy decisions or practice changes. However, such evidence-based public health practice is low across various settings. Only 58% of the health programs in the United States of America were found to run based on the latest evidence [22]. Thus, challenges and gaps exist in transferring the evidence from public health research to practice.

Some examples of public health research and its influence on policy decisions are provided below. The landmark Kangra Valley study, which established the effectiveness of iodization of salt to control goiter, has been one of India's earliest public health research projects that influenced the policy. The National Goitre Control Program was launched based on the findings from this study, and later India adopted universal salt iodization as the official policy to manage iodine deficiency disorders [23]. The iconic Framingham Heart Study is one of the longest, population-based cohort studies ever since 1948. Framingham is a town located in Massachusetts, USA. The study contributes to a newer understanding of cardiovascular disease risk from time to time, based on which diagnostic and treatment guidelines are revised [24]. Population-level research on the impact of tobacco on human health and inequities has been the greatest advocate for global and national tobacco control regulation and laws. Multiple studies on various public health aspects of epidemiology and tobacco control interventions in the last 70 years have made drastic changes in tobacco policy and practices of the politicians, policy makers, governments, public health practitioners, and the general public [25,26].

Conclusion

Evidence-based public health practice is the norm that is linked with the generation of evidence using public health research. Public health research must adopt a standard and robust methodology to precisely measure and report the findings. Public health professionals must have the skills to conduct research, collate the evidence, and communicate the findings to various target groups. The best available evidence should be identified and reviewed as part of evidence synthesis to make informed and cost-effective policy or practice decisions in public health.

References

[1] Public health leadership society principles of the ethical practice of public health. 2002.

[2] Barrett DH, Ortmann LW, Brown N, DeCausey BR, Saenz C, Dawson A. Public health research. In: Public health ethics: cases spanning the globe; 2016. p. 285−318.

[3] Mathur R. National ethical guidelines for biomedical and health research involving human participants. 2017. New Delhi.

[4] Beauchamp TL, Childress JF. Principles of biomedical ethics. Edicoes Loyola; 1994.

[5] Wallack L. Building a social justice narrative for public health. Health Educ Behav 2019;46:901−4. https://doi.org/10.1177/1090198119867123.

[6] World day of social justice 2020. DISD; n.d. https://www.un.org/development/desa/dspd/international-days/world-day-of-social-justice/2020-2.html. (Accessed 20 November 2022).

[7] The principle of reciprocity: how can it inform public health and healthy public policies? National Collaborating Centre for Healthy Public Policy; n.d. https://ccnpps-ncchpp.ca/the-principle-of-reciprocity-how-can-it-inform-public-health-and-healthy-public-policies/. (Accessed 20 November 2022).

[8] Dawson A, Jennings B. The place of solidarity in public health ethics. Publ Health Rev 2012;34:1−15. https://doi.org/10.1007/BF03391656.

[9] Hulley SB. Designing clinical research. Lippincott Williams & Wilkins; 2007.

[10] Booth A, Noyes J, Flemming K, Moore G, Tunçalp Ö, Shakibazadeh E. Research guides: public health: research questions. BMJ Glob Health 2019;4. https://doi.org/10.1136/BMJGH-2018-001107.

[11] Develop SMART objectives. n.d. https://www.cdc.gov/publichealthgateway/phcommunities/resourcekit/evaluate/develop-smart-objectives.html. (Accessed 20 November 2022).

[12] Quantitative and qualitative methods for public health. n.d. https://sphweb.bumc.bu.edu/otlt/MPH-Modules/PH717-QuantCore/PH717-Module1B-DescriptiveStudies_and_Statistics/PH717-Module1B-DescriptiveStudies_and_Statistics9.html. (Accessed 20 November 2022).

[13] Isaacs A. An overview of qualitative research methodology for public health researchers. Int J Med Public Health 2014;4:318. https://doi.org/10.4103/2230-8598.144055.

[14] Shorten A, Smith J. Mixed methods research: expanding the evidence base. Evid Based Nurs 2017;20:74−5. https://doi.org/10.1136/eb-2017-102699.

[15] Thiese MS. Observational and interventional study design types; an overview. Biochem Med (Zagreb) 2014;24:199. https://doi.org/10.11613/BM.2014.022.

[16] Malhotra S, Zodpey SP. Operations research in public health. Indian J Public Health 2010;54:145. https://doi.org/10.4103/0019-557X.75737.

[17] Andrade C. Sample size and its importance in research. Indian J Psychol Med 2020;42:102. https://doi.org/10.4103/IJPSYM.IJPSYM_504_19.

[18] Rosenstock IM, Hochbaum GM. Some principles of research design in public health. Am J Public Health 2010;100:1861. https://doi.org/10.2105/AJPH.100.10.1861.

[19] Dickert N, Sugarman J. Ethical goals of community consultation in research. Am J Public Health 2005;95:1123−7. https://doi.org/10.2105/AJPH.2004.058933.

[20] Viergever RF, Hendriks TCC. The 10 largest public and philanthropic funders of health research in the world: what they fund and how they distribute their funds. Health Res Policy Syst 2016;14:12. https://doi.org/10.1186/s12961-015-0074-z.

[21] Public health writing guide. n.d. https://publichealth.charlotte.edu/sites/publichealth.charlotte.edu/files/media.

[22] Dreisinger M, Leet TL, Baker EA, Gillespie KN, Haas B, Brownson RC. Improving the public health workforce: evaluation of a training course to enhance evidence-based decision making. J Public Health Manag Pract 2008;14:138−43. https://doi.org/10.1097/01.PHH.0000311891.73078.50.

[23] Pandav CS, Yadav K, Srivastava R, Pandav R, Karmarkar MG. Iodine deficiency disorders (IDD) control in India. Indian J Med Res 2013;138:418.

[24] Mahmood SS, Levy D, Vasan RS, Wang TJ. The Framingham Heart Study and the epidemiology of cardiovascular diseases: a historical perspective. Lancet 2014;383:999. https://doi.org/10.1016/S0140-6736(13)61752-3.

[25] Thomas S, Fayter D, Misso K, Ogilvie D, Petticrew M, Sowden A, et al. Population tobacco control interventions and their effects on social inequalities in smoking: systematic review. Tob Control 2008;17:230. https://doi.org/10.1136/TC.2007.023911.

[26] Warner KE. The role of research in international tobacco control. Am J Public Health 2005;95:976. https://doi.org/10.2105/AJPH.2004.046904.

Chapter 3

Evidence-based public health practice

Rizwan Suliankatchi Abdulkader
Indian Council of Medical Research-National Institute of Epidemiology (ICMR-NIE), Chennai, Tamil Nadu, India

Evolution of EBPH

The 1970s and 80s saw a significant change in how medicine was practiced in the western world. The new field of Evidence-Based Medicine (EBM) was quickly becoming the standard for delivering health to individual patients. Even in the face of criticism that it downplayed the art of medicine and physician expertise and tried to formalize all treatment decisions, EBM stood the test of time by denouncing ineffective and sometimes outright harmful medical practices. The success of EBM led to its gradual adoption by other fields, such as psychology, nursing, dentistry, nutrition, and others. It was only a matter of time before public health also adopted evidence-based approaches to implementing public health interventions. Since then, Evidence-Based Public Health (EBPH) has become a common vocabulary in the teaching and practice of public health [1].

Definition of EBPH

The broad concepts behind the idea of EBPH originate from the defining pillars of EBM, i.e., the use of science to decide effective interventions and community participation in decision-making. While EBM is applied to individual patients in the clinical setting, EBPH is applied to communities and whole populations. Although different definitions of EBPH exist in the literature, they all have one thing in common: emphasizing the application of science-based evidence to improve population health. The most widely used definition given by Brownson et al. states that "EBPH is the development, implementation, and evaluation of effective programs and policies in public health through the application of principles of scientific reasoning, including systematic uses of data and information systems and appropriate use of program planning models" [2]. This definition lacked the inclusion of the

Principles and Application of Evidence-Based Public Health Practice
https://doi.org/10.1016/B978-0-323-95356-6.00015-X

41

community in the decision-making process. So Kohatsu and colleagues gave an updated definition overcoming this, stating that "EBPH is the process of integrating science-based interventions with community preferences to improve the health of populations" [3]. No matter what definition we use, the underlying principles of using established scientific evidence in conjunction with community values and preferences for improving population health remain paramount. This underpinning is probably best exemplified by the illustration of the domains of decision-making in EBPH given by Satterfield et al. [4]. These domains include (a) best available research evidence, (b) taking into account population characteristics like needs, values, beliefs, and preferences, (c) available resources and expertise, and (d) within the context of the environment and organization (Fig. 3.1).

A point to note here is that we need to have the best available evidence, not the best one. Unlike in clinical medicine, randomized trials for the effectiveness of many interventions may not be readily available in public health. In such cases, the best available evidence must come from observational studies; more often than not, there may be no studies. This limitation will be ever-present in public health, probably best exemplified by the public health responses initiated during the COVID-19 pandemic, when very little was known about the agent or the disease. However, the resilience of public health systems lies in modifying our strategies as and when more evidence becomes available and constantly evolving as the population needs change.

FIGURE 3.1 Domains of evidence-based public health. *Adapted from Satterfield JM, Spring B, Brownson RC, Mullen EJ, Newhouse RP, Walker BB, et al. Toward a transdisciplinary model of evidence-based practice. Milbank Q 2009;87:368—90. https://doi.org/10.1111/j.1468-0009.2009. 00561.x.*

Why should EBPH be practiced?

In real-world settings, interventions to improve population health are often taken without systematic scientific processes: political whims, populist pressures, lobbying, and urgency guide which interventions get funded and implemented. The danger of using such processes is that these interventions may fail to produce the desired effect or may even harm the communities. In addition, they also carry the opportunity cost of not implementing programs with known effectiveness. Examples of such programs include the school-based education programs for obesity and physical activity in the United States of America, the failure of the compulsory family planning programs, fear-based awareness campaigns for control of HIV/AIDS, and the public toilets campaign to end open-defecation in India. On the other hand, we can see how EBPH-based interventions have helped population health. One of the striking examples is the vaccination programs to prevent infectious diseases in children. With the strong support of scientific evidence, childhood vaccination programs have prevented the deaths of 36 million under-five children in the past two decades in low and middle-income countries (LMICs) [5]. Other success stories include the control of HIV/AIDS through improved public awareness campaigns, increased uptake of barrier contraception among high-risk groups, and effective drug regimens like HAART for HIV control and prevention. Evidence for the effectiveness of these interventions is generated through carefully and systematically designed studies which provide the confidence required by the implementers.

The argument for using EBPH wherever and whenever possible is getting stronger. The alternative, to try what we "think" will work, has no place in modern public health practice. However, public health practice faces several challenges in implementing EBPH. Increasingly, funding institutions and the informed public require evidence before implementing public health programs. Therefore, it becomes imperative for policymakers to use the best available evidence for implementing population-level health programs.

One of the strongest arguments in favor of the EBPH movement is its ability to reduce health inequities across all levels of development. This consequence is even more relevant in LMICs, where a large percentage of the population depends on public health services to meet their health needs. When evidence is generated with an all-inclusive approach, focusing on the needs of those who occupy the bottom rungs of society, the programs based on them will be well equipped to reduce the health disparities between the 'haves' and 'have-nots' [6].

Levels of evidence in EBPH

Evidence in the field of public health comes in a variety of flavors. Unlike clinical settings where study designs for evidence fall into neat categories like randomized controlled trials and cohort or case-control studies, evidence for public health programs do not, and most often, cannot come from randomized controlled trials. In order of their merit, four different levels of evidence have been suggested in literature: (a) evidence-based, (b) effective, (c) promising, and (d) emerging. Evidence-based programs are time-tested and supported by systematic reviews of several independent studies from more than one group. If "evidence-based," the traditional hierarchy of evidence (used in EBM) can be applied for classification to grade the recommendations [7]. Effective programs are based on individual studies or technical reports that have undergone peer review. Promising programs arise from government reports or conference presentations that have not undergone formal peer review and can potentially be a low-cost and wide-reaching intervention. Emerging programs are in the pipeline/being conducted and seem to have face validity based on pilot studies or privately funded research.

Other classification types also exist. For example, the United States Community Preventive Services Task Force uses six domains to evaluate the strength of evidence [8]. These include execution, design suitability, number of studies, consistency, effect size, and expert opinion. Based on these, evidence is classified into strong and sufficient pieces of evidence (recommended), insufficient but supported by expert opinion (recommended based on expert opinion), insufficient (not recommended), and strong evidence of harm or ineffectiveness (discouraged).

Who should practice EBPH?

EBPH should be practiced by everyone who has a role in population health, including public health students, researchers, program managers, policy-makers, politicians, and local community leaders. Researchers are unique among these because they not only use EBPH but also generate the evidence required to practice EBPH. Studies have shown that less than half of the personnel involved in public health are likely to have formal training or degree. Therefore, principles of EBPH must be taught through in-service training and continuous skill development programs for public health personnel.

Tools that are required to generate and implement EBPH

There are several tools at the disposal of practitioners of EBPH [9]. Systematic reviews and metaanalyses are the workhorses of the EBPH movement. Other sources of evidence include public health surveillance data, evidence-based guidelines by public health organizations, health technology assessments, health impact assessments, economic evaluations, quasi-experimental studies such as interrupted time series, before-after studies, and, last but not least, community-based participatory research methods.

Systematic reviews can quickly produce much-needed evidence on public health programs and interventions. Specifically, many international organizations, such as the Campbell Collaboration, the Joanna Briggs Institute, and the Cochrane Collaboration, have been set up for this sole purpose.

Framework to implement EBPH

Brownson and colleagues have devised a seven-stage nonlinear process which guides a user on how to implement EBPH [10]. The process mirrors EBM in that both begin with asking a well-framed research question to address.

Formally, the steps include (1) community assessment, (2) quantifying the issue, (3) developing a concise statement of the issue, (4) determining what is known through the literature, i.e., literature review, (5) developing and prioritizing program and policy options, (6) developing an action plan and implementing the interventions, and finally, (7) evaluating the program or policy and circling back to community assessment.

Practically, one should begin with a clear and focused question to answer. Then, conduct a careful and systematic literature review to see what has been done and summarize the information into actionable programs. While looking for different types of studies, we need to be aware of the types and hierarchies within. There are three types of evidence to consider: (1) burden of disease or etiological studies, (2) studies comparing the effectiveness of interventions, and (3) translation or implementation research conducted in real-world settings. After considering the type of evidence, one should proceed to assess the level of evidence as discussed above. Then the interventions are ranked from best to least suited. A community needs assessment with local stakeholders, community members, and leaders should be conducted to ensure that community beliefs and preferences are considered before planning the implementation. Once a program has been finalized for implementation, it must be customized to the community by considering community needs. This part is usually done by a community-based participatory approach, where the community members provide inputs at all levels of the research activity. The program should be implemented with a built-in monitoring and evaluation framework, in order to provide constant feedback to the program managers and improve the program in the long run. The program must be flexible to adapt to the population's changing needs and the problem in question.

In case no context-specific programs can be found in the literature, one must think of devising a primary implementation or operational research. Once field-tested, this program can be put through the same steps described above. However, one thing to keep in mind is that since the program managers are the ones who will be designing the implementation research, they can start involving the community early in the decision-making process.

Let us discuss two real-world case studies highlighting the various principles that led to the failure/success of public health programs focused on drug abuse.

Case study 1

In the United States, drug addiction became a major health problem in the mid-1900s. In response to this problem, the Comprehensive Drug Abuse Prevention and Control Act of 1970 and Comprehensive Crime Control Act of 1984 were passed by the US congress to control the availability/prohibition of drug substances and criminalize the possession of controlled substances. Drug abuse was considered "public enemy number one." These policies, along with others, lead to the mass incarceration of a large proportion of young African-American males, huge cost to the taxpayer for maintaining the prisons, proliferation and easy availability of more dangerous drugs in the streets, and increasing burden of drug addiction, drug overdosages, and related deaths. This "war on drugs" leading to the implementation of a program for five decades that was unsuccessful can be considered a massive failure in the comprehensive understanding of a public health problem [11].

Case study 2

Portugal dealt with the drug abuse epidemic using decriminalization and a public health reorientation approach. In this approach, "victims" were referred to a "dissuasion board" where qualified members decided on what further action was needed. The possible options could be no action, referral to treatment, or a fine. Many studies have documented the effects of the Portuguese decriminalized approach on the drug abuse epidemic. The effects are more positive, where there was a reduction in illicit drug use among users and adolescents, reduced number of offenders in the criminal justice system, reduced amounts of drugs seized by enforcement authorities, and reduced drug prices. In short, this policy of harm reduction as a solution to tackle the drug problem has been described as one that is based on "humanism, pragmatism, and participation" [12].

From these two case studies, we can see that for similar public health problems, diagonally opposite approaches have been implemented with marked differences in effectiveness. Sometimes, even with evidence of no effectiveness or outright harm, it might be difficult to change course, as is currently evident in the case of the "US War on Drugs" approach.

Challenges in the adoption and implementation of EBPH

As alluded to earlier in the chapter, EBPH has far greater challenges in implementation when compared to EBM.

First and foremost is the lack of scientifically rigorous randomized controlled studies on which EBM guidelines usually rely. Public health interventions do not easily lend themselves to be evaluated by RCTs due to cost, ethical considerations, or time required to establish effectiveness. However, they rely on cross-sectional studies, quasi-experimental studies like before—after studies, trials with historical controls, or natural experiments (like implementing a new TB diagnostic program in a community) with posthoc evaluations. And these studies are not always available for every intervention that we might wish to implement. Even when available, they may not have the context in which we wish to implement them. Much emphasis has been placed on rigorous evaluations of public health interventions which come with their levels of uncertainty, not to mention falling for the trap of inferring causation from an association in such designs. Secondly, the programs we wish to implement for the larger public good do not form a singular intervention but rather a package of strategies. Evaluating such a complex set of strategies poses a unique set of challenges especially in assessing what worked and what did not work. Thirdly, the public health workforce comprises professionals from diverse backgrounds. Though there are merits to having a wide range of expertise in the public health team, they do not tend to have uniform training in the public health discipline. This diversity poses challenges in how each member with a different background perceives the proposed program and decides whether it is worth implementing or not. Fourthly, unlike EBM, where the decision affects a few patients in a clinical setting, EBPH has the potential to change the outcomes for entire communities. So, the risk of implementing an ineffective or harmful program looms large in the minds of the implementer. Finally, before any intervention or program is rolled out under the umbrella of EBPH, the community must be fully engaged in the decision-making process. Communities being diverse and representing the views of various sections of society in a program may become too complex and untenable. Nevertheless, we do see programs being introduced with very little community involvement or stakeholder consensus.

Future of EBPH

Building on what is available now, EBPH will likely benefit greatly from the explosion of data and information available for various public health issues. As novel technologies such as artificial intelligence and its components like machine learning, deep learning, and others in data processing and analysis emerge, more can be done to answer prevailing questions of what and how public health interventions work. There is an increasing demand for future programs to address health disparities. As more studies become available globally, policymakers will benefit from the free and open sharing of programmatic experiences, data, and technical reports. Adapting to local contexts can be facilitated by sharing the technical know-how in open-source platforms as a mandate of publicly funded research. EBPH, when implemented across all walks of public health management, will ensure transparency in the decision-making process, enforce accountability, and ensure that the public good is always at the forefront. Every member of the public health workforce should be trained in the essentials of EBPH, regardless of their background, nature of work, level of authority, and role in program implementation.

References

[1]   Sheridan DJ, Julian DG. Achievements and limitations of evidence-based medicine. J Am Coll Cardiol 2016;68:204−13. https://doi.org/10.1016/j.jacc.2016.03.600.

[2]   Brownson RC, Fielding JE, Maylahn CM. Evidence-based public health: a fundamental concept for public health practice. Annu Rev Public Health 2009;30:175−201. https://doi.org/10.1146/annurev.publhealth.031308.100134.

[3]   Kohatsu ND, Robinson JG, Torner JC. Evidence-based public health: an evolving concept. Am J Prev Med 2004;27:417−21. https://doi.org/10.1016/j.amepre.2004.07.019.

[4]   Satterfield JM, Spring B, Brownson RC, Mullen EJ, Newhouse RP, Walker BB, et al. Toward a transdisciplinary model of evidence-based practice. Milbank Q 2009;87:368−90. https://doi.org/10.1111/j.1468-0009.2009.00561.x.

[5]   Li X, Mukandavire C, Cucunubá ZM, Echeverria Londono S, Abbas K, Clapham HE, et al. Estimating the health impact of vaccination against ten pathogens in 98 low-income and middle-income countries from 2000 to 2030: a modelling study. Lancet 2021;397:398−408. https://doi.org/10.1016/S0140-6736(20)32657-X.

[6]   Brownson RC, Haire-Joshu D, Luke DA. Shaping the context of health: a review of environmental and policy approaches in the prevention of chronic diseases. Annu Rev Public Health 2006;27:341−70. https://doi.org/10.1146/annurev.publhealth.27.021405.102137.

[7]   Burns PB, Rohrich RJ, Chung KC. The levels of evidence and their role in evidence-based medicine. Plast Reconstr Surg 2011;128:305−10. https://doi.org/10.1097/PRS.0b013e318219c171.

[8]   U.S. Department of Health and Human Services. Evidence-based clinical and public health: generating and applying the evidence. In: Secretary's Advisory Committee on National Health Promotion and Disease Prevention Objectives for 2020; 2010. p. 32. https://health.gov/sites/default/files/2021-11/Committee%27s Report on Evidence-Based ClinicalandPublicHealth-GeneratingandApplyingtheEvidence.pdf.

[9]   Jacobs JA, Jones E, Gabella BA, Spring B, Brownson RC. Tools for implementing an evidence-based approach in public health practice. Prev Chronic Dis 2012;9:E116. https://doi.org/10.5888/pcd9.110324.

[10] Brownson RC, Baker EA, Deshpande AD, Gillespie KN. Evidence-based public health. 3rd ed. Oxford University Press; n.d.

[11] Transform Drug Policy Foundation. The war on drugs: options and alternatives. 2013. http://fileserver.idpc.net/library/The-war-on-drugs_Options-and-alternatives.pdf. [Accessed 6 November 2022].

[12] Moreira M, Hughes B, Storti CC, Zobel F. Drug policy profiles—Portugal. Drug Policy Profiles 2011:28. https://www.emcdda.europa.eu/attachements.cfm/att_137215_EN_PolicyProfile_Portugal_WEB_Final.pdf. [Accessed 6 November 2022].

Chapter 4

Applying evidence-based strategies for public health preparedness and emergency management

Vinayagamoorthy Kalaiselvi[1] and Jaya Prasad Tripathy[2]
[1]Centre for Public Health, Punjab University, Chandigarh, Punjab, India; [2]Department of Community Medicine, All India Institute of Medical Sciences (AIIMS), Nagpur, Maharashtra, India

Introduction

Public health practice includes delivering preventive, promotive, curative, and rehabilitative healthcare services at normal times and during emergencies. Though management of public health emergencies or disasters is integral to public health practice, it remains a neglected area across countries. Public health emergency and disaster terminologies are interchangeably used and nearly carry the same definitions. The growing threats of public health emergencies, especially due to globalization, need to be predicted and prevented in time.

Definition of public health emergency

Public health emergency is defined by New Jersey Emergency Health Powers Act (EHPA) in 2005 as
"An occurrence or imminent threat of an occurrence that:

(a) is caused or is reasonably believed to be caused by any of the following:
  (1) bioterrorism or an accidental release of one or more biological agents;
  (2) the appearance of a novel or previously controlled or eradicated biological agent;
  (3) a natural disaster;
  (4) a chemical attack or accidental release of toxic chemicals; or
  (5) a nuclear attack or nuclear accident; and

Principles and Application of Evidence-Based Public Health Practice
https://doi.org/10.1016/B978-0-323-95356-6.00004-5
49

(b) poses a high probability of any of the following harms:

    (1) a large number of deaths, illnesses, or injuries in the affected population;

    (2) a large number of serious or long-term impairments in the affected population; or

    (3) exposure to a biological agent or chemical that poses a significant risk of substantial future harm to a large number of people in the affected population" [1,2].

The World Health Organization's Revised International Health Regulations, adopted in 2005, define a public health emergency of international concern (PHEIC) as "[A]n extraordinary event which is determined ... to constitute a public health risk to other States through the international spread of disease and to potentially require a coordinated international response" [3]. According to the RAND Corporation, public health emergency is an event "whose scale, timing, or unpredictability threatens to overwhelm routine capabilities" [4,5]. In this background, the public health emergency is not the same for all types of conditions, people, settings, countries, and time periods.

Types of disasters [6]

Emergencies/disasters are broadly classified as follows:

1. Simple and compound (Table 4.1)
2. Natural and man-made (Table 4.2)

TABLE 4.1 Nature and the difference between simple and compound disasters [6].

Simple disasters	Compound disasters
A simple disaster does not destroy the community's physical infrastructure and has no impact on the adjacent communities. Therefore, the rescue operation can be carried out by the local and provincial resources.	A compound disaster destroys the structure of the community. In a compound disaster, the damage is heavy, including damage to vital physical infrastructure and fatalities; therefore, national and international organizations are involved in rescue operations.
e.g., A train accident involving a few passengers	e.g., An earthquake causing mass fatality in a heavily populated area

TABLE 4.2 Nature and the difference between natural and man-made disasters.

Natural disaster	Man-made disaster
Natural disasters are caused by a natural phenomenon	Man-made disasters are caused due to human negligence or man-made hazards
It can be grouped as follows based on the cause: • Topological (e.g., landslides, earthquakes, floods, avalanches) • Meteorological (e.g., cyclones, storms, hurricanes, heat waves, snow storms) • Biological (e.g., disease epidemics)	It is grouped as follows: • Armed conflict (e.g., war) • Accidents (transportation accidents, mine disasters, nuclear or chemical leaks, collapse of buildings) • Warfare: Use of chemical, biological, nuclear, or radiological weapons • Large-scale human displacement (refugees) • Disasters that happen in human settlements (e.g., pollution, antimicrobial resistance)

Evolution of public health preparedness and emergency management [5]

Outbreaks of infectious diseases globally, such as the Severe Acute Respiratory Syndrome coronavirus 1 and 2 and other disasters in the recent past, have established the significance of reinforcement of public health systems to defend people from natural and man-made threats. To fulfill these needs, several activities and programs have been established in national and global settings; one such approach is integrating emergency management with public health practice and vice versa.

Public health emergency management (PHEM) is a new and growing field that is based on the specialized knowledge, skills, and systematic concepts of public health and emergency management that are essential for the successful management of emergencies with significant health implications. PHEM is given importance after the September 11, 2001, event in the United States of America and the anthrax outbreak. It led to the widespread adoption of the National Incident Management Systems to guide emergency response and management of events of an emergency nature [7]. The International Health Regulations (IHR) was revised to protect the populations from public health threats in the WHO member countries and advised them to strengthen the emergency preparedness and response activities. Aligned with this initiative, the Centers for Disease Control and Prevention hosts a Public Health Emergency Management Fellowship program. The fellowship program trains existing public health experts on understanding and use of principles and concepts of PHEM [8]. In 2014, the Global Health Security Agenda called for strengthening the national Public Health Emergency Operation Center (PHEOC), among other elements of PHEM [9]. The World Health Organization, in 2015, proposed an outline for developing and managing a PHEOC which integrates traditional public health services into an emergency management model [10]. PHEOC includes the following:

1. Prevention and mitigation of hazards
2. Implementing public health surveillance programs
3. Establishing institutional and technical capacities (laboratories, clinics, rapid response teams)
4. Enhancing readiness by planning for and stockpiling response resources
5. Enhancing environmental health programs
6. Engaging communities
7. Training staff and validating plans

Despite these global health initiatives, the Ebola outbreak in West Africa, the Zika virus outbreak in the Americas and elsewhere, and the recent COVID-19 pandemic reemphasized the urgent need to strengthen the PHEM capacities for better preparedness and management of future emergencies. COVID-19 has revealed that the current level of preparedness is insufficient to manage public health emergencies affecting the major population of the world.

The Delhi Declaration-Emergency Preparedness in the South-East Asia Region (SEAR) in 2019, the Five-year Regional Strategic Plan 2019–2023, the Risk Communication Strategy for public health emergencies in the WHO SEAR 2019–2023, and the Asia–Pacific Strategy for Emerging Diseases and Public Health Emergencies (APSED III): Advancing implementation of the IHR (2005) demonstrate ministerial-level political commitment in implementing IHR for health security in the SEAR.

Principles and components of public health preparedness and emergency management

PHEP definition [5]

"Public health emergency preparedness (PHEP) is the capability of the public health and health care systems, communities, and individuals to prevent, protect against, quickly respond to, and recover from health emergencies, particularly those whose scale, timing, or unpredictability threatens to overwhelm routine capabilities. Preparedness involves a coordinated and continuous process of planning and implementation that relies on measuring performance and taking corrective action" [5,11]. The definition includes all hazards or disasters that overwhelm the routine capabilities of the public health system.

Principles of PHEP

PHEP should be combined as a part of daily public health practices and built upon existing systems. PHEP response should have essential building block capabilities with functions that can be adapted to small and large events. Further, it should have integrity, responsibility, and transparency and engage the community and other stakeholders. PHEP must involve prevention, mitigation, and recovery activities and response to events. Further, it should have the capability and capacity to ensure readiness along with exercises to improve the health and strengths of the communities.

Elements of PHEP

Each aspect of PHEP is made up of several action-oriented elements which have to be achieved by the communities in order to respond to an emergency. The elements are categorized into three groups: preplanned and coordinated rapid-response capability, expert and fully staffed workforce, accountability, and quality improvement [5].

Preplanned and coordinated rapid-response capability

It focuses on evaluating the community to recognize and fulfill the breaches in planning, recognizing, and informing accountable parties of their roles in a rapid response operation, followed by the capacity to promptly implement public health functions.
   "It includes 11 elements:

1. Health risk assessment: Find the risks and susceptibilities because the occurrence of a PHE is based on the preexisting characteristics and strengths of the target population.
2. Legal climate: Recognize and pay attention to legal and accountability barriers to efficiently monitor, avoid, or react to a public health emergency.
3. Roles and responsibilities: Explain, allocate, and check responsibilities in every sector, at various levels of the government and with every individual and ensure group integration.
4. Incident Command System (ICS): An integrated ICS is used to enhance decision-making and response-ability at every level.
5. Public engagement: Encourage active participation of the public in public health emergency preparedness by educating and organizing them.
6. Epidemiology functions: Develop and maintain a surveillance system for possible environmental, radiological, toxic, or infectious threats.
7. Laboratory functions: Maintain and expand the systems to test for possible threats

8. Countermeasures and mitigation strategies: Mitigation strategies that target the community (e.g., isolation and quarantine, social distancing) and countermeasure dissemination approaches should be developed and improved when applicable.

9. Mass health care: Build and maintain the competence to deliver healthcare amenities in a large scale.

10. Public information and communication: Build and maintain the competence for quick communication of precise and reliable facts to the public by socially apt means.

11. Robust supply chain: Ascertain vital resources for public health emergency response and develop a supply chain for the prompt delivery of the resources" [5].

Expert and fully staffed workforce

It focuses on the necessity to develop and sustain a competent workforce that can perform under stressful situations, which can be achieved by establishing leadership skills among public health leaders.

1. Operations-ready workers and volunteers: Improving the capacities and competencies of human health resources at all levels to combat a public health emergency.

2. Leadership: Institutionalizing public health leadership skills (e.g., arranging resources and connecting with the public).

Accountability and quality improvement

It emphasizes conducting public health events, assessing performance, providing incentives, and safeguarding the financial responsibility of the systems.

1. Testing operational capabilities: Rehearse, conduct analysis, and enhance public health emergency preparedness by conducting public health events, training, and exercises when apt.

2. Performance management: Establish a performance appraisal or audit and fix accountabilities linked with responsibilities.

3. Financial tracking: Keep records of the resources and make sure to track resources and ensure sufficient and apt repayment by maintaining a financial system.

Public health emergency preparedness and response capabilities

The Public Health Emergency Preparedness and Resource (PHEPR) management is linked with 15 capability standards under six domains that improve the capacity of the public health system for emergency preparedness and response at all levels. The six identified domains are community resilience, incidence management, information management, countermeasures and mitigation, surge management, and biosurveillance [12]. Of the 15 capabilities, nine are core and foundational capabilities of PHEPR and the remaining six are cross-cutting capabilities (Table 4.3).

TABLE 4.3 The domains and capabilities of public health emergency preparedness and response [12].

Domain	Capability	Core capability	Cross-cutting capability
Community resilience	Community preparedness	Yes	
	Community recovery		Yes
Incident management	Emergency operations coordination	Yes	
Information management	Emergency public information and warning	Yes	
	Information sharing	Yes	
Countermeasures and mitigation	Medical countermeasure dispensing and administration	Yes	
	Medical materiel management and distribution	Yes	
	Nonpharmaceutical interventions		Yes
	Responder safety and health	Yes	
Surge management	Fatality management		Yes
	Mass care		Yes
	Medical surge		Yes
	Volunteer management		Yes
Biosurveillance	Public health laboratory testing	Yes	
	Public health surveillance and epidemiological investigation	Yes	

Predictions, mitigation, preparedness, management, and rehabilitation [7]

The emergency management cycle best describes the interface between public health and emergency management. The phases of the emergency management cycle define the roles and abilities of an emergency management system.

Mitigation emphasizes decreasing the loss due to hazard or risk and monitoring predicted loss; mitigation activities may be executed before, during, or after an event, e.g., vaccination, surveillance, and other activities to decrease outbreaks of infectious diseases.

Preparedness activities are carried out before an event occurs and focus on developing or sustaining workforce and infrastructure along with preparation, training, and exercises needed to ascertain gaps and improve emergency response, e.g., developing and maintaining emergency response plans, warning systems, manpower, and information technology set-up like surveillance and reporting systems.

Response to an emergency is based on the hazard that has the risk of affecting routine functions or capacities. It can include the following activities in general.

- Intersectoral coordination of public health response functions
- Data collection and analysis of epidemiologic, laboratory testing, and other data
- Sharing information with various stakeholders
- Communication of emergency information and recommendation with the target audience
- Coordinated execution of regulatory actions like the delivery of healthcare services.

Recovery or rehabilitation activities are carried out during and after the response and include activities to regain or adjust to "new" usual circumstances after an event. They are the activities aimed to bring about a systematic transformation of response-related activities to usual public health programs, activities to restore or reinforce health systems, or observing longstanding consequences such as mental or behavioral health problems in affected populations after a public health emergency.

Role of various government and other stakeholders in risk mitigation and management in public health emergencies

PHEM or disaster management includes coordinated activities undertaken by various organizations to attain a shared objective. It includes a combination of resources and efforts from people, agencies, and organizations within and outside the state/government.

Role of the state/government

The principal concern of the state is the welfare of its citizens. It plays a main role in providing sufficient food, education, hygiene, and health services, which can be attained using the resources and the state's power to formulate policies and implement them. In order to reduce liability to emergencies, prediction, risk analysis, and reduction have to be part of the development programs of the state, both at micro- and macrolevels. Further, as part of emergency preparedness, the existing resources of the state should be made a part of the humanitarian response.

The government is responsible for providing assistance to people affected and the major objectives are as follows:

- To assist in the coping of the line dept., local administration, and the community
- To deliver care and resources to people in disaster.
- To arrange for a structured, methodical, flexible method for dealing with the emergency.
- To spread genuine facts in an apt, timely, precise, and beneficial method.
- To build fast and lasting support plans for susceptible people after an emergency
- Safeguarding dynamic involvement of government, community, volunteers, and nongovernment organizations (NGOs) and with maximum exploitation of manpower, supplies, and funds during an emergency.

Extraordinary executive powers are sanctioned through various public health laws passed at national and subnational levels like the epidemic disease act, during an emergency.

Role of local administration

The local administration undertakes preparations and ensures readiness to handle disasters. For example, precautionary measures like operational control rooms, emergency shelters, and health services. The government officers of various sectors are explained their duties and trained to engage in preventive and preparative actions to fight disasters by working together with the local administration. The administrative or executive officer will control the relief procedures, such as food distribution, free kitchen, etc., carried out by civil society organizations (CSOs), NGOs, and voluntary organizations. The CSOs/NGOs can start further long-term programs in affected areas following appropriate permissions from the Board of Revenue/Special Relief Commissioner under the suggestion of the Government.

Coordination with civil society organizations, NGOs, and CBOs

Disaster management is achieved in rural and marginalized areas by the coordinated activity of NGOs and CBOs. NGOs, through their access to information and advanced technology, form and activate the CBOs to distribute information, create awareness, bring social mobilization, and build the capacity of the community.

Public health preparedness and emergency management during the COVID-19 pandemic [13,14]

The COVID-19 pandemic has questioned the ability of subnational, national, and global communities in preparedness and response to a public health emergency. Several strategies have been put forward to manage the pandemic, the effectiveness of which is based on the coordination of the existing health system, and the population targeted. Key findings of public health preparedness and emergency management across the globe during the COVID-19 pandemic are as follows:

- *COVID-19 response*: There was an initial delay in taking several essential decisions by the international and national agencies during the early stage of the COVID-19 pandemic: issuing a caution of the human transmission of the virus; declaration of public health emergency of global concern; insisting on travel restriction imposition; and use of face mask to contain the virus spread [15]. In January 2020, when the pandemic became widely recognized, most governments were sluggish in identifying its significance and responding promptly. Countries like Singapore, Taiwan, and Hong Kong had a rapid response to the outbreak by choosing strategies to suppress the spread of the virus, like screening at airports, stringent contact tracing, and quarantine of contacts of confirmed cases and travel restrictions that resulted in low mortality. On the contrary, European countries like Italy, Spain, and the United Kingdom nearly failed to predict the impact of the COVID-19 pandemic and delayed implementing available preventive strategies to stop the transmission of the virus [16].
- *International cooperation*: There was insufficient synchronization among the countries over several issues: delay in warning about the outbreak of COVID-19, its nature of transmission in the early stages, and inconsistencies in the containment measures across the countries and globally; several countries failed to adopt evidence-based control measures; the low- and middle-income countries (LMICs) were poorly funded and there was disproportionate dispersal of essential supplies like a mask, screening kits, drugs, pharmaceutical devices, and vaccines; nonavailability or inaccurate data on infections, severity, deaths, variants, and health system responses; and the absence of international and national safety nets to defend the vulnerable population. There were indeed positive aspects in the

coordination between countries, like public–private partnerships for developing the vaccines quickly, financial aid from the World Bank, and other health agencies like the Global fund to support the COVID-19 response of countries, and innovations in healthcare delivery.

- *Widespread political ineffectiveness*: A notable pandemic-related event is the carelessness of some powerful political figures. In the early stages of the pandemic, a number of national leaders made reckless remarks, disregarded scientific facts, and needlessly endangered lives to maintain the economy [16,17]. The health outcomes of the pandemic were influenced to a great extent by the national policy of the countries. Further, to foster confidence and mobilize efficient responses, strong leadership practices and public engagement are crucial.
- *Role of media in information sharing*: A devastating environment causes a lack of trust in health officials. It encourages the fact that personal opinions are valid compared to the best scientific evidence due to the quick dissemination of false information and a lack of adequate vision. For instance, the false promotion in the media of dangerous, experimental treatments, such as hydroxychloroquine and ivermectin, led to needless public chaos and a shortage of drugs [18]. A study on YouTube videos about COVID-19 established the presence of false information in about 27.5% of the videos studied [19].
- *Public attitudes toward pandemic control:* Across the globe, there were significant differences in how people felt about the pandemic. Several factors like culture, educational attainment, social media, and the statements of local leaders influenced public perceptions. Social media significantly influenced the public's perception of social and public health initiatives. The lack of space for in-depth debate on social media made it easier for conventions to form quickly, which frequently invalidated the need for changes to recommendations and public health and social policies in response to the new findings on the virus. An example of the public attitude toward COVID-19 measures is the protest against COVID-19 restrictions and the antivaccine movements that has led to vaccine hesitancy.
- *Community engagement*: During public health emergencies, community involvement like involving the local leaders and collaborating with community members to personalize messaging and campaigns is essential since most of the nonpharmaceutical interventions targeting viral transmissions, like mask use, hand washing, and social distancing, are based on a feeling of social responsibility among communities. High-income nations with robust national health systems where healthcare was integrated with public health systems that support clinical healthcare have performed better in combating COVID-19 and preserving routine healthcare delivery. Likewise, LMICs had better responses since they are built upon their past experiences of epidemics and utilized public health resources like

community health workers to establish a rapport with the communities and for screening and contact tracing.

- *Health service delivery:* Countries worldwide have used three primary strategies to quickly scale up their healthcare set-up: building new healthcare accommodations, remodeling public spaces, and converting current healthcare facilities to accommodate COVID-19 patients. China constructed two field hospitals dedicated to COVID-19 in 2 weeks time-span [20]. Further, to accommodate COVID-19 patients, elective surgeries were canceled, home isolation was advocated for mild to moderate cases of COVID-19, and telemedicine services were used to treat non-COVID cases in some countries.
- *Health workforce:* The disproportionate case burden of COVID-19 across countries has challenged the resiliency of the health system due to several factors: a shortage of manpower and uneven distribution of the healthcare staff, inadequate supply of protective gear, screening kits, inadequate training of the workforce, and poor mental health. In order to address these issues, different strategies were adopted by countries, like virtual training followed by supervised work of nursing and medical students in managing COVID-19 cases, incentives, and others.

Challenges and barriers to public health preparedness and emergency management [21]

Lack of community engagement

A community's vulnerability to local hazards depends on the context in which it is located, which depends on the area's physical topography and climatic circumstances. It also establishes the material resources that the community has access to. Similarly, the local financial resources are determined by the socioeconomic character of the town. For instance, even if vulnerable people are aware, they live in a disaster-prone area, a poverty trap condition may prohibit them from moving to a safer location [21]. Variables including socioeconomic status, race and ethnicity, language, and place of residence influence people's capacity to prepare for emergencies and respond to specific situations [22–24].

Community-level influences, including the demographic composition of a community, the effect of migration, and the level of urbanization, also impact disaster management. Evidence and practice should be contextualized to the local setting, needs, and community knowledge to make it easier for agencies to implement interventions and for the community to adopt them. Increasing community ownership for disaster risk reduction initiatives and building trust makes everything easier, all depending on community engagement [21].

The project should be made known to the community to develop the trust attained by social mobilization. However, even though it is widely acknowledged that community involvement is crucial, problems were found, including low levels of community comprehension that restrict participation in disaster management planning and policy. Due to the community's limited understanding, ideas for policies and programs are not very helpful and, if ignored, may upset them [21].

Part of this attitude of taking risks stems from the socioeconomic constraints that prevented some tribes from relocating. This is accompanied by a significant amount of fatalism and decreased risk awareness. Fatalism may be prevalent, influencing not only practitioners but also policymakers and the general public, and it can prevent people from participating in catastrophe planning and mitigation efforts [21]. Certain demographic groups' responses to and comprehension of preventive messages are also influenced by behavioral traits like self-efficacy and perceived norms, previous views, readiness to follow advice, and faith in the government [25−29].

Problems with the system

The disaster management system behaves and functions according to the procedures already in place and the way system is structured. Some challenges in the disaster management system are coordination issues, problems with defining roles and responsibilities, fragmented plan and approaches, problems with communication channels, problems with the hierarchical structure of authority, and issues with how the various agencies coordinate their actions [21]. The allocation and availability of resources for disaster management is always an issue since it is difficult to calculate or predict the impact of a disaster as each disaster is different.

Political barriers

The following factors contributed to the failure of disaster management: Insufficient political ownership, lack of coordination across operations, commitment and policy direction, divergent institutional goals, and a lack of personal ownership or relevance among individuals. The lack of governmental stability and priority and the political concentration on constitutional issues could be the reasons for the lower priority given to disaster management [21].

Old laws and regulations: The existing disaster management act may be outdated and unworkable because it primarily focuses on rescue and relief. For example, the initial Epidemic Disease Act of India was promulgated in 1897 and after nearly 120 years, it has been updated [30]. At the same time, the main elements of the act must include readiness, mitigation, and adaptation.

Inadequate political ownership: Inadequacies in the legal framework or its enforcement on crucial issues such as building regulations and codes result from a lack of political ownership. Insufficient government capability is also one of the reasons why existing legislation may be difficult to enforce.

Ineffective coordination: Lack of coordination will cause one agency to act one way, while another works another course in the same area, and occasionally duplication may happen. If this is not managed correctly, the resource will be inefficiently used and some people may be denied aid.

Lack of harmonization among multiple stakeholders: Agendas of the numerous external actors, agencies, and organizations are not coordinated with those of the national government. Due to the diversity of interested parties, the disaster management system is less driven by evidence-based decision-making and more by synthesizing many domestic and international stakeholder agendas.

Knowledge barriers

There is a lack of evidence-based disaster management information across settings. Information gaps hamper the development and use of efficient catastrophe reduction strategies because of a lack of demand for evidence in the first place, difficulties accessing the evidence, and issues collecting the evidence base. In order to generate interest and demand, there is a need to raise evidence-based awareness of information and show stakeholders its importance. Evidence generation needs to be integrated into the routine systems and as an audit after managing the disaster.

The definitions of the evidence and the relevant experts should be crystal clear. To make an informed decision free from conflicts of interest, the meaning of the evidence base should be the same for all parties involved. Various stakeholders' varying capacities and understandings made translating the evidence into practice difficult. Due to insufficient learning procedures, the systems should integrate organizational learning into the framework. It takes time to develop the capacity within the implementing partner.

Staff turnover and a lack of qualified human resources in this field are the reasons that make it difficult to retain organizational memory. Institutionalizing learning is challenging because the workforce constantly switches between tasks. The level of stakeholder organizations' openness to sharing knowledge is another obstacle to the spread of best practices [21].

Barriers to preparedness

Inconsistent and unsustainable funding is a barrier to emergency preparedness (notably outbreak preparedness) [31]. Financial resources are essential for prevention and reaction, logistical systems, medical supplies necessary for survival, and devoted, qualified, and equipped human resources for crises. Funding for prioritizing research before, during, and after public health emergencies has been shown to be currently distributed across numerous funding agencies. It is inconsistent and does not support the advancement of high-quality research and the long-term development of research expertise [32].

Early access to financing at the operational level is still essential to promptly prepare for an impending risk of epidemic, respond to alerts, and guarantee that any spread is restrained [31]. Contingency finance is related to immunization, risk communication, and the preparation of the health workforce. Besides this, collaborations between practitioners and policymakers could guarantee that research fills crucial knowledge gaps and assist in closing the gap between research and practice. Incentives and a solid and ongoing commitment to workforce capacity development for PHEPR researchers and practitioners will be necessary. It will be a method for incorporating evidence into public health decision-making to create an evidence-based policy [33]. Without efforts to coherently, transparently, and rigorously synthesize and evaluate PHEPR research, it seems likely that the practitioners may continue to use empirical and easy-to-use strategies without any scientific basis, which may cause more harm rather than benefit.

Implementing coordinated, evidence-based responses in a public health emergency is complicated because of the time constraints, high levels of scientific uncertainty and political pressure, and potential absence of indisputable evidence, as the COVID-19 pandemic taught us. It is required to more clearly characterize the mechanisms via which interdisciplinary evidence feeds into decision-making processes and the context in which these mechanisms function to increase the influence of evidence on decision-making for public health emergency preparation [34]. WHO emphasizes the need for consistent, and widely used tools and methods for developing disaster preparation plans using predetermined indicators [31].

Role of communication during public health emergencies

The responses to various disasters show that public awareness and warning are among the essential emergency preparedness skills. Effective communication is necessary for excellent emergency management because it lowers risks, encourages the adoption of preventative measures, lessens the adverse effects of disasters on mental health, and speeds up recovery after a crisis [35,36].

Public health authorities can plan for strategic communication efforts in future emergencies by understanding communication practices at various stages of public health emergencies. There is also a need for clear criteria for evaluation techniques and study designs to create reliable results for the different kinds of inquiries.

To strengthen the evidence and gather more information on long-term trends and how communication behaviors and their effects are likely to change over time, longitudinal studies rather than cross-sectional ones with a large enough sample size for the generalization of the findings may be beneficial in this situation. Effective planning during emergencies can benefit from experimental studies like RCTs that compare the effects of various communication messages on multiple outcomes. Still, it is challenging to conduct this research practically because so many external factors could affect the results and make it difficult to conclude.

As a result, some have argued for more innovative alternatives, including time-series designs, to address the difficulties in researching extensive communication efforts [37]. Another strategy that could be used to produce data to forecast the outcome of a specific intervention is simulation. Qualitative research approaches can assist in explaining why and how practices might be effective or ineffective, which may help explain study results or guide the design of interventions. They can also help develop theories that can be empirically tested [32].

The dissemination of information to the public is just as important as making decisions based on facts. It is crucial to disseminate reliable information that can be rapidly and easily supplied to the public during an emergency. For the first time in history, social media and technology are being used widely to keep people connected, informed, and productive during the COVID-19 pandemic [34]. The same technology, on the other end, facilitates and magnifies infodemics, which continued to jeopardize efforts to control the pandemic and weaken the global response. Avoid infodemics by promptly disseminating accurate information, supported by research and evidence, to all communities, focusing on high-risk populations. At the same time, use fact-checking and other techniques to stop the spread of false information while protecting the right to free speech.

It is crucial because communication enables information interchange between the public and the health authorities. The health authorities should remember that the information delivered should be of high quality and understandable to the general population. People ought to be able and willing to use that information daily. It should fulfill the informational needs of the public during health crises. Information should be spread so that, whether people actively seek it out, they still learn what they need to protect their health and engage in daily activities.

Evaluation of evidence to inform public health preparedness and emergency management practices

GRADE: level of confidence in the evidence generated from Reviews of Qualitative Research and Grading of Recommendations Assessment, Development and Evaluation (GRADE-CERQual) [38,39].

The certainty of the evidence (COE) is assessed using the GRADE approach, which ranks bodies of proof according to eight distinct areas. As opposed to the majority of previous frameworks, GRADE modifies the COE based on the caliber and risk of bias of the studies included in the analysis rather than establishing explicit quality standards for research inclusion. Additionally, GRADE makes transparent, evidence-based recommendations in the form of guidelines, considering evidence other than that linked to effect (e.g., feasibility, acceptability). GRADE is used by numerous international review and recommendation bodies, and the procedures are regularly improved. GRADE has recently been modified to evaluate qualitative evidence (GRADE-CERQual) [40].

Evidence-based decision criteria according to GRADE [40]

- *The priority of the problem* can be ascertained by considering the population affected by the intervention and the importance of the desired and unfavorable outcomes.
- *Certainty of the evidence* evaluated utilizing the GRADE domains.
- *Balance of benefits and harms* based on determining whether one considerably outweighs the other after weighing the magnitude of both desired and undesired impacts.
- *Acceptability* considers the opinions of those who stand to gain (or lose) and the timing of the benefits, negative consequences, and costs.
- *Values/preferences/outcome importance* reflects on the significance of the health outcomes for individuals affected, their variability, and any lingering doubts.
- *Equity* can be assessed by recognizing equity as the desired outcome, evaluating equity-relevant results, comparing the severity of effects on disadvantaged and advantaged groups, comparing the baseline risk for underprivileged populations, and assessing relevance to poor people and situations.
- *Resource use* evaluates the need, quantity, and alternate resources available, identifies resource use issues that may be significant to stakeholders, and cost-effectiveness.
- *Feasibility* examines the intervention's viability, significant obstacles to implementation that need to be considered, and supply side capabilities.

How to review PHEPR evidence [40]

1. Decide on the review topic while considering stakeholder feedback and available literature on gaps and priorities.
2. Create the essential review questions and the analytical framework.
3. Look up studies in the gray and peer-reviewed literature and request them from stakeholders.
4. Use criteria for inclusion and exclusion.
5. Sort the evidence into several methodological streams (qualitative studies, after-action reports, case reports, quantitative research, comparative, noncomparative, modeling studies, and descriptive surveys).
6. Use and modify current tools for quality assessment of specific studies based on research design.

7. Combine the body of evidence into methodological streams and use the correct grading scale
8. To calculate the final COE, consider evidence of influence from other streams (such as modeling, mechanical, and qualitative evidence) and agree or disagree with results from quantitative research investigations.
9. Combine data from several methodological streams to fill up the PHEPR Evidence to Decision framework and determine implementation factors.
10. Create implementation advice or practice recommendations.

Developing and expanding the evidence for public health preparedness and emergency management

Research in the field of PHEPR is fragmented and mainly limited to academic exercises within the walls of universities. Though there is increasing evidence of PHEPR reported globally, especially in the last decade and after the SARS-CoV-2 pandemic, it is predominantly from selected developed countries [41,42]. Further, the PHEPR research evidence is available only for certain capabilities among the 15 listed capabilities [43].

PHEPR research funding goes up after one disaster and declines as time elapses since the last event. Poor research infrastructure and workforce capacity could be attributed to the lack of continuous funding for PHEPR research. Thus, significant strides are needed to increase the quantum of evidence base and its quality for PHEPR practices. The following strategies may be helpful in generating evidence on PHEPR:

1. Developing a National PHEPR research platform: This will provide academic researchers, implementers, and policymakers a common platform to support collaborative research efforts to address the evidence gaps through sustained research funding, effective leadership, and high-quality PHEPR rigorous research
2. Recognize PHEPR science as a subspecialty of the broader public health field
3. Ensure adequate and stable funding and infrastructure to support PHEPR research
4. Develop guidance on research methods, good research practices, evaluation approaches, and reporting guidelines to improve the quality of PHEPR research
5. Capacity building of PHEPR researchers and practitioners
6. Culture of the evidence-based process to inform PHEPR decision-making: Evidence generated by PHEPR research will be useful to PHEPR practitioners only if its translational value can be realized and incorporated into public health practice. A transparent methodology of identifying, systematically reviewing and evaluating PHEPR evidence is necessary to formulate consensus guidelines for PHEPR practices.

Bridging the gap between research and practice in the context of public health preparedness and emergency management

In this chapter, we have extensively discussed the need and methods to generate evidence for PHEPR practices. An equally important yet neglected aspect is the dissemination, translation, and implementation of the evidence to policy or practice. A major barrier to the uptake of evidence-based practices is the disconnect between the policymakers/practitioners and the researchers, who work on their research agenda within respective academic circles without effective collaboration with the decision-makers.

A coordinated implementation science approach is needed to ensure that the evidence from PHEPR research achieves broad reach and becomes the standard of practice of the target audience. To achieve this, we propose the following:

1. Creating a priority agenda for research and implementation with active engagement not only from the researchers but also from the implementing partners and policymakers
2. Engage the decision makers and practitioners right from the beginning when the research agenda is being formulated and make them active partners in the process of evidence generation
3. A dissemination plan should be part of every research proposal
4. Partnering with public health professional organizations or national associations to disseminate evidence-based practices
5. Evidence briefs should be written in simple layman's language with clear, actionable points for the policymakers
6. Incentivize the public health agencies to evaluate new or adapted practices. Embed program evaluations into routine operations to understand whether evidence-based practices worked, under what conditions, and at what cost with what impact and consequences

Conclusion

PHEPR should be recognized as a subspecialty of the broader public health discipline. The PHEPR field is relatively young and has been evolving rapidly over the past 2 decades. The COVID-19 pandemic has highlighted critical evidence gaps and underscored the importance of a robust Public Health Emergency Response and Management system backed by sound evidence and evidence-based practice. Research must be embedded within the PHEPR system, which must be integrated with the routine public health system. A common National PHEPR research platform is proposed to provide a common forum for academic researchers, practitioners, and policymakers to address the evidence gaps through sustained research funding, effective leadership, and high-quality PHEPR rigorous research. Besides generating credible evidence, translation of evidence into PHEPR practice is needed through effective collaboration between academia and policymakers. PHEPR research must be routinely used to manage any public health emergency. The effect in the newer setting needs careful documentation and dissemination.

References

[1] Loretta W. Senate, no. 1048. State of New Jersey. 217th legislature. 2016.

[2] Senate and General Assembly of the State of New Jersey. Emergency health powers act. New Jersey: Senate and General Assembly of the State of New Jersey; 2005.

[3] World Health Organization. International health regulation (2005). 2nd ed. 2005. p. 1—82. http://apps.who.int/iris/bitstream/handle/10665/43883/9789241580410_eng.pdf;jsessionid=1 3A152EB513E6BC55F55B99AB17398A9?sequence=1. [Accessed 9 February 2023].

[4] RAND Corporation. RAND panel identifies key components of public health emergency preparedness. RAND; 2007. News Releases, https://www.rand.org/news/press/2007/04/05. html. [Accessed 9 February 2023].

[5] Nelson C, Lurie N, Wasserman J, Zakowski S. Conceptualizing and defining public health emergency preparedness. Am J Public Health 2007;97:S9. https://doi.org/10.2105/AJPH.2007.114496.

[6] Rutherford W, De Boer MGJ. The definition and classification of disasters. Injury Br J Accid Surg 2013;15:10—2. https://doi.org/10.1017/CBO9781107050990.013.

[7] Rose DA, Murthy S, Brooks J, Bryant J. The evolution of public health emergency management as a field of practice. Am J Public Health 2017;107:S126. https://doi.org/10.2105/AJPH.2017.303947.

[8] Center for Preparedness and Response. Public health emergency management fellowship. 2021. Emergency Operations, https://www.cdc.gov/cpr/eoc/emergencymanagementfellowship. htm. [Accessed 9 February 2023].

[9] Katz R, Sorrell EM, Kornblet SA, Fischer JE. Global health security agenda and the international health regulations: moving forward. Biosecur Bioterror 2014;12:231—8. https://doi.org/10.1089/BSP.2014.0038.

[10] World Health Organization. Framework for a public health emergency operations centre. 2015. p. 1—80. [Accessed 9 February 2023].

[11] Maddock JE. Preparing public health for the unexpected. Am J Public Health 2018;108:S348. https://doi.org/10.2105/AJPH.2018.304611.

[12] Centers for Disease Control and Prevention. Public health preparedness capabilities: national standards for state and local planning | state and local readiness. CDC; 2021. p. 1—176. https://www.cdc.gov/cpr/readiness/capabilities/index.htm. [Accessed 12 February 2023].

[13] Sachs JD, Karim SSA, Aknin L, Allen J, Brosbøl K, Colombo F, et al. The Lancet Commission on lessons for the future from the COVID-19 pandemic. Lancet 2022;400:1224—80. https://doi.org/10.1016/S0140-6736(22)01585-9.

[14] Haldane V, de Foo C, Abdalla SM, Jung AS, Tan M, Wu S, et al. Health systems resilience in managing the COVID-19 pandemic: lessons from 28 countries. Nat Med 2021;27:964—80. https://doi.org/10.1038/S41591-021-01381-Y.

[15] ET Online. Pandemic numbers hidden—delayed warning, hidden numbers: panel slams China, WHO for acting too slow on Covid-19. The Economic Times; 2021.

[16] Sjölander-Lindqvist A, Larsson S, Fava N, Gillberg N, Marcianò C, Cinque S. Communicating about COVID-19 in four European countries: similarities and differences in national discourses in Germany, Italy, Spain, and Sweden. Front Commun 2020;5:97. https://doi.org/10.3389/FCOMM.2020.593325/BIBTEX.

[17] Sumit G, Dorothy C, Elizabeth JK, Elize Massard da F, Fundação Getulio V, Salvador Vázquez del M, et al. 5 leaders who badly mishandled the COVID-19 pandemic. US News; 2021.

[18] Schellack N, Strydom M, Pepper MS, Herd CL, Hendricks CL, Bronkhorst E, et al. Social media and COVID-19—perceptions and public deceptions of ivermectin, colchicine and hydroxychloroquine: lessons for future pandemics. Antibiotics 2022;11:445. https://doi.org/10.3390/ANTIBIOTICS11040445/S1.

[19] Li HOY, Bailey A, Huynh D, Chan J. YouTube as a source of information on COVID-19: a pandemic of misinformation? BMJ Glob Health 2020;5. https://doi.org/10.1136/BMJGH-2020-002604.

[20] Cai Y, Chen Y, Xiao L, Khor S, Liu T, Han Y, et al. The health and economic impact of constructing temporary field hospitals to meet the COVID-19 pandemic surge: Wuhan Leishenshan Hospital in China as a case study. J Glob Health 2021;11:5023. https://doi.org/10.7189/JOGH.11.05023.

[21] Lee ACK. Barriers to evidence-based disaster management in Nepal: a qualitative study. Publ Health 2016;133:99—106. https://doi.org/10.1016/J.PUHE.2016.01.007.

[22] McGough M, Frank LL, Tipton S, Tinker TL, Vaughan E. Communicating the risks of bioterrorism and other emergencies in a diverse society: a case study of special populations in North Dakota. Biosecur Bioterror 2005;3:235—45. https://doi.org/10.1089/bsp.2005.3.235.

[23] Aburto NJ, Pevzner E, Lopez-Ridaura R, Rojas R, Lopez-Gatell H, Lazcano E, et al. Knowledge and adoption of community mitigation efforts in Mexico during the 2009 H1N1 pandemic. Am J Prev Med 2010;39:395—402. https://doi.org/10.1016/j.amepre.2010.07.011.

[24] Wray RJ, Becker SM, Henderson N, Glik D, Jupka K, Middleton S, et al. Communicating with the public about emerging health threats: lessons from the pre-event message development project. Am J Public Health 2008;98:2214—22. https://doi.org/10.2105/AJPH.2006.107102.

[25] Horney JA, MacDonald PDM, Van Willigen M, Berke PR, Kaufman JS. Individual actual or perceived property flood risk: did it predict evacuation from hurricane Isabel in North Carolina, 2003? Risk Anal 2010;30:501—11. https://doi.org/10.1111/j.1539-6924.2009.01341.x.

[26] Plough A, Bristow B, Fielding J, Caldwell S, Khan S. Pandemics and health equity: lessons learned from the H1N1 response in Los Angeles county. J Publ Health Manag Pract 2011;17:20—7. https://doi.org/10.1097/PHH.0b013e3181ff2ad7.

[27] Meredith LS, Eisenman DP, Rhodes H, Ryan G, Long A. Trust influences response to public health messages during a bioterrorist event. J Health Commun 2007;12:217—32. https://doi.org/10.1080/10810730701265978.

[28] Hilyard KM, Freimuth Prof VS, Donald director M, Supriya K, Sandra CQ. The vagaries of public support for government actions in case of a pandemic. Health Aff 2010;29:2294—301. https://doi.org/10.1377/hlthaff.2010.0474.

[29] Paek HJ, Hilyard K, Freimuth V, Barge JK, Mindlin M. Theory-based approaches to understanding public emergency preparedness: implications for effective health and risk communication. J Health Commun 2010;15:428—44. https://doi.org/10.1080/10810731003753083.

[30] Government of India. The epidemic diseases (amendment) ordinance, 2020. New Delhi: Ministry of Law and Justice; 2020.

[31] World Health Organisation. A strategic framework for emergency preparedness. 2017.

[32] Calonge N, Brown L, Downey A. Evidence-based practice for public health emergency preparedness and response: recommendations from a national academies of sciences, engineering, and medicine report. JAMA 2020;324:629–30. https://doi.org/10.1001/JAMA.2020.12901.

[33] European Centre for Disease. Prevention and Control (ECDC). The use of evidence in decision-making during public health emergencies; 2019. https://doi.org/10.2900/63594.

[34] World Health Organisation. Managing the COVID-19 infodemic: promoting healthy behaviours and mitigating the harm from misinformation and disinformation; 2020. https://www.who.int/news/item/23-09-2020-managing-the-covid-19-infodemic-promoting-healthy-behaviours-and-mitigating-the-harm-from-misinformation-and-disinformation. (Accessed 11 December 2022).

[35] Young ME, Norman GR, Humphreys KR. Medicine in the popular press: the influence of the media on perceptions of disease. PLoS One 2008;3. https://doi.org/10.1371/journal.pone.0003552.

[36] Rubin GJ, Potts HWW, Michie S. The impact of communications about swine flu (influenza A HINIv) on public responses to the outbreak: results from 36 national telephone surveys in the UK. Health Technol Assess 2010;14:183–266. https://doi.org/10.3310/hta14340-03.

[37] Hornik RC. Public health communication: evidence for behavior change. Taylor & Francis; 2008.

[38] Guyatt G, Oxman AD, Akl EA, Kunz R, Vist G, Brozek J, et al. GRADE guidelines: 1. Introduction—GRADE evidence profiles and summary of findings tables. J Clin Epidemiol 2011;64:383–94. https://doi.org/10.1016/j.jclinepi.2010.04.026.

[39] Lewin S, Glenton C, Munthe-Kaas H, Carlsen B, Colvin CJ, Gülmezoglu M, et al. Using qualitative evidence in decision making for health and social interventions: an approach to assess confidence in findings from qualitative evidence syntheses (GRADE-CERQual). PLoS Med 2015;12. https://doi.org/10.1371/journal.pmed.1001895.

[40] Evidence-based practice for public health emergency preparedness and response. National Academies Press; 2020. https://doi.org/10.17226/25650.

[41] Sweileh WM. A bibliometric analysis of health-related literature on natural disasters from 1900 to 2017. Health Res Policy Syst 2019;17. https://doi.org/10.1186/s12961-019-0418-1.

[42] Lin T, Qiu Y, Peng W, Peng L. Global research on public health emergency preparedness from 1997 to 2019: a bibliometric analysis. Disaster Med Public Health Prep 2022;16:153–62. https://doi.org/10.1017/DMP.2020.206.

[43] National Academies of Sciences Engineering and Medicine. Evidence-based practice for public health emergency preparedness and response. 2020. https://doi.org/10.17226/25650.

Section II

Case studies of public health practice

Chapter 5

Public health practice and clinical practice

Kathiresan Jeyashree

Indian Council of Medical Research-National Institute of Epidemiology (ICMR-NIE), Chennai, Tamil Nadu, India

Public health practice and clinical practice—The definitions

As described in Chapter 1, Winslow defined public health as the science and art of preventing disease, prolonging life, and promoting health and efficiency through organized community effort for the sanitation of the environment, the control of communicable infections, the education of the individual in personal hygiene, the organization of medical and nursing services for the early diagnosis and preventive treatment of disease, and the development of the social machinery to insure everyone a standard of living adequate for the maintenance of health [1]. Public health physicians, a subset of all public health practitioners, juggle roles as a clinician who focuses on health promotion, prevention, cure and rehabilitation; a manager of health services who plans, implements, and monitors various community health programs; a researcher who generates knowledge from the community s/he works in to inform global and local action; and an adviser who advocates for the health of the community with decision makers [2,3]. A clinician, on the other hand, is seen as a person delivering a spectrum of healthcare services, usually more specialized and targeting the individual's disease condition.

Public health practice and clinical practice—The dichotomy

The Radcliffe line between clinical and public health practice is sadly established quite early in the medical education. As a graduate enters medical school, this divide is covertly and overtly reinforced. Public health is accorded a step-motherly treatment both by graduates during their education and clinicians during their practice. Public health is dangerously and falsely regarded quite often as less scientific or rigorous compared to clinical medicine. The general population also is unable to appreciate the utility and necessity of

Principles and Application of Evidence-Based Public Health Practice
https://doi.org/10.1016/B978-0-323-95356-6.00017-3
73

public health measures mainly due to its lesser visibility and lesser perceived direct utility to an individual. They are more accustomed to the picture of public health officials as people who are to be reached out for complaints regarding civic nuisance (garbage disposal, loud noise, and others) and of public health physicians as people for vaccination and awareness creation.

The basic differences between clinical and public health practice are presented in Table 5.1.

Healthcare from different points of view

It helps to look at the conundrum from different points of view—the public health practitioner, the clinician, and the individual.

The clinician

The clinician looks at the individual's disease or health issue with prime focus on alleviation of symptoms and cure. There is limited time, resources, and expertise available to address the social determinants and more distal determinants of the health condition. This approach is gratifying both to the patient and the practitioner as the results, in most, if not all, cases are obvious and quick and the direct requirements of the individual are specifically addressed. The dialogue is direct between the practitioner, his team, and the individual. The team involved in coordination and decision-making is predominantly the medical and paramedical team with an increasing involvement of communication and managerial experts. The principles guiding the care are the standard treatment guidelines and medical ethics as relevant to the treatment of an individual. Larger attributes of equity, affordability, accessibility, and appropriateness are not automatically factored into the solutions. Though there are exceptions, there is a significant involvement of the individual in every stage of the solution and the individual is the direct recipient of benefits and adverse effects. The cost is borne by the Government, the insurance systems, and in countries where these two have minimal role in health care and there is a predominant private health sector, the patient.

Public health practitioner

The public health practitioners, on the other hand, seldom interact with the individual. The community is their working ground from where they identify problems to be addressed and where they implement and evaluate the solutions to the problems.

The problems are of larger scope and the solutions influencing many more lives—human, animal, and plant. Their priorities are determined by what affects many, what can be fixed with available resources, and which solutions are efficient, economical, equitable, and inclusive. The idea is to fix fundamental

TABLE 5.1 The basic difference between clinical and public health practice.

Domains	Clinical medicine	Public health
Core disciplines	Medical and surgical specialties and super/ subspecialties, allied health sciences	Medical and surgical specialties, allied health sciences, epidemiology, social and behavioral sciences, management, economics, political sciences, public health engineering, biostatistics, demography, environmental and occupational sciences, nutrition, information technology
Primary focus on (beneficiary)	Individual Family	Community (local and global) Family Individual
Primary focus on (type of care)	Cure (focused and specialized care) rehabilitation	Health promotion, prevention, cure, disability prevention, and rehabilitation
Status among doctors and medical students as preferred career	Highly preferred	Not as much preferred
Emphasis in teaching and assessment during undergraduate and postgraduate	High, from the first year of medical education	Not as much as clinical medicine and offered in siloes rather than integrated
Financial implications of care	Cost borne by the individual mostly	Significant proportion of cost borne by the government
Team	Doctors, nurses, paramedics	Doctors, social scientists, economists, paramedical staff based in community, engineers, environmental and occupational scientists, nutritionists, and others
Impact on individual	Immediate, direct, and high	May vary from no direct benefit to long term huge
Impact on population	No or minimal	Slower, long term, huge
Research	Based in hospitals	Based primarily in community but also in health centers/ hospitals

determinants or roots of problems rather than just the problems or the symptoms of the problems. This will yield richer dividends. For example, public health interventions focusing on obesity prevention are expected to impact several diseases associated with obesity. The results in public health take longer, if not forever, not be direct and may not single out an issue for fixing. The individual may not feel the direct impact of the public health practitioner's work and the returns are less palpable at the individual level. The effects are not easily visible or comprehensible to the individual. I quote a senior colleague who told me at the very beginning of my public health career, "You will never be able to point out to a person and say that there walks a person whose life I saved by preventing diphtheria by vaccination." You can make a difference in the lives of a larger number of people though it may not be immediately apparent or obvious.

Individual

The individual looks at his/her health situation, symptoms as a problem requiring a solution, preferably a quick, effective, and affordable one. The roots of the problem may lie within their bodies or immediate vicinity or larger community, which may also require action at different levels. But they do not generally factor in larger and more distant determinants of the current health issue. For example, a person suffering from pulmonary tuberculosis (TB) is concerned about medication, its side effects, and probably transmission to household members. S/he is not probably concerned about indoor air pollution, malnutrition, neighborhood prevalence of TB, and airborne infection control measures in the DOTS center or other determinants of TB. They are generally content with direct fixes to their disease condition. There may be elements of follow-up and prophylaxis for some conditions, but adherence is usually poor after initial recovery. To pay for the care, the individual looks up to the government, insurance companies, or one's own pockets.

Let us consider the example of a 55-year-old female with a fracture of femur sustained after a fall inside her home and see the expectations and perceptions of the three stakeholders (Table 5.2).

Need for integration of public health and clinical practice

Public health and clinical medicine are increasingly and inevitably interdependent [4]. The gains for the individual and community are maximized when both disciplines complement each other and work together. The benefits and downside for these two ideologies to come together have been made crystal clear during the recent COVID-19 pandemic. Overflowing hospitals with patients queued on the roads for oxygen and the precious lives lost highlighted what can ensue even with state-of-the-art specialized care with ineffective public health measures [5]. Resistance to vaccination and other control measures to prevent transmission of infection and the differential magnitude and impact of the pandemic across nations and different subsections of the population highlighted the importance of the community in the context of an individual's health.

TABLE 5.2 The points of view of a public health practitioner, clinician, and individual toward a health problem.

	Public health practitioner	Clinician	Individual
Problem	Organizing care of the victim; determinants of the fracture in the elderly; prevention of future falls in elderly, pain, disability, loss of quality of life. Health promotion orientation in home (settings-based approach)	Pain, fracture femur, course of treatment, surgical complications, surgical outcome	Pain, mobility, restore activity, and social life as before; cost of treatment, loss of wages
Roots	Old age, postmenopausal, gender distribution of household roles, elderly nutrition, underlying medical disease, elderly friendly home or healthcare facility, elderly social support and caregiving, elderly health insurance and pension, lack of awareness of elderly care	Old age, underlying medical disease, vision, proprioception, stability, postmenopausal, low calcium, and Vit D3	Old age, slippery floor
Solution(s)	Fixation and referral for complications Possible points of prevention—calcium supplementation for postmenopausal women Prevention of future falls Strengthening the elderly care under public health Periodic screening or assessment for early detection of signs Home visits for elderly living alone Elderly medical care Social support for elderly Funds from government and NGOs	Stabilization and fixation of the fracture Management of surgical or postoperative complications Physiotherapy	Pain relief, mobility Affordable, accessible, appropriate care

In the COVID-19 example, let us consider two extreme hypothetical scenarios: how would the scenario have been had we dissed public health actions and relied on clinical medicine alone. The piling up of patients at hospitals and overwhelming the entire hospital network, which translates to compromised quality of care, poorer outcomes for patients, and higher expenditure for the system and the individual. Let us consider the alternate extreme scenario with a defunct clinical practice and exclusive reliance on public health action. That leaves us with prevented infections due to measures to cut down transmission; but the cases that do occur inevitably will be deprived of appropriate, up-to-date clinical care, and consequent poorer treatment outcomes. The message is that if either of these two strategies fails, the patient and community get suboptimal benefits leading to greater human and financial suffering.

There are numerous benefits for the community and the individual when these two disciplines intertwine and complement each other. When functioning optimally, these two disciplines can help reduce each other's burden. For example, an effective physical activity promotion program for adolescents is going to reduce the number of obese adults at risk of acute coronary syndrome and thus the burden on the clinician and health facilities. Similarly, timely and effective treatment of TB will reduce the number of infective people who are actively spreading the infection in the community.

To understand health-related behavior and choices and improve effectiveness of interventions

There is a need for the clinician to understand the disease or health state in its social, environmental, and political context. This shall enable them to see the milieu in which health-related choices are being made by the patient/individual. Such an in-depth understanding is not optional but essential to prescribe the optimal care for the person and better outcomes. Environment in which health-related choices are made determines the adherence to prescribed interventions and their effectiveness. Prescribing tobacco cessation services to an adolescent requires not just medical and pharmacological knowledge but a keen understanding of the social, political, and economic issues that influence tobacco use behavior. Without this understanding, the adolescent becomes another entry in the hospital register. Such care offered in siloes is not beneficial to the patient. Trying to treat a diarrheal episode in an under-five child without understanding the source of water and food in the community where the child lives will amount to treating symptoms and not the cause. This is not to undermine the value of clinical medicine in saving lives, but a reminder to focus as much on disease prevention and health promotion also.

To understand and shorten the pathway to care

The clinician sees the tip of the iceberg—the person who fell ill and is manifesting symptoms. The layers lying beneath are those not reporting to a clinic, those without symptoms, and those with risk factors, all of whom are potential patients. A good number of them, if addressed early, may result in avoiding the development of the disease. For example, the cardiology centers diagnose and treat patients reporting symptoms of myocardial infarction. For every such patient, there are a 100 who are unable to reach a clinic or are incubating risk factors waiting to evolve into a cardiac disease. It is essential that the clinician understands that this section is also in need of attention and by meeting their needs we can ultimately reduce the disease burden to a greater extent than by treating the patients with manifest disease alone.

To understand and establish equity in clinical care

The clinical practice can attempt to establish equity in the services delivered through their hospitals/clinics across socio-economic strata, gender, race, orientation, and other social divides. There is a general uncertainty about how these are within the spheres of concern for a physician given that their role in determining the current health state of the patients is far distal and removed from the time of patient-clinician interaction. There is a moral responsibility for clinicians to strive to deliver equal care to all patients, through systemic changes. Training in inclusive care should begin from undergraduation for a fundamental understanding of pervading inequities in determinants of health, health states, health care, and disease outcomes. One can imbue equity in the policies and strategies at the clinical care establishments by providing inclusive care, recruitment of healthcare staff to be representative, development of selective strategies for the care to reach the more vulnerable, addressing social determinants of health and illness, inclusive representation in research, etc. Focused leadership, financial commitment, partnership with community organizations, and assessment of level of equity of care delivered in clinical establishments will help achieve equity in clinical practice [6].

How to facilitate the integration between clinical and public health practice?

Integration of training

Training of postgraduates in all specialties and on-the job training of clinicians about the role of social determinants of health will equip them to design and deliver comprehensive care to the individual and community. Routine assessment of social determinants among all patients and employment opportunities for public health practitioners within the hospitals will mainstream public health within a set-up, which is usually exclusively for clinical medicine. Similarly, training of clinical medicine practitioners in some, if not all, core competencies of public health as epidemiology, social and behavioral sciences, and health promotions will improve the comprehensiveness of care offered to the patients they treat.

Linkage of public health and clinical medicine systems by programs of each nested within the other is highly desirable [7]. The example of the base of the Barts Health NHS Trust in Royal London hospital is an example of public health activities hosted within a hospital system. Their work focuses on developing care pathways and analyzing inequalities in access, referrals, and outcomes according to deprivation, advising on trust polices on inequalities, and managing operational public health services such as smoking cessation and screening [8].

Integration of implementation

While there are numerous opportunities and grounds on which public health and clinical medicine can collaborate, we highlight a few such avenues with examples. The clinician must be able to link the patient to the public health programs that s/he may benefit from. This puts the patient under an umbrella of care being offered as part of the system, in addition to the individualized tailor-made care at the clinician's hands as well. For example, in the Tamil Nadu state of India, free care is offered and there are also numerous benefits for the mother and child from the state-run program for reproductive and child health including, but not limited to, cash benefits, regular monitoring for fetal and maternal well-being, assistance for transport to an institution for delivery, follow-up during postnatal care, and a well-knit referral system. A mother who is registered under this program is followed through till the complete immunization of the child and this care is not dependent on any individual caregiver but runs on an auto mode. A mother who is in the private health system consulting a clinician on the other hand receives specialized and tailor-made care. The patient pays out-of-pocket toward meeting the cost of the care. The state has now mandated registration of all pregnancies with the public health system and this registration id is required for the issue of birth certificate. The system now is a rich common database for understanding the profile of pregnancies, complications, childbirth outcomes, and immunization. The information from such systems provides rich insights for informing policy and program for maternal and child health.

The model for population-based screening for noncommunicable diseases followed by linkage to specialist care and follow-up in public health facilities is another example of effective integration of public health and clinical practice. The "*Makkalai Thedi Maruthuvam*" model in Tamil Nadu, India, has employed this model of taking screening and primary health care to the doorstep of people while establishing effective referral linkages for clinical care and follow-up. One of the major areas for improvement to facilitate this integration is to '*catch them young.*' It is interesting to study the experiment at the New Mexico School of Medicine where 'Health Equity Principles of Public Health' course is taught as the first course to medical students even before they are introduced to anatomy [9]. The program was designed to teach students to address disparity, promote health equity, and address the root causes of disease. Assessment of undergraduates and postgraduates on disease prevention and public health competencies will also enable them to see the role of linking clinical, basic sciences, and public health competencies in patient care [10].

Integration of monitoring

At the level of implementation, an integration of the directorates of implementation of public health and clinical practice will also facilitate dialogue and liaison between these two approaches to health care. This will enable them to complement each other to maximize the benefits for the community and the individual. In the state of Tamil Nadu in India, there are three apex directorates that oversee heath namely, the Directorate of Public Health and Preventive Medicine which is responsible for all primary health care level facilities and services, Directorate of Medical Services that controls the secondary level healthcare facilities (up to the level of district hospitals), and the Directorate of Medical Education which oversees all tertiary and teaching healthcare facilities. There is a close dialogue and collaboration between these three directorates to plan and deliver health at community level and facility level.

Challenges in integration of public health and clinical practice

While these examples are those where both the community and individual benefit, there might be situations where one benefits more than the other. As in Geoffrey Rose's hallmark description of the 'prevention paradox', where the measure that benefits the community is not offering much to the individual as a benefit [11]. For example, the TB sensitive policy approaches target the determinants of TB like socioeconomic status, malnutrition, etc. The benefits to the community at large are higher than that of patients with TB themselves. A salt reduction program in the community may benefit the community by reducing the mean quantity of salt consumed in the community but may not translate into a benefit for an individual in terms of the reduction in his salt consumption and risk of developing hypertension (Fig. 5.1).

There might also be situations where one is compromised in the interest of the other. For example, consider the infringement into individual sovereignty by some public health measures. The asymmetrical development of specialized care within the ivory towers of hospitals focusing on individuals, mostly the privileged, deprives the majority even of essential healthcare services.

Way forward for the integration

Newer health threats have nudged us into remembering that the greatest health achievements of human race were achieved when public health and clinical care came together to complement and bolster each other toward better health for all. Moving forward, there are higher hopes for integration of clinical medicine and public health which is further facilitated by advances in

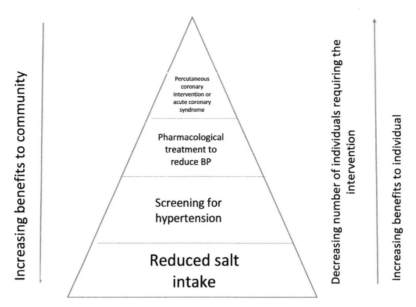

FIGURE 5.1 Effect of public health and clinical medicine approaches on the incidence of hypertension and coronary artery disease.

information technology. These advances have opened limitless opportunities for data sharing and generating evidence to inform action. The future is replete with these opportunities for integration which, if capitalized correctly by able leadership, professional training, and political commitment, will promise better health for all.

References

[1] Winslow CE. The untilled fields of public health. Science January 1920;51(1306):23–33.
[2] Shewade H, Jeyashree K, Chinnakali P. Reviving community medicine in India: the need to perform our primary role. Int J Med Public Health 2014;4(1):29.
[3] Singh A. What is (there) in a name. Indian J Community Med 2004;29(4):151–4.
[4] Frieden TR. Shattuck lecture: the future of public health. N Engl J Med October 2015;373(18):1748–54.
[5] Zhang Y, Olufadewa II, Adesina MA, Ekpo MD, Akinloye SJ, Iyanda TO, et al. Lessons from the coronavirus disease 2019 (COVID-19) pandemic response in China, Italy, and the U.S.: a guide for Africa and low- and middle-income countries. Glob Health J March 2021;5(1):56–61.
[6] Mate KS, Wyatt R. Health equity must be a strategic priority. NEJM Catalyst Innovations in Care Delivery 2017. Available from: https://catalyst.nejm.org/doi/full/10.1056/CAT.17.0556. [Accessed 8 November 2022].
[7] Maher D, Ford N, Gilmore I. Practical steps in promoting synergies between clinical medicine and public health. Clin Med April 2017;17(2):100–2.

[8]   Middleton J. From the President. Faculty Public Health Ebulletin; September 2016 (139). Available from: https://us12.campaign-archive.com/?u=9e818256890cafc10f8039758&id=147 96cd87e&e=[UNIQID].

[9]   Geppert CMA, Arndell CL, Clithero A, Dow-Velarde LA, Eldredge JD, Eldredge JP, et al. Reuniting public health and medicine: the University of New Mexico School of Medicine Public Health Certificate. Am J Prev Med October 2011;41(4 Suppl. 3):S214−9.

[10] Schapiro R, Stickford-Becker AE, Foertsch JA, Remington PL, Seibert CS. Integrative cases for preclinical medical students: connecting clinical, basic science, and public health approaches. Am J Prev Med October 2011;41(4 Suppl. 3):S187−92.

[11] Rose G. Strategy of prevention: lessons from cardiovascular disease. Br Med J Clin Res Ed June 1981;282(6279):1847−51.

Chapter 6

Effect of public health research on policy and practice

Punam Bandokar[1] and Madhur Verma[2]

[1]*Indian Council of Medical Research-National Institute of Epidemiology (ICMR-NIE), Chennai, Tamil Nadu, India;* [2]*Department of Community and Family Medicine, All India Institute of Medical Sciences (AIIMS), Bathinda, Punjab, India*

Introduction

Historically, public health practice was based on trial and error method of using empirical or expert knowledge and skills. Over time, the emphasis was placed on documenting the various observations and intervention testing. Scientific journals became the place of documentation and mode of dissemination. The methodology, i.e., the quality of research conduct, is getting refined and better reported nowadays, including improved access to research evidence. Thanks to the open access policy and information technology, which facilitate access to the available evidence from anywhere in the world.

Worldwide, billions of dollars are spent annually on health research to improve understanding of health and disease [1]. More than half of the grants are sanctioned for basic science research, followed by clinical research. Public health research receives the lowest proportion of grants. As a result of all these research works, at least one million research papers are published yearly in scientific journals. However, the translation of this research into policy or practice remained low over decades. Iain et al. call these "avoidable waste in the production and reporting of research evidence." Translation of research or the impact of research is getting popular recently, and most funding agencies and government organizations seek clarification before sanctioning any project.

Public Health Research needs to change or modify the prevailing policy or practice based on the findings. It is the primary principle of practicing evidence-based public health. The influence of public health research on policy or practice at local, sub-national, national, and global levels and strategies used for linking and influencing the same needs greater focus. Public health professionals play a vital part in policy development and implementation, disseminating the results to facilitate action, developing collaborations

Principles and Application of Evidence-Based Public Health Practice
https://doi.org/10.1016/B978-0-323-95356-6.00003-3

and fund allocations, and encouraging resource efficiency through evidence-based public health interventions.

Definition of research effect or impact

Research impact is one of the critical phenomena checked from every corner of the world, i.e., by researchers, funders, government, policymakers, policy implementors or program managers, developmental organizations, members and practitioners of scientific communities, and the general public. The researcher will be primarily concerned about the name and impact factor of the journal where the research manuscript is published and the number of citations received for their published work. The number of citations indirectly indicates the impact of the research on further research. The funders, government, and developmental organizations are primarily concerned about the effect on population health and its scalability, including cost-effectiveness. For practitioners, it is about efficient service delivery, quality of care, patient or population satisfaction, and improved health outcomes.

Research impact is phrased differently across settings as 'effect,' 'influence,' 'benefits,' 'translation,' 'uptake,' 'scale-up,' 'cost-effectiveness,' 'diffusion,' and others. There are more than 100 definitions documented for 'research impact,' which could be due to varied stakeholders. Alla et al. grouped all these definitions as (a) impact with a demonstrable contribution or change or benefit to society and economy, (b) bibliometric definition, and (c) use-based definition [2]. The United Kingdom's Research Excellence Framework defined the research impact as "an effect on, change or benefit to the economy, society, culture, public policy or services, health, the environment or quality of life, beyond academia." [3] It is the most commonly used definition. Based on a systematic review, the revised definition defines research impact as "a direct or indirect contribution of research processes or outputs that have informed (or resulted in) the development of new health policy/practices, or revisions of existing health policy/practices, at various levels of governance (international, national, state, local, organizational, health unit)." Though the revised definition includes the change in any policy or practice, measuring the indirect contribution of the research is difficult in any setting.

Policy and practice

At the most basic level, a policy is a principle of action codified and adopted by the decision authority. The approach is a law, regulation, procedure, administrative action, incentive, or voluntary practice of governments and other institutions. It can be formulated and applicable to global, national, subnational, and local levels. Policies can influence health in many different sectors. The policies developed by one sector may apply to another since no sector is independent. Within public health, policy development includes advancing and implementing public health laws, regulations, or voluntary procedures that influence systems development, organizational change, and individual behavior to promote health and well-being. Traditionally, policy development and implementation is a cyclical process, i.e., agenda setting, policy formulation, adoption, implementation, evaluation, and maintenance [4].

The policy implementation process and its maintenance are documented to be known as practices. Practices are the way laws/policies are implemented. They may include formal procedures, but often they result from organizational culture and habits that have accumulated over time. Practices are reviewed occasionally to determine whether they conform to the nation's mission, philosophy, policies, and international obligations. There is a minute yet substantial difference between policy and practice, and it should be clear to anyone who wants to advocate for change. The difference becomes evident if we have an idea, a problem, or a suggestion; knowing if it is a policy or a practice will ensure that we go to the right people in the concerned domain who can work with us to make that change. To be an effective advocate, we should know who owns what in the decision-making structure of a region.

Knowledge translation (KT) is of paramount importance for formulating a policy to be practiced. The World Health Organization (WHO) defines KT as "the exchange, synthesis, and application of knowledge by relevant stakeholders to accelerate the benefits of global and local innovations in strengthening health systems and improving people's health." [5] Along with Pan American Health Organization (PAHO), WHO has visioned a world where the available scientific evidence in public health research is utilized by the policymaker and other stakeholders in policy-making to be translated into practice by the people to improve their health. To bridge this know-do gap, PAHO Knowledge Management, Bioethics, and Research Department plays a crucial role in PAHO/WHO and other American countries [6]. In developing countries, the Ottawa Model of Research Use framework is suggested as a valid approach to bridge the know-do gap [7]. It focuses on context-specific modeling of KT strategies according to the barriers and the supports found. It also emphasizes measuring the improved health outcomes by applying the research findings in the given setting.

Models of research translation to policy or practice

Research's impact on policy or practice is complex. It is a complex process and warrants more resources than needed for research, which is straightforward. It takes years for translation into policy or practice. Morris et al. reported 17 years lag period for translating clinical research into clinical practice. The same for public health research will be longer due to the complexity of the health issues dealt with by public health and the involvement of various stakeholders.

Gentry mentions different models by which research can be translated into policy in their study [8]. A brief description is attempted here. Linear model: where new knowledge directly drives the new policy. Clinical research is more linear, especially in changing clinical practice. For example, a new drug found efficacies through a clinical trial in controlling blood pressure will quickly shift the prescription practice of clinicians as it is available in the market. The relevant change in national guidelines may take some time though the same is not valid for public health research since conducting randomized control trials and attributing one intervention to the effect are pretty tricky. Further, end users like clinicians in clinical research are not directly involved in changing the policy, which is under the policy or decision makers.

Cyclical/circular model: where a linearly developed policy generates new knowledge gaps, and working further on it may have further policy formulation. It follows the traditional policy cycle from agenda setting to implementation, evaluation, and maintenance. *Interactive, diffusion/sedimentation, and dialogue models:* research is considered one-factor influencing policy, and the dialogue model develops on this, saying that this interaction will generate knowledge. *Translation framework model:* where the study not only serves as an evidence generator but also have interdisciplinary and intersectoral collaboration, which influences the development and uptake of policy.

Research and policy/practice change: Is there a disconnect? [9]

The aim of research, policy, and practice is the betterment of the masses at large. But it is observed that each of these systems only sometimes functions mutually toward their goals (Fig. 6.1). Many practices are not a direct result of the research, and often research is not conducted on issues that could be more important practically. Sometimes the research findings need to be communicated better to the policymakers so as to be incorporated into policy and practice. There needs to be more connection between the three systems. To understand this severance, we should be cognizant of the work cycle of the systems.

Grossly, it depicts how each system finds a solution to an identified problem in a given context. The steps followed by each niche resemble the other but are confounded with the disintegration. Some of the identifiable disconnects are enlisted in the following Table 6.1.

We shall understand the reasons for these disconnects, which will guide us to recommend relevant solutions so that these three vital domains can be collaborated well to achieve benefits and near our aim of improved public health.

  i. **Relevance:** The problem identified depends on the perception of each niche. Policymakers find it relevant to political ideology and public

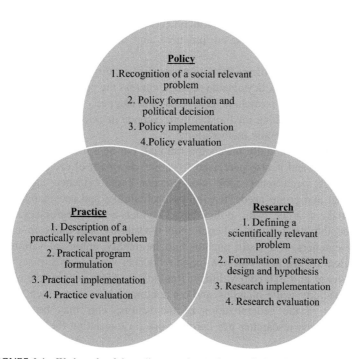

FIGURE 6.1   Work cycle of the policy, practice, and research domains of public health.

TABLE 6.1 Stepwise disconnections between the policy, practice, and research cycles of public health.

Steps in the work cycle	Disconnections
1. Problem recognition	**a.** Relevance **b.** Policy agenda setting **c.** Status
2. Formulation of policy, practice, and research	**d.** Formal power in policy **e.** Goals **f.** Evidence **g.** Legitimacy **h.** Value of theory and practice **i.** Work attitude
3. Implementation	**j.** Adjustment during pilot
4. Evaluation	**k.** Lifespan **l.** External versus internal validity **m.** Public accountability

opinion, practitioners to individual needs, and researchers to a knowledge gap and theory. This social, practical, and scientific relevance may not overlap.

ii. **Policy agenda setting:** Practitioners may sometimes exhibit indirect influence in local policy agenda setting, whereas the political group beholds that authority. Researchers have no say but can put a topic on the political agenda by grabbing media attention. Agenda or priorities need to be set following systematic methodology.

iii. **Evidence:** essentially means the rational, but here it is analogous to the relevance and is also perceived from the viewpoint of each domain. For instance, practitioners look upon practice-based evidence, namely individual and group needs, feasibility, profitability, and acceptability. Policymakers argue with policy-based evidence like legitimacy, public acceptability, political salience, public visibility, and pubic immediacy. Researchers would still anchor onto rationality, empirical validity, and logical precision as research-based evidence.

iv. **Internal and external validity:** Researchers often concentrate on internal validity while planning and conducting, which is an essential determinant of the generalization of the results. This may be secondary because the researcher has hardly work on translating the findings into policy or practice.

v. **Different capacity levels:** Stakeholders in policy making and implementation will have further research and implementation capacities. Hence, understanding and bringing consensus on the problem would be far-fetched.

vi. **Formal power in policy:** Practitioners and researchers are secluded from the formal power of decision-making during policy formulation and are dominated by political decisions.

vii. **Goals:** The policy and practice goals are often difficult to measure, whereas research goals need to be more detailed.

viii. **Legitimacy:** Practice and research primarily focusing on individual approaches may create political unrest as some may feel it interferes with their personal life. The ventured environmental dimension should be defined to sustain a legitimate linkage between research and policy or practice.

ix. **Value of theory and practice:** It is to be argued based on the educational background and academic experience applied in the work cycle.

x. **Work attitude:** The stereotypic images (formed by the work attitude) of scientists as thinkers, practitioners as doers, and policymakers as bureaucrats hamper the collation across the domains. This calls for acceptance of differences in powers, working style, and knowledge of another language to understand the evidence and other influencing factors.

xi. **Adjustment during the pilot:** Technically called interim adjustment, are strongly discouraged in research, but some "muddling through" between steps two and three is permissible in policy and practice implementation, guided by the advances learned in the policy and practice-based evidence.

xii. **Lifespan:** The difference in the dynamic and the lifespan of the work cycles hinders the cohabitation of the niches.

xiii. **Public accountability:** Researchers, practitioners, and policymakers are duty-bound to communicate to the population the process and the outcomes of any research or policy with due transparency. This needs to be improved due to the scientific jargon, length of documents, and ambiguous formulations. The majority of the publications have exclusive usage within the domains.

To overcome these disconnections, awareness of these clinches is necessary. Mutual respect, face-to-face interactions, personal relations, and collaboration will boost the identification of relevant public health problems and synthesize effective remedies to tackle the goon. Relevant news media publications will bridge the prevailing gap due to scientific and ambiguous vocabulary used in international journals and policy documentation. Public health professors and managers of practice institutions endeavoring as social entrepreneurs will have a significant impact on policy-making. A measurable performance indicator that yields mutual benefits should be coined for all the collaboration to begin.

Operational research: A successful example in public health research translation

Operational research (OR) is a critical research method used in public health research focusing on global operations and implementation of healthcare service delivery. It is defined as a "search for knowledge on interventions, strategies, or tools that can enhance the quality, coverage, effectiveness, or performance of the health system, specific health services, or disease control programs in which research is conducted." [10] It has guided tuberculosis (TB) and HIV programs across countries and led to policy and practice change based on evidence. The evidence generation is quick since most of the OR uses existing secondary data based on which the policy or practice change decision is taken. The International Union Against Tuberculosis and Lung Disease, in collaboration with Medecins Sans Frontieres, provides OR training "Structured Operational Research Training Initiative (SORT IT)" under the banners of the WHO. The participants, mainly from low-and middle-income countries (LMICs), are trained in OR, i.e., from identifying program-relevant research questions, developing it as a protocol (contact program), data collection/extraction (distant program), writing the manuscript (contact program), dissemination meetings, and assessing the research impact on policy or practice. Two experts who were already trained in OR guide and provide hands-on training to one participant until publication and assessment of the effects of a research protocol.

Guiding principles and determinants of successful OR [11]

Identifying and understanding the policy priority or program objective is one of the main guiding principles in OR. If no policy priority is available, the same needs to be deliberated and created through the systematic use of evidence. Understand the issues related to policy priorities and their implementation or program objective and plan research on the same. Since the research is conducted in and around the policy priorities or program, there is a high chance of translation in addition to the availability of resources and support. Since most ORs are conducted in the routine program setting, it reflects the practical scenario and eases future implementation or fine-tuning. Based on the experience from conducting successive SORT IT courses and impact assessments, the training group identified various determinants that facilitate the conduction and successful translation of the OR in public health practice [12]:

1. Research question relevant to policy priority or program: Problem identification should be made in collaboration with the domain experts as they better understand which problem is timely and vital. Direct relevance to policy priority or program easily engages the decision-makers or program managers and links the resources, including funding.

2. Collaboration or partnership: OR-based analysis is often complex and interdisciplinary. Collaboration or association with the policymaker or program manager facilitates their ownership and responsibility. Hence, the uptake of the research will be better. It will improve the relationship between the academic researcher and the policy maker or program manager. Their involvement as investigators in the OR will also build research skills directly or indirectly. Similarly, collaboration with development or nongovernmental organizations is equally essential, facilitating the conduction and translation of OR.

3. Simplicity and replication: OR should be simple enough in conduction and data requirement, and at the same time, it should be detailed enough to capture all the aspects of the problem under the radar. Robust methods must be adopted with adequate quality control to measure outcomes accurately.

4. Build the research capacity in the program: It is one of the crucial aspects of any program to train the staff on research methodology. However, such training will help them to understand, undertake, and publish ORs which will impact the policy or practice. Organizations' critical mass of trained researchers will bring a positive culture and increased research outputs.

5. Identification of critical findings and dissemination: The sophisticated OR-based analysis should be expressed as simple one or two sentences with infographics understood by the decision-makers. The work needs to be published in peer-reviewed medical journals, which are usually referred to by decision-makers. Publication in local newspapers or as a newsletter is also essential to disseminate the findings among people.

6. Regular evaluation of implementation: The policy or practice could be changed based on the initial OR, which needs evaluation regularly. New ORs need to be proposed based on the evaluation findings and further identification of issues.

Effect or impact of OR

The SORT IT training group follows the trained participants after the publication of the OR to assess the research impact periodically. They evaluated the impact of OR at the international, national, subnational, local or organization, or hospital level after 18 and 36 months, verified with documentary evidence or through triangulation. Their success rate in translation remained more than 60%, i.e., at least 60% of the research publications on OR (derived from the SORT IT course) had a research impact on policy or practice.

Reasons for no effect or impact on policy or practice

The common reasons for failure to make an impact are (a) short of time to expect an impact, (b) movement to a new organization, (c) no dissemination meeting, (d) minimal quantum of evidence and others. Here we enlist some examples of effect OR on policy or practice related to the prevention and management of TB [10]:

1. The number of follow-up sputum smears among multidrug resistant TB has been changed from two to one based on the research findings of Kumar et al. [13].

2. TB patients are assigned a directly observed treatment (DOT) provider and not given an option of self-supervision in Fiji. This policy change occurred in response to OR-based research stating that TB outcomes were more successful than self-administered treatment [14].

3. There was a scaling up of TB drug sensitivity testing (DST) facilities in the national policy following a study finding by Dave et al. 2015, in India. It said that the treatment outcomes improved in TB patients with the option of DST at diagnosis [15].

4. Ciobanu et al. in their study, concluded that incentive provided to TB patients improved treatment success rates and needs to be continued in Moldova. This led to the financial shift from national budget sources and global funds to incentivize TB patients [16].

5. The designing of an information system for the hospital, which will improve the communication between the hospital and laboratories, was undertaken following an OR-based finding by Nanjebe et al. Uganda, in treating opportunistic infection among HIV-infected patients. It was found that only a tiny proportion of culture reports were found in the patient's chart and hence had less influence on the physician's choice of antibiotics [17].

Specific ORs change the policy or practice in real time, i.e., even before the completion of the research based on interim findings or sometime before the publication. One example was bidirectional testing for TB and diabetes mellitus, i.e., screening for TB among patients with diabetes and vice versa, tested in India. During the conduction of this OR based on the interim report, the decision maker integrated the bidirectional screening for TB-diabetes with the national program. Immediately, the necessary logistics and revision of reporting format were done to start the services. In addition to the determinants of research impact previously mentioned, the multicentric nature and alarming findings are important reasons for such quick translation of research to policy or practice [18].

Case studies on research impact on policy and practice beyond tuberculosis

1. The Swiss Programme for Research on Global Issues for Development (r4d Programme), initiated by the Swiss National Science Foundation and the Swiss Agency for Development and Cooperation, was conducting a randomized controlled trial (RCT) on improving the HIV care cascade in Lesotho: toward 90-90-90—a research collaboration with the Ministry of Health-Lesotho, wherein they had published their protocol to introduce antiretroviral therapy (ART) on the day of diagnosis in rural communities in Lesotho; WHO and International AIDS Society approached the team based on the publication and were very much interested in results of the study. WHO and the other committees took up the recommendation of same-day ART initiation [19].

2. De Angelis et al. used their multiparameter evidence synthesis (MPES) approach to estimate the HIV burden. This approach has been adopted in the national HIV prevention program for England, the Halve It campaign, and Terence Higgin's "It starts with my campaign." In collaboration with Biostatistics Unit, the European organization also use the MPES approach to estimate the HIV burden in other countries like Netherlands and Poland. It is also used to estimate the severity of flu in the United Kingdom and United States of America and hepatitis B prevalence [20].

3. Fall-prevention programs for older ambulatory community dwellers: Gillespie et al. in 2001, in their RCT, showed weak evidence that evaluated a

single-element intervention. However, the providers weighed up screening the "at-risk" elderly and followed up with the intervention. U.S. Center for Disease Control and Prevention has adopted a multifactorial approach in many programs to prevent falls in the elderly [21].

4. Home management of fever among under-five children was tested by Dodowa Health Research Centre, Ghana, in 2000−09. The study showed that the home-based management of fevers with antimalarial only was more cost-effective than antimalarial plus antibiotics. This supported the continuation of the existing national strategy of home management of fevers among children using antimalarial only for fevers and prescribing antibiotics only when there are respiratory signs of pneumonia [22].

Case studies on research impact on policy and practice management of the COVID-19 pandemic

The world has been hit hard by the COVID-19 pandemic; vaccination is the most effective preventive measure to control the pandemic. We will now briefly state a few strategies adopted for improving the COVID-19 vaccine coverage, which is research based.

(a) Scaling and mass vaccination for COVID-19 was one of the essential and quick impacts following the safety and efficacy studies across countries. The rapid development of a vaccine and vaccination of the global mass has never been documented in history.

(b) Booster doses—Administration to healthcare workers and individuals with comorbid conditions was initiated on the recommendation of the National Technical Advisory Group as, lately, 97% of the cases were infected with Omicron. It was based on findings of the Indian SARS-CoV-2 Genomics Consortium that undertakes surveillance of cases [23].

(c) Vaccine ambassadors—research has shown that people tend to get vaccinated if they think most people around them are vaccinated [24]. Endorsement of vaccination by peers, especially when they were vaccinated, either through social media or by discussing the risk of contracting the disease and the usefulness of vaccination, influences the decision [25]. Some examples of such strategies successfully utilized to increase the COVID-19 vaccination uptake by the masses are Motivate, Vaccinate, and Activate Campaign and community-level implementation [26] in San Francisco; Philly Teen Vaxx: Encouraging Peer Vaccination and Vaxx-it-ball in Philadelphia.

(d) Medical provider vaccine standardization integrates vaccination and default options in regular medical practice. More so, a standardized announcement made by the clinician in the treatment room, can impact vaccine uptake [27]. These strategies were used in Partnership Standardizes COVID-19 Vaccination practice in provider offices in Arizona, a state in the southwestern United States. This has also reduced the missed opportunities for vaccination [28].

(e) Motivational interviewing—A pilot study [29] showed a 15% increase in a mother's intention to vaccinate her child and a 7% and 9% increase in vaccination coverage at 7 months and 2 years, respectively. Another study showed improved human papilloma virus (HPV) vaccination completion among adolescent girls [30]. During the COVID-19 pandemic, such strategy was utilized in the innovative notification and motivation interviewing strategy to improve the COVID-19 vaccination in Pennsylvania.

(f) Combating misinformation—it is of utmost importance to burst the misinformation and the disinformation around any issues that are important in promoting public health. These can form a difficult barrier to overcome regarding vaccine acceptance. Research supports fact-checking [31], debunking [32], and refuting a falsehood [33] as effective tools to overcome these barriers. To overcome these barriers, "physicians use TikTok to address vaccine misconceptions" in Florida. The Tennessee department of health's division of disparities elimination formed a COVID-19 health disparity task force to combat the misconception. The team created connections with small business owners, faith leaders, and other key partners to share correct information regarding COVID-19 vaccination and to disrupt the false information.

Case studies where public health research failed to impact the policy or practice

1. WHO developed a framework for the prevention and treatment of Malaria during pregnancy in African regions. It recommended using intermittent preventive treatment in pregnancy (IPTp) with sulfadoxine-pyrimethamine. Malawi, Kenya, Uganda, Tanzania, and Zambia were the first five countries to adopt IPTp as the national policy. The effective delivery of this initiative would be a joint venture of the reproductive health program and the malaria control program, wherein antenatal attendance will be the point of interaction between the two, in addition to IPTp. WHO recommends the use of insecticidal bed nets for the effective management of malaria and anemia. But the framework could have been a success in Zambia. The study cited numerous reasons for the failure of the policy [34].

- The most important of them was that the evidence was based on only one drug (sulfadoxine-pyrimethamine), to which resistance developed subsequently. Also, the policy recommendations were made based on RCT conducted outside Zambia, i.e., Kenya and Malawi
- High lag time from adopting policy and creating a guideline almost took 6 years from 1998 (WHO recommendation) to 2004 (release of

guideline). Also, the translation of the documents from English to other languages hampered the acceptance of the policy. All these issues created a kind of uncertainty regarding the policy recommendation

- Lack of stakeholder commitment and consensus
- Adverse publicity on sulfadoxine-pyrimethamine usage among health workers and the pregnant mother was detrimental, as the headlines flashed figuring a child with Stevens—Johnson syndrome along with the tag line "sulfadoxine kills." Numerous media articles were circulating in the region reiterating similar adverse findings.
- Other probable reasons—Artesunate combination therapy was used as the first line of treatment of uncomplicated malaria in other regions, and sulfadoxine-pyrimethamine was seen as a failed drug. Also, it failed to make the patient feel better as it did not have an antipyretic effect, unlike chloroquine. Since malaria transmission was often asymptomatic, the pregnant women did not appreciate the need for the second dose. A weak linkage between the malaria prevention and reproductive health programs, lack of sustainable funds, limited human resources, and inadequate monitoring systems also played a crucial role in the failure of the policy to be transmitted into practice

2. India's total sanitation campaign was launched in 1999 by the GOI to improve the quality of life in rural areas and increase sanitation coverage. As a joint state and federal government collaboration, it was a community-led, people-centered, and incentive-based program. Initially, there was an increase in sanitation coverage from 22% in 2001 to 31% in 2011, but the program eventually failed due to low political priority, inadequate monitoring, corruption, and other reasons [35].

Summary

Translation of research findings into policy or practice change is a paradoxical event influenced by numerous factors, ranging from identifying the relevant problem, the research setting, communicating the findings, policy formation, the actual practice change in the community, and collaboration or engagement of decision makers. Research-based evidence is a strong base for policy change with numerous success stories, but it is often questionable due to a single intervention and ideal study setting. OR conducted in routine programmatic settings showed a better impact on policy or practice, especially as it is a boon for resource-limited settings like LMICs. Stakeholder and multi-domain collaboration is the cornerstone of successful policy formulation or research translation, but the scope is broader if research managers endeavor as social entrepreneurs.

References

[1] Chalmers I, Glasziou P. Avoidable waste in the production and reporting of research evidence. Obstet Gynecol December 2009;114(6):1341—5. https://doi.org/10.1097/AOG.0b013e3181c3020d.

[2] Alla K, Hall WD, Whiteford HA, Head BW, Meurk CS. How do we define the policy impact of public health research? A systematic review. Health Res Policy Syst December 2, 2017;15(1):84. https://doi.org/10.1186/s12961-017-0247-z.

[3] Higher Education Funding Council for England—GOV.UK. Available from: https://www.gov.uk/government/organisations/higher-education-funding-council-for-england. (Accessed 17 February 2023).

[4] European Geoscience Union. The policy cycle. Policy. Available from: https://www.egu.eu/policy/cycle/. (Accessed 17 February 2023).

[5] World Health Organization. Rapid response knowledge translation mechanisms to translate evidence into public health policy in emergencies rapid response knowledge translation mechanisms to translate evidence into public health; 2006. Available from: https://apps.who.int/iris/handle/10665/341972. (Accessed 17 February 2023).

[6] Pan American Health Organization, World Health Organization. Knowledge translation for health decision making. Available from: https://www3.paho.org/hq/index.php?option=com_content&view=article&id=14477:knowledge-translation-for-health-decision-making&Itemid=0&lang=en#gsc.tab=0. (Accessed 17 February 2023).

[7] Santesso N, Tugwell P. Knowledge translation in developing countries. J Contin Educ Health Prof 2006;26(1):87—96. https://doi.org/10.1002/chp.55.

[8] Gentry S, Milden L, Kelly MP. Why is translating research into policy so hard? How theory can help public health researchers achieve impact? Publ Health January 2020;178:90—6. https://doi.org/10.1016/j.puhe.2019.09.009.

[9] Jansen MW, van Oers HA, Kok G, de Vries NK. Public health: disconnections between policy, practice and research. Health Res Policy Syst December 31, 2010;8(1):37. https://doi.org/10.1186/1478-4505-8-37.

[10] Tripathy JP, Kumar AM, Guillerm N, Berger SD, Bissell K, Reid A, et al. Does the structured operational research and training initiative (SORT IT) continue to influence health policy and/or practice? Glob Health Action January 6, 2018;11(1):1500762. https://doi.org/10.1080/16549716.2018.1500762.

[11] Brandeau ML. Creating impact with operations research in health: making room for practice in academia. Health Care Manag Sci December 24, 2016;19(4):305—12. https://doi.org/10.1007/s10729-015-9328-0.

[12] Walley J. How to get research into practice: first get practice into research. Bull World Health Organ June 1, 2007;85(6):424. https://doi.org/10.2471/blt.07.042531.

[13] Kumar RS, Kumar AMV, Claassens M, Banurekha VV, Gomathi NS, Venkatesan P, et al. Number of sputum specimens during treatment follow-up of tuberculosis patients: two or one? Public Health Action December 21, 2013;3(4):304—7. https://doi.org/10.5588/pha.13.0049.

[14] Narayan N, Viney K, Varman S. Comparison of tuberculosis treatment outcomes by method of treatment supervision in the Fiji Islands. Public Health Action September 21, 2014;4(3):174—8. https://doi.org/10.5588/pha.14.0020.

[15] Dave P, Vadera B, Kumar AMV, Chinnakali P, Modi B, Solanki R, et al. Has introduction of rapid drug susceptibility testing at diagnosis impacted treatment outcomes among

previously treated tuberculosis patients in Gujarat, India? PLoS One April 13, 2015;10(4):e0121996. https://doi.org/10.1371/journal.pone.0121996.

[16] Ciobanu A, Domente L, Soltan V, Bivol S, Severin L, Plesca V, et al. Do incentives improve tuberculosis treatment outcomes in the Republic of Moldova? Public Health Action October 21, 2014;4(Suppl 2):59–63. https://doi.org/10.5588/pha.14.0047.

[17] Nanjebe D, Wilkinson E, Reid T. Is bacterial culture and sensitivity testing of value in treating infections in HIV+ patients in Uganda? (Unpublished manuscript)

[18] Kumar AMV, Satyanarayana S, Wilson NC, Chadha SS, Gupta D, Nair S, et al. Operational research leading to rapid national policy change: tuberculosis-diabetes collaboration in India. Public Health Action June 21, 2014;4(2):85–8. https://doi.org/10.5588/pha.14.0012.

[19] Erismann S, Pesantes MA, Beran D, Leuenberger A, Farnham A, Berger Gonzalez de White M, et al. How to bring research evidence into policy? Synthesizing strategies of five research projects in low-and middle-income countries. Health Res Policy Syst December 6, 2021;19(1):29. https://doi.org/10.1186/s12961-020-00646-1.

[20] Cambridge Biomedical Campus. Case studies. Available from: https://cambridge-biomedical.com/discoveries/case-studies/. (Accessed 17 February 2023).

[21] Marks R, Allegrante JP. Falls-prevention programs for older ambulatory community dwellers: from public health research to health promotion policy. Sozial-und Präventivmedizin/Soc Prev Med June 1, 2004;49(3):171–8. https://doi.org/10.1007/s00038-004-3040-z.

[22] INDEPTH Network. Translating research into policy and practice-examples from health and demographic surveillance sites. p. 1–7. Available from: http://www.indepth-network.org/pec/translating_research_into_policy_and_practice.pdf.

[23] The Hindu. Genome consortium advises booster shots. COVID-19; 2022. Available from: https://www.thehindu.com/news/national/covid-19-genome-consortium-insacog-advises-booster-doses-be-considered/article37813524.ece. (Accessed 17 February 2023).

[24] Quinn SC, Hilyard KM, Jamison AM, An J, Hancock GR, Musa D, et al. The influence of social norms on flu vaccination among African American and White adults. Health Educ Res December 1, 2017;32(6):473–86. https://doi.org/10.1093/her/cyx070.

[25] Bronchetti ET, Huffman DB, Magenheim E. Attention, intentions, and follow-through in preventive health behavior: field experimental evidence on flu vaccination. J Econ Behav Organ August 2015;116:270–91. https://doi.org/10.1016/j.jebo.2015.04.003.

[26] Marquez C, Kerkhoff AD, Naso J, Contreras MG, Castellanos Diaz E, Rojas S, et al. A multi-component, community-based strategy to facilitate COVID-19 vaccine uptake among Latinx populations: from theory to practice. PLoS One September 20, 2021;16(9):e0257111. https://doi.org/10.1371/journal.pone.0257111.

[27] Brewer NT, Hall ME, Malo TL, Gilkey MB, Quinn B, Lathren C. Announcements versus conversations to improve HPV vaccination coverage: a randomized trial. Pediatrics January 1, 2017;139(1). https://doi.org/10.1542/peds.2016-1764.

[28] The room where it happens: primary care and COVID-19 vaccinations | Commonwealth Fund. Available from: https://www.commonwealthfund.org/publications/2021/jul/room-where-it-happens. (Accessed 17 February 2023).

[29] Gagneur A. Motivational interviewing: a powerful tool to address vaccine hesitancy. Canada Commun Dis Rep April 2, 2020;46(04):93–7. https://doi.org/10.14745/ccdr.v46i04a06.

[30] Dempsey AF, Pyrznawoski J, Lockhart S, Barnard J, Campagna EJ, Garrett K, et al. Effect of a health care professional communication training intervention on adolescent human papillomavirus vaccination. JAMA Pediatr May 7, 2018;172(5):e180016. https://doi.org/10.1001/jamapediatrics.2018.0016.

[31] Zhang J, Featherstone JD, Calabrese C, Wojcieszak M. Effects of fact-checking social media vaccine misinformation on attitudes toward vaccines. Prev Med April 2021;145:106408. https://doi.org/10.1016/j.ypmed.2020.

[32] Chan MS, Jones CR, Hall Jamieson K, Albarracín D. Debunking: a meta-analysis of the psychological efficacy of messages countering misinformation. Psychol Sci November 12, 2017;28(11):1531−46. https://doi.org/10.1177/0956797617714579.

[33] Bode L, Vraga E. The Swiss cheese model for mitigating online misinformation. Bull At Sci May 4, 2021;77(3):129−33. https://doi.org/10.1080/00963402.2021.1912170.

[34] Crawley J, Hill J, Yartey J, Robalo M, Serufilira A, Ba-Nguz A, et al. From evidence to action? Challenges to policy change and programme delivery for malaria in pregnancy. Lancet Infect Dis February 2007;7(2):145−55. https://doi.org/10.1016/S1473-3099(07)70026-9.

[35] Hueso A, Bell B. An untold story of policy failure: the Total Sanitation Campaign in India. Water Pol December 1, 2013;15(6):1001−17. https://doi.org/10.2166/wp.2013.032.

Chapter 7

Public health policy for social action to ensure better population health

Kalaiselvi Selvaraj

Department of Community and Family Medicine, All India Institute of Medical Sciences (AIIMS), Madurai, Tamil Nadu, India

Introduction

The 'Health in All Policies' approach and commitment toward sustainable development goals (SDGs) emphasize the role of social determinants of health to improve and maintain the health of individuals and the population. The recent World Social Report cautions that growing inequality in the world can jeopardize the progress of internationally committed development goals such as SDGs [1].

Health is a social good. Often, most health outcomes result from situations prevailing in sectors other than health. The health system acts as the shock absorber for other sectors' involvement. Hence now the time has come to realize the role of other partners and make entry points to have a strong collaboration with non-health sectors such as education, transport, engineering, social welfare, food, information technology, and other sectors. The decision-makers involved in health-related decision-making should try to collaborate successfully with other sectors. There could be several approaches to make such collaboration depending on the available resources, needs, and expertise. Social action through better public health policies is one of the important approaches to addressing the social determinants of health.

Social actions are often better understood in the name of community participation and collective actions. The following quote from a famous sociologist from the United States, Mr. Charles Tilly, best captures collective action, "people acting together to pursue their common interests." [2] Community Action for Health (CAH) is "collective efforts by communities directed toward increasing control over the determinants of health and thereby improving health" [3].

Principles and Application of Evidence-Based Public Health Practice
https://doi.org/10.1016/B978-0-323-95356-6.00009-4

Improving public health is not about money or targets. It's about attitude and shared vision. The only resource you will ever need is the community itself. The widely applied social action theory better explains the phenomena under social actions related to health [4]. The social action theory conceptualized by Ewart et al. has proposed factors to achieve desired health action or behavior at three levels. According to this theory, contextual factors influence self-change process, impacting the desired health behavior or action [4].

Social actions in health include community activism for healthcare reform, reducing disparities to render social justice, and public safety. It could be in specific areas like physical disability and rehabilitation, AIDS, cancer, and others. The health-related social actions and movements could be better linked and understood with health activism. Health activism is defined as "taking responsibility for individual health, working to improve health conditions for a group and making efforts to change and improve policies for a large group of people" [5]. Hence, such health activism needs policy-level decisions in various sectors since it deals with every aspect of daily living and the equal distribution of resources to achieve health.

Public health policy refers to statements or formal positions issued by a government or its departments that often suggest the accomplishment of a goal or purpose. Programs that embody policy may not be recognized as policy in several instances. For instance, providing healthy food choices in an industrial environment could be considered an action rather than a policy, even though all these actions emerge from collective local occupational policy intended to provide a healthy workforce. In recent decades, diseases like tuberculosis reflect the cumulative actions of the socio-cultural and political environment and have been set as an indicator to assess progress in development goals. In short, the prevalence of certain health conditions such as maternal mortality, infant mortality, malnutrition, and diseases like tuberculosis reflects the "causes of the causes," which need to be tackled through social actions.

Role of public policies or social action on population health

Apart from biological factors and direct healthcare system—related factors, several social constructs play a significant role in health. Some of these determinants act directly to influence health, and some determinants affect health by acting as an intermediary determinant in the pathway. Hence globally, there is a strong felt need to adopt the 'Health in All Policies' approach.

Prof. Amartya Sen, the Nobel Laureate, quotes the health outcome as a product of various dimensions of functioning and its agency [6]. However, public health experts are comfortable dealing with biomedical aspects currently despite the fact that the approach has shown limited success in improving the health and well-being of the population. Though it sounds great, the health system and other related sectors have faced significant challenges in implementing the stated plans to achieve population health in reality.

In the situation of various emerging socio-political, economic, and environmental factors that endanger life, failure to create social mechanisms will be a significant threat to inclusive development. There is an urgency to have this social mechanism in the world, especially in low- and middle-income countries (LMICs), where the health system is already overwhelmed with existing health conditions. Through a pathway of greater knowledge, collective efficacy, and demands for greater rights, some marginal groups stood with policymakers to leverage their rights and identity. They ensured access to housing facilities and other basic amenities and financial security. Social actions, which are collective efforts of common people, policy advocates, civil society organizations, research, technical experts, and policy communities, can effectively portray the issue for further deliverable actions. The schematic diagram depicts how social actions can influence policy at various levels through different methods (Fig. 7.1) [7].

The social actions influence public policy at all four stages: agenda setting (initial placement and positioning of issue), technical design of the policy reform (policy design), legislative follow-up, and passage of the proposed

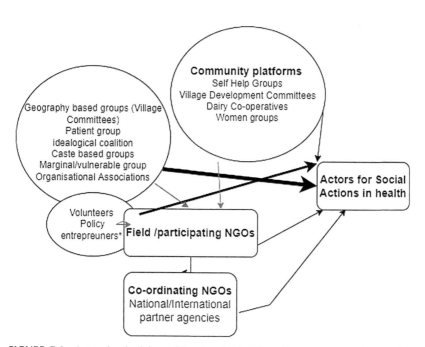

FIGURE 7.1 Actors involved in social action for health; policy entrepreneurs/communities include community volunteers, technical experts, researchers, and activists.

political bill (adoption), and implementation of adopted policy (policy implementation).

Pathway of social action to improved health

The method or pathway of social action to improved health is very complex. Social actions facilitate positioning of a public health problem by converting the material view into a social constructionist view. The social construction view triggers the action toward policy reform. The policy reform could be at the micro- or macro level, modulated by various social actions, including public hearings, advisory forums, consultative meetings, lobbying, and other modern social action methods (Fig. 7.2).

The process and methods of social action leading to population health may vary for every setting depending on the local need and available resources. However, broadly raising awareness and empowering the community, regulatory methods, fiscal methods, and creating enabling environment are the common methods of social action (Table 7.1).

Some countries like Sri Lanka, Cuba, and Costa Rica are examples of successful implementation of primary care. These countries made progress in health and social indicators and achieved this success, not from health sector efforts alone. These countries focused on areas of social determinants such as essential food security, improved female education, and improved access to safe water and sanitation. These countries have invested in improving access to health through social health insurance policies, enhanced occupational safety through labor laws, mobilizing civil society participation for health, and equity-driven health care to reach the more vulnerable and poorer.

Case studies on the impact of public policies or social action on population health

Public policies or linked social action impact the population health directly and indirectly. This varies from setting the policy agenda to developing and implementing macroeconomic policies to bringing transparent governance. Further, it varies from dealing with the whole population well-being to control of specific diseases or health issues.

Role of social actions in public health policy agenda setting

In the context of recent transformation, several health systems issues have competing priorities with each other. In resource constraint countries (LMIC), the health system is struggling to achieve its goal. For instance, diseases like HIV/AIDS could get the attention of policymakers and donor agencies, which is reflected by around one-third of external agency support being attracted to HIV/AIDS alone though the mortality due to HIV/AIDS alone is <5%.

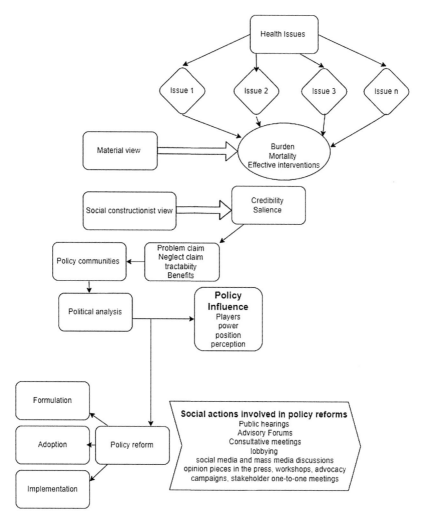

FIGURE 7.2 Process involved in portraying the issue from material to social construction through social actions for policy reforms. *The author's own framework is adapted from Jeremy Shiffman's paper on the social explanation for the rise and fall of global health issues.*

Despite the huge number of deaths and morbidities, diseases like undernutrition, pneumonia and noncommunicable diseases could not receive that attention.

In countries like India, where more than 80% of healthcare expenditure comes from out-of-pocket payments, neglecting health issues and forgone healthcare becomes a systemic problem. To counter this, many communities have devised some financial risk protection and resource-generating mechanisms. Community-based health insurance scheme (CBHI) is one such

TABLE 7.1 Methods followed in social actions to produce implementation change in health.

Environment change	Behavior modifications
Poverty elimination and employment Dealing with proximal risk factors/social determinants of health Community development actions Change service delivery sites	Raise awareness • Conduct educational meetings • Informing local opinion leaders • Use of mass media Increase demand
Laws and regulations	**Fiscal measures**
Redistribution of health-related resources and power Eliminate monopoly and increase the access	Access to new funding could involve new uses of existing money, cost-sharing, and raising private funds. These funds are used to support necessary actions such as purchasing materials, arranging transport, paying fees, or reimbursement of incurred expenditure
Social movement	**Policy measures**
Community mobilization and empowerment Equitable distribution of resources People's collective responsibility	Supporting social policies

successful effort that received close attention from stakeholders from developing countries. Under this, community members formed a group based on their poor socio-economic status or belonging to marginal communities. These low-income families enroll voluntarily with a minimum premium starting from INR 20/person/year to less than INR 1000/family/year. Premium is paid to any credible NGOs working for that community development for a long time. Depending upon the scheme's design, health care was delivered directly through NGOs or empaneled private providers. During a health crisis, the family benefits to access health care without out-of-pocket payment, or the incurred health expenditures are reimbursed. In the Indian context, several successful CBHI schemes benefitted the tribal community and women from poor households. Following CBHI examples from India are often quoted as best examples: (i) Self-Employed Women's Association—from Gujarat (India) benefitting 535,000 families of women SEWA members, (ii) Karuna trust—benefitting around 278,000 people from scheduled caste and tribal community in Narsipur block, Karnataka (India), (iii) Yeshaswini—gives health insurance coverage to tenant farmers in Karnataka who are part of cooperative society for in-patient services [8−10].

Collective efforts through these schemes have led to improved access and affordability to health services which is otherwise deprived in these communities, and the utilization of healthcare services is also found to be more in these families. Under these schemes, the enrolled families could face a health crisis without much financial hardship.

Impact on social determinants of health

Several initiatives like subsidies or scholarships for education, agricultural pro-policies, and social security schemes like tax exemption for the vulnerables have increased social mobility. Actions that are targeted to improve social mobility are proven to impact health in many ways. Table 7.2 summarizes how social mobility promotion affects health improvement.

Considering the heavy reliance on social determinants of health, Laverack quotes health workers to play an intermediary role in enabling the community through their communication and facilitating community actions [12]. India's most extensive flagship program of Accredited Social Health Activists (ASHAs) is also an example of how social contribution can result in better healthcare access and outcome. However, ASHAs are being increasingly used in clinical practice. Again, this trend has created a vacuum in addressing social determinants of health and advocacy from the community [4].

Impact of social actions on health governance and accountability

The following few examples reflect how the public policies which had the scope of social actions mediated to achieve health-related outcomes by ensuring governance and accountability.

Facility level

The initiative by the National Health Mission of the Indian context, Rogi Kalyan Samiti (RKS) [Patient Welfare Committee], has demonstrated the impact of collective social actions on health. RKS is a local governance mechanism comprising representatives from the community like village representatives, elected members, healthcare providers, civil society members, and representatives from the administrative division. This committee is mandated to improve the amenities required for patient care in health facilities, including developing infrastructures, ensuring supplies, and maintaining cleanliness. In addition, the committee is responsible for arranging transport, directing the resources to match the local needs, fostering contributions and donations for health care in an accountable manner, seek interim solutions like the contractual appointment of health staff, and making decisions related to user fees, etc.

As this mechanism has a wide scope to capture community felt needs, well-functioning RKS facilities could provide patient-centric care with more satisfied healthcare attendees. Often, these structures could act as an effective

TABLE 7.2 Examples of impact of public policies or social actions on social determinants of health.[a]

Problem	Health impact pathways	Policy approaches	Social actions	Example scenarios
Education Ill health is the cause of 200–500 million lost school days. Low-resourced and unsafe teaching environment causes many tragic incidents and hampers teacher and student health. School dropout is inversely related to immunization coverage in various countries.	More school years, especially among females, are associated with reduced incidence of infant and child mortality and undernutrition. In recent years, the health sector's contribution to shaping the school curriculum inculcates hygiene, health promotion, oral health, and nutrition. Some recent initiatives such as health-promoting schools, child-friendly school initiatives, and school health nutrition reflect it.	Policies enhancing access and quality of education are crucial to achieving health outcomes. In developing countries, improving school infrastructure through potable water and toilets has reduced school dropouts. Free mid-day meals in schools have the potential to reduce school dropouts and improve school attendance and child's nutrition. Schools play a significant role in protecting against physical and sexual abuse, emotional ill-treatment, neglect, and exploitation. Health policies should create strong entry points for activities such as vaccination, health education, health literacy, and nutrition supplementation. Health points near the school	The recent initiative of the Ayushman Bharat (National Health Protection Scheme, India) Ambassador under the multisectoral plan to control NCDs gives a promising scope to set a promotive health environment through student-led approaches. This initiative is targeted to improve healthy eating, promote physical activity, control substance use, enhancing emotional	Several integrated community and school-based interventions have contributed to raising self-esteem and channelizing adolescents addicted to substances or violence into productive activities such as sports. Intervention in the name of a 'stepping stone' in South Africa gave appropriate exposure to learners to equip them with negotiating life skills. As a result, the country has seen a decline in sexually transmitted illnesses, risky behaviors, and gender-based violence in adolescents. Free education for girls till higher studies Fellowship/Scholarship for the first graduate of the family to complete higher studies.

The behavior adopted at the school age is a determining factor for the future incidence of various noncommunicable diseases. Education attainment is also associated with better disease control states such as glycemic and blood pressure control. In the era of emerging NCDs, having a self-reliant population in self-care could be achieved through education alone.	provide health services, and speech and language therapy which has improved school attendance.	intelligence. This scheme also gives a scope by guiding the school children to become change agents in their community to set a health promotive environment in their villages. Since historical times, school teachers have acted as great resources to identify health conditions (vision/hearing impairment, disabilities etc.) among school children and refer them for appropriate care.	
Housing Due to unplanned rapid urbanization, millions of people worldwide live in	Lack of connectivity to services and public utilities from the	Legislating housing standards, promoting affordable housing by leveraging tax and subsidies, social housing schemes, and	Community awareness campaigns to build fire safety	In some cities like Philadelphia, US, citizens' forums were instrumental in revising the housing safety standards to

Continued

TABLE 7.2 Examples of impact of public policies or social actions on social determinants of health.[a]—cont'd

Problem	Health impact pathways	Policy approaches	Social actions	Example scenarios
houses with poor living standards and unsafe environments. In urban slums, more than one-third of income goes for house rent which compromises the other needs, especially health. Forced evictions and forced resettlement are increasing and negatively affecting health. More than 3% of the global disease burden is attributed to indoor air pollutants alone.	house leads to poor educational attainment, loss of job security, and neglect/delay in health-seeking. Housing is a direct agent in causing multiple diseases such as infestations with parasites, vector-borne diseases, respiratory illnesses due to overcrowding, homicide, fire, accidents, and injuries. This housing is the source of exposure to various indoor biological and chemical agents,	rental voucher schemes Rent control policies Special slum-based healthcare services	measures, criminal act preventive measures, awareness generation on basic sanitation and hygiene and resource generation for public utilities.	protect from injuries such as electrical/fire-related hazards. Some of the experiences of including community voices while drafting a housing master plan have resulted in better implementation of housing and health-related interventions in the area.

	including indoor air pollutants. An overcrowded environment, in the long run, can compromise the quality of life, thereby the mental health issues as well			Despite the clean cooking fuel policy, the Indian states differ in the use of clean fuels. A national sample survey-based analysis shows that better use of clean fuel in Tamil Nadu compared to Odisha which was hypothecated to governance, and other social determinants of clean fuel use.
Energy Globally, more than 1.3 billion people could not avail of electricity, and about 2.7 million could not have access to clean fuels, and the problem is more concentrated in African and developing Asian countries.	Exposure to traditional fossil fuels leads to indoor air pollution, thereby causing respiratory and cardiovascular disorders, burns, injuries, and fire accidents The energy issue also carries gender and equity dimensions, as the poorer primarily rely on fossil fuels. Women spent most of their time collecting wood,	Increase access to a renewable energy source (wind energy, solar power), regulatory approach for safety standards, construction standards to match with energy efficient measures, fuel choice and replacement vehicle policy, subsidy and loan for clean fuel options, and eco-labeling policies	Raising awareness and information sharing are promoting the use of energy-efficient appliances. Dissemination of health risks due to fossil fuel Comply with regulatory and safety standards	

Continued

TABLE 7.2 Examples of impact of public policies or social actions on social determinants of health.[a]—cont'd

Problem	Health impact pathways	Policy approaches	Social actions	Example scenarios
	leading to compromises in education and income opportunity. Lack of affordability to health access which in turn created by lack of income opportunities due to scarcity of energy-efficient resources			
Transport: Road traffic injuries. The 10th common cause of death. It consumes nearly 1.5% of GDP for its management. A family from an area with poor transport access spends 25% –35% of total family income on it. Due to this, fewer	Direct: accidents and injuries Indirect: air and noise pollution, congestion Geographic access and access to public transport are the primary	Regulation of road infrastructure standards, vehicle safety standards, driver licenses, training, and safety regulations such as seat belts, helmets, forbidding drunken driving. Enabling an environment for pedestrians to promote walking	Improving the non-motorized forms of transport (bicycles and pedestrians) as an alternative to the motorized vehicle. Expansion of public transport	The voucher scheme adopted in the state of West Bengal, India, aims to provide access to transport services for marginalized pregnant women to reach the facility at the time of delivery. Under this scheme, the woman gets reimbursement for the financial expenditure

resources are left for food, health, and other expenditures of the family. Often transport accounts for 20%–35% of expenditures incurred for health matters. However, most social security measures and health insurance schemes do not provide coverage for transport.	influencers for maternal and child healthcare utilization. Most home deliveries are the result of poor access to transport services. Extended travel time, which compromises involvement in other activities such as family time and physical activity. Resource-poor healthcare settings deviate from other competing health emergency priorities such as conducting delivery. Road constructions that are friendly to pedestrians and cyclists have influenced the reduction of	and cycling. Adequacy of transport Subsidiaries or special vouchers	and reducing car usage has reduced the fatalities due to road traffic injuries and reduced the traffic congestion. Some of the transport fare waiving schemes for vulnerable like women and the elderly have increased their mobility to a great extent and, in turn, increased their access to health care and social services. During several environmental impact assessment processes, stakeholders from the community have raised their voices to halt the project, which can endanger the	incurred for the vehicle. Various Indian states provide free transport services to patients with cancer and disability. Recently, the state of Tamil Nadu and Punjab (India) has been implementing free local transport for women to empower and help them reach various public utilities, including health facilities. Initiatives like car-free days or restriction of personal vehicle use with restricted parking time and tax are helping in overall vehicle movement. It, in turn, helped to reduce air pollutants and hospitalization

Continued

TABLE 7.2 Examples of impact of public policies or social actions on social determinants of health.[a]—cont'd

Problem	Health impact pathways	Policy approaches	Social actions	Example scenarios
	obesity and noncommunicable diseases related mortality in urban settings.		equity and health of the people.	
Poverty and social protection Chronic poverty harms health by restricting health access and deterring earning opportunities. The vicious cycle of poverty and adverse health impacts never ends.	Social vulnerability to poverty-related causes results in the intergenerational transmission of poor nutritional status, loss of social cohesion, and conflict behaviors.	Globally policies such as old age pension, the benefit for the disabled, income assurance for the unemployed, maternal health protection scheme, cash/cash in kinds transfers for the vulnerable, equity-driven legislation, food subsidies, employee's insurance schemes or injury benefits have been implemented to ensure social protection. Apart from the direct impact, such as food security, these social protection schemes lead to better school attendance, human development potential, and improved health care utilization.	Income security measures and non-conditional benefit transfers act as gateways to facilitate access to several other amenities. Rural employment guarantee acts and disaster relief funds such as crop security are income-ensuring measures to improve the livelihood of rural marginalized and gender equity in India. These measures protect the people by bringing down dangerous coping behaviors such as sending the children to employment, neglecting health care, and slashing expenditure on basic amenities.	

Health services challenges and social protection measures	Direct payment or lack of affordability hinders access to health care.	Reduction of out-of-pocket payments in the form of levy of user fees, risk pooling, and reimbursement through health insurance schemes have increased healthcare utilization such as immunization, institutional deliveries, etc.	In Cambodia, an initiative of UNICEF, in the name of health equity funds, included community members, local health authorities, and local NGOs. This initiative acted as a third-party payer to purchase health care for the poor. Access improved as this initiative took care of the money for the needy to buy health care. Further, this committee was also responsible for identifying the beneficiary who needs this fund the most. India's largest flagship social health insurance program for the informal workforce, "Rashtriya Swasthya Bima Yojana" is in line to increase the access to health care among the marginalized casual laborer and reduce the direct payment at the time of health service delivery. India's health system initiative under Janani Suraksha Yojana entitled to transfer money for the women who give birth in the institute. As a result, institutional delivery has increased by two-three folds which impacted on reducing the maternal mortality, especially in states where the health indicators are poor. This scheme also led to the overall increase in antenatal and postnatal services, which were highly neglected areas before the scheme's implementation.

[a]Examples in this table are from WHO reports on social determinants of health—Sectoral briefing series [11].

forum to address patient redressal, transparent governance, accountability, and a pivotal force to ensure quality assurance in health facilities [13,14].

Along similar lines, social audits, which people jointly carry out with healthcare administrators for issues related to maternal and infant deaths, identified the gaps, and facilitated in implementing the potential solutions. These social audits act as a tool for awareness generation, creating accountability, transparency, and monitoring and evaluation mechanism for the services related to maternal and child health issues [15].

Community level

India's largest flagship program National Rural Health Mission launched 'Community Action for Health (CAH)' at its inception. Under this, there is a three-way partnership between healthcare providers, healthcare managers, and community members. In the community, members are proposed to monitor the access, service availability, such as the presence of healthcare providers, drugs, medical negligence, and staff behavior, and quality of the services. Till 2018, more than 0.2 million villages had implemented this CAH. Though there is minimal documentation on its impact, the available and empirical evidence suggests that the role of community and civil societies was instrumental in improving governance and accountability in the health sector [16]. In Maharashtra (India), since 2010, three rounds of community-based monitoring of public services, including health sector roles, were conducted through a team of representatives from Panchayat Raj Institution (District administrative committee). The primary aim of this exercise was to check whether the services proposed in the district action plan had reached the grassroots level.

The following box summarizes some of the significant improvements produced by community members as a part of the CAH community monitoring: Box 7.1.

Through self-help group

In Gujarat, trained local women run the community resource hub under the women-run society in the name of Shakti Seva Kendras (SSKs). The SSK centers are the initiatives of women who work in the informal sector. These centers help the community to raise awareness of health issues, health facilities, social security schemes, and ways to achieve them by acting as a knowledge resource hub. As a result, in the state of Gujarat, among the vulnerable where these SSK centers are available, the coverage on utilization of social security schemes like pension schemes, maternity conditional cash transfer schemes (Janani Suraksha Yojana [JSY]), and others have improved. Similarly, the knowledge of public health services also improved [17].

BOX 7.1 The list of achievements of community action for health in India

- Nonfunctional health subcenters and laboratory facilities are revived to get their functions.
- Due to community pressures, new health worker posts were sanctioned, and vacant posts were filled.
- The number of OPD attendees doubled after this community monitoring.
- The frequency of auxiliary nurse midwife (ANM) visits improved
- Several health facilities significantly improved institutional delivery and immunization coverage.
- Vulnerable populations and adolescents are represented in conveying their needs and getting it done.
- Conditional cash transfers had regular disbursement to beneficiaries without any delay.
- Prescription and outside purchase of drugs stopped. Provisions were made to purchase the drug under untied funds of RKS
- The contribution of the successfully functioning Village Health Sanitation and Nutrition Committee (VHSNC) is enormous from efforts to make drinking water safe and facilitate the building of toilets.

Social action toward specific health issues or disease control

Community members have contributed to various social actions, from mere participation in consultative and substantive participation to the level of structural involvement in the health system. Historically social activities have been associated with health outcomes and intermediary health-related determinants such as access, equity, utilization, quality, and responsiveness to a specific disease or health condition. Because of these valuable outcomes of community involvement globally, there has been a steady growth in community participation, especially in primary care [18].

Infectious diseases

The recent COVID-19 pandemic has given rise to several unpredictabilities. With the perception of living in a risky society, the administrative or executive agencies enforced quite challenging public health interventions. Most communities faced the problem of continued COVID-19 spread, low supplies, conflict with authorities in cooperating with measures for strict social isolation, and social distancing measures.

Despite the world having faced the spread of COVID-19, many case studies demonstrated effective transmission control in a short duration. Here is an example from a Chinese rural community of how the collective and

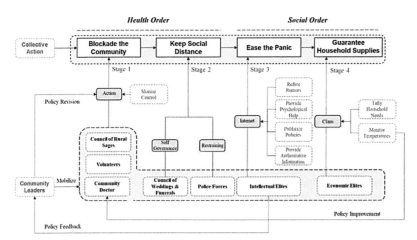

FIGURE 7.3   Collaborative social action for control of COVID-19 transmission in Quanjingwan.

collaborative partnership from different community forums had effectively resulted in effective control of COVID-19 transmission in the community (Fig. 7.3).

The Quanjingwan community is a rural village in Hubei province catering to 115 households and 765 residents, including 63 migrants who recently moved from Wuhan. This village is 200 km away from Wuhan city, the epicenter of COVID-19. The village committee and Council of Rural Sages [group of villagers with prestigious positions such as retired, military employees, and literates] had come together to work for common agenda on pandemic control. However, they had various power conflicts before this. This group blocked the outsiders' entry by blocking all seven intersections in the village by building mud walls or stone piles for villagers. Similarly, large meetings and weddings/funerals were restricted or canceled with the help of local volunteers. Also, this team ensured that no one entered the village and the police force ensured that there was no deviation from the social distancing measures. Inside the village, a community doctor screened everyone for COVID-19 and suspected COVID-19 patients, including the people who returned from Wuhan, were put in isolation rooms. The community doctor took care of these isolated patients through daily check-ups. Commodities, including food supplies, were distributed uninterruptedly to the isolated patients through villagers, and garbage was disposed of carefully by the assigned staff from the village. Thus, the COVID-19 transmission inside the village was kept under control.

As there was a threat of running out of supplies and commodities, village leaders approached the economically better off members of the community to take the responsibility of getting a pass and arranging supplies in bulk to distribute among villagers. Individual coordinators, assigned for each clan (street-like structures), calculated the actual estimates of required supplies for their clan. Thus, a team of community representatives ensured the health order, and another group of community members ensured the social order. To alleviate fear and share up-to-date, relevant, credible evidence, intellectual elites from the village actively disseminated knowledge through various social media platforms [19]. Importantly, the role of the health department or doctor was minimal and the community predominantly managed the whole situation on their own.

Noncommunicable diseases and risk factors

A case study from the Aboriginal Australian community demonstrated how social actions influenced the noncommunicable disease−related outcome, which is considered challenging with individual-based strategies [20]. Based on the successful NCD project from the neighborhood area, community members approached the healthcare provider who had designed the project. Following the meeting, they established a community-based project named 'Looma lifestyle project.' In this project, community members assigned representative people as diabetes workers, store managers, and sports and recreation officers.

Community members actively involved the whole community in helping the patients with diabetes in various aspects like diabetes education, healthy cooking classes, hunting trips, and store tours to identify healthy food choices. In addition, community members helped them improve the food quality, ensure the availability of whole meal flour and bread, and provide healthy food options (for example, replacing butter with margarine). Community members revived several sports teams and organized mega-sports events to increase physical activity. Also, several walking groups were formed in the community. In addition, the community contributed a vehicle to support the project and ensured a tobacco ban in public buildings. As a result, the community achieved a significant reduction in proportion of people consuming sugary drinks and not doing physical activity. Community members who had diabetes and were part of this intervention area significantly decreased their blood glucose levels.

Several case studies demonstrated the community's collective role in closing the existing alcohol outlet and their firm stand against establishing newer alcohol outlets in their community [21]. In India, several community-based organizations, especially women's self-help groups, have fought to achieve this.

Maternal health recognized as a policy priority

Compared to 1970, maternal health status, as indicated by the maternal mortality ratio, has seen a significant reduction in most countries, including the LMICs, which are edging toward achieving maternal health-related goals. The improvement in biomedical science alone cannot take credit for it. It is a result of civil society's role and efforts of several social forums to identify maternal death as a violation of human rights and position this issue as a developmental and hence a political priority. For example, 'Jan Swasthya Abhiyan (JSA),' an Indian circle of the People's Health Movement, conducted several public hearings in collaboration with the National Human Rights Commission. JSA shared the evidence on gaps identified in maternal healthcare service delivery and excerpts from public hearings formally with technical support from several academic agencies to the government. Media was also roped in for further support to set the context, highlight the issue, and attract the attention of policy stakeholders. The consequence of these reactions made the National Health System Resource Centre a knowledge and guiding resource for the health system in India. It recommended that federal and state health system functionaries strengthen the health system to provide quality maternal services in the country [22].

The success story of India's declining maternal mortality ratio is not the sole contribution of any advancement in biomedicine. The increase in the institutional delivery rate in tandem with strengthening the health system enabled this success (Fig. 7.4). Changing the preference of delivery place from home to institutional delivery was not a simple behavior change. The community health workers (ASHAs) facilitated the receptiveness to approach the health system. They educated the family regarding the services available and also escorted these pregnant women to promote their ANC visits, delivery, and immunization.

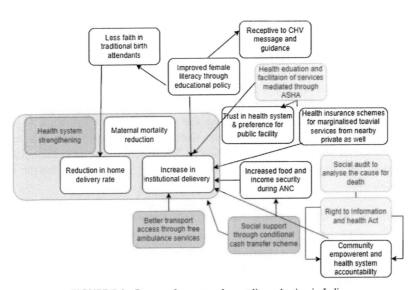

FIGURE 7.4   Reasons for maternal mortality reduction in India.

In India before 2005, more than 60% of deliveries did not happen at the institutional level. The reason could not be entirely attributed to the preference of the people. The reason also includes the fact that many people could not travel to those far distant areas in time. Thus, the sequential effect of introducing free transport access through 108 ambulance services (toll-free number: 108) in most states has helped women reach the facility for institutional delivery. This mechanism also helped the health system with timely referral and management of complications at the earliest at the nearest well-equipped health facility. Above all, the conditional cash transfer under JSY for institutional delivery significantly changed people's preferences. In addition to the JSY scheme, various state-led maternal cash assistance support has helped poorer mothers experience a better quality of life during pregnancy by ensuring adequate rest without losing wages. This schematic diagram depicts how the public health policies addressing social factors contributed to saving lives in India from maternal mortality [23].

Social movement as part of social action for health

Social actions have not only helped to improve health outcomes but also to ensure social justice and equity as well. Since the movement on "Health For All," there has been a heavy emphasis given to the people's health movement.

In the context of neoliberalism which increased health inequities, the way forward to achieve health equity is through the peoples' health movement:

Though recent decades have created better opportunities to prioritize and achieve health outcomes, the inequalities that could be addressed through available means have also been growing. Often the positive ray/scope in the form of a solution for social inequalities is identified through social movements. The role of social movements in health was first witnessed during cholera epidemic, which led to the first public health act in Britain [6]. Some of the important interventions which improved health access include people's movement to get antiretroviral drugs for HIV/AIDS, access to healthy foods, and implementation of an alcohol ban in the community.

New York City food movement to bring about health-promoting school

Adopting healthy food habits can play a significant role in combating the recent increasing trend of non-communicable diseases. Inculcating healthy food practices among children was found to have a sustainable impact. Along these lines, a New York parent-led advocacy group called "New York City healthy school food alliance" in collaboration with parents, school authorities, chefs, not-for-profit lobbies, and politicians to provide healthy food as well as culinary, and nutritional education at schools. Also, this group organizes several awareness campaigns and promotes mindful eating in schools, in addition to educating students about kitchen and garden, to realize the value of food among students [24].

New Zealand's prostitutes collective to partner with health system

New Zealand prostitutes collective (NZPC) is a group that emerged during the inception of the HIV spread (1987). This rights-based forum is an organization for the sex workers by the sex workers and has stood firmly against the stigma experienced by sex workers. With a strong commitment to political dialogue, they were instrumental in decriminalizing sex work and legalizing and improving the occupational safety of migrant sex workers in New Zealand. To make this change, they were in constant consultation with the Ministry of Health and the police department. As a result, NZPC was seen as the first resource group when policy reforms were discussed regarding sex work [25].

A similar example is the "Sonagachi community intervention project" from India, an initiative to empower sex workers against HIV spread. Under this project, through peer education and building social capital as a locus of change, sex workers were able to negotiate their health-related decisions. They did advocacy with a multidisciplinary team to influence power [26].

Trade unions negotiating healthy workplace

Until 1970, safety in the workplace was not that favorable to employees. The collective labor workforce, which emerged as the labor movement during the industrial revolution, is an example of how social actions transformed health among the occupational group. These labor movements ensured policy decisions to commit to workplace safety, equal treatment, welfare activities, compensations, and learning opportunities among industrial workers.

Role of Treatment Action Campaign in South Africa

Many theoretically proven biological and medical interventions, including HIV-related preventive interventions, failed because of lack of community mobilization.

South Africa runs the largest antiretroviral treatment program in the world. Several challenges faced by the program and the success stories demonstrate the role of social actions in health. In 1998, South Africa was the biggest victim of HIV, with thousands of new HIV transmissions and 900−1000 deaths daily. Despite the availability of ARV drugs which reduced the death rate by 80%, the government refused to make the drugs available to people for political and economic reasons. The annual cost of ARV drugs per patient came out to be US$15,000/year. At the same time, some countries where generic ARV drugs were used, it cost around US$350/year. However, due to pharmaceutical companies' patent and intellectual property rights, access to ARV drugs was denied to the poor for a decade. Treatment Action Campaign formed by some activists, including some medicos, constantly fought through several campaigns with mass social movements and legitimate social mobilization. As a result, South Africa provided access to generic ARV drugs, bringing down the cost to US$150/year/patient. Apart from these social movements, 'Trained HIV literacy educators' provided HIV health literacy through massive campaigns throughout the country. Due to these two strategies of social action, namely legitimate social movement and community empowerment, South Africa evidenced more than 10-fold reduction in death rate and transmission of newer infections [27].

Role of social actions in conflict and internally displaced population

In the context where people are internally displaced within the country, they lose access to health care and affordability. Moreover, due to their temporary unorganized settlement and lack of basic amenities, they become simultaneously vulnerable to infectious diseases, psychological morbidities, and internal conflicts.

In these settings, the role of volunteers and collective actions by community members or nongovernmental organizations played a significant role in bringing back normalcy in their life. In recent times, in the Democratic Republic of Congo, where more than 80,000 people, who were marginalized due to internal displacement, managed to handle the health issues through community efforts. Collective social actions led to the empowerment of the people to seek early care and provide primary care within the community. With the support of donor agencies and implementers like Médecins Sans Frontières, these communities have managed health issues through a referral from higher centers. In this way, collective social actions have made this vulnerable, marginalized community self-reliant in access to health care [28].

Similarly, healthcare access efforts invested in social actions in providing income security through a 100-day employment guarantee program (Mahatma Gandhi National Rural Employment Guarantee Act-MGNREGA) positively impacted the infant feeding and nutritional status in communities. These determinants enhanced a child's survival through a complex pathway.

Factors that can undermine social actions

Though social actions have contributed to achieving health outcomes in varied ways, especially in the context of having high access to technology, several challenges can undermine its role. Factors such as individualism, lack of solidarity, social capital, lack of mutual support from other social support agencies, and vested interest in privatizing public goods had been reported to be threats while considering social actions to achieve health outcomes.

The social actions or movements were often considered troubles or political threats. Hence, the funding agencies do not support such activities, mostly to scale up or make them sustainable. Reactive actions rather than proactive actions make the solutions nonsustainable. Often, solutions are contextualized; hence, the universal approach of the transformative approach may not benefit all. Moreover, retaining the same level of social cohesion and sustenance is challenging.

Some of the processes followed, such as demanding rights, confronting gatekeepers and opinion leaders, mediation, and removing community members from advocacy, have led to different paths that hindered real success.

Conclusion

Public policy and linked social action must address the structural or intermediary social determinants of health. Though policy decisions or reform is one of the social action approaches, the programmatic, regulatory, and community development approaches need to be efficiently used. The COVID-19 pandemic expedited and facilitated collaboration with people outside the health sector. Such established and better-performing collaborations and social actions need continuation and expansion to other areas for better health and well-being. The national and subnational multisectoral action plan and its effective implementation can facilitate the "Health in All Policies" and address most of the social determinants of health.

References

[1] Department of Economic and Social Affairs (United Nation). World social report 2020. Inequality in a rapidly changing world. 2020. p. 216. https://www.un.org/development/desa/dspd/wp-content/uploads/sites/22/2020/02/World-Social-Report2020-FullReport.pdf. [Accessed 30 December 2022].

[2] Tilly C. Popular contention in Great Britain. New York: Routledge; 2005. p. 1758–834.

[3] Nutbeam D, Kickbusch I. Health promotion glossary. Health Promot Int 1998;13:349–64. https://doi.org/10.1093/heapro/13.4.349.

[4] Ewart CK. Social action theory for a public health psychology. Am Psychol 1991;46:931–46. https://doi.org/10.1037/0003-066X.46.9.931.

[5] Laverack G. Health activism. Health Promot Int 2012;27:429–34. https://doi.org/10.1093/heapro/das044.

[6] López Barreda R, Robertson-Preidler J, Bedregal García P. Health assessment and the capability approach. Global Bioeth 2019;30:19–27. https://doi.org/10.1080/11287462.2019.1673028.

[7] Shiffman J. A social explanation for the rise and fall of global health issues. Bull World Health Organ 2009;87:608–13. https://doi.org/10.2471/BLT.08.060749.

[8] Devadasan N, Ranson K, Van Damme W, Acharya A, Criel B. The landscape of community health insurance in India: an overview based on 10 case studies. Health Policy 2006;78:224–34. https://doi.org/10.1016/j.healthpol.2005.10.005.

[9] Reshmi B, Sreekumaran N, Unnikrishnan B. A systematic review of community-based health insurance programs in South Asia. 2016.

[10] Joshi R, Pakhare A, Yelwatkar S, Bhan A, Kalantri SP, Jajoo UN. Impact of community-based health insurance and economic status on utilization of healthcare services: a household-level cross-sectional survey from rural central India. 2020. p. 74—82.

[11] World Health Organization. Social determinants of health sectoral briefing series. 2011.

[12] Loewenson R. Activism for health. Lancet 2013;381:2157. https://doi.org/10.1016/s0140-6736(13)61428-2.

[13] Adsul N, Kar M. Study of Rogi Kalyan Samitis in strengthening health systems under national rural health missionvol. 38. District Pune, Maharashtra; 2013. https://doi.org/10.4103/0970-0218.120157.

[14] Sector Wide Approach to Strengthening Health (SWASTH) Bihar. Rogi Kalyan Samitis help health care facilities become patient friendly. 2015.

[15] Puri M, Lahariya C. Social audit in health sector planning and program implementation in India. Indian J Community Med 2011;36:174—7. https://doi.org/10.4103/0970-0218.86515.

[16] Society for Community Health Awareness Research and Action (SOCHARA). Annual report 2014—15. Community action for health. 2015. https://doi.org/10.1001/jama.1954.02940510024012.

[17] Lahariya C, Roy B, Shukla A, Chatterjee M, De Graeve H, Jhalani M, et al. Community action for health in India: evolution, lessons learnt and ways forward to achieve universal health coverage. WHO South East Asia. J Public Health 2020;9:82—91. https://doi.org/10.4103/2224-3151.283002.

[18] Bath JJW. Impact of community participation in primary health care: what is the evidence. Aust J Prim Health 2015;21:2—8.

[19] Shu Q, Wang Y. Collaborative leadership, collective action, and community governance against public health crises under uncertainty: a case study of the quanjingwan community in China. Int J Environ Res Public Health 2021;18:1—12. https://doi.org/10.3390/ijerph18020598.

[20] Rowley KG, Daniel M, Skinner K, Skinner M, White GA. Effectiveness of a community-directed 'healthy lifestyle' program in a remote Australian Aboriginal community. 2000.

[21] Conway K. Booze and beach bans: turning the tide through community action in New Zealand. Health Promot Int 2002;17:171—7. https://doi.org/10.1093/heapro/17.2.171.

[22] Jat TR, Deo PR, Goicolea I, Hurtig AK, San Sebastian M. The emergence of maternal health as a political priority in Madhya Pradesh, India: a qualitative study. BMC Pregnancy Childbirth 2013;13. https://doi.org/10.1186/1471-2393-13-181.

[23] Chatterjee P. Tackling social factors to save lives in India. Bull World Health Organ 2011;89:708—9. https://doi.org/10.2471/BLT.11.031011.

[24] Hunter College New York City Food Policy Corner. Food justice; n.d. https://www.nycfoodpolicy.org/this-parent-led-advocacy-group-is-working-to-change-nyc-school-food/.

[25] Global Network of Sex Work Projects. New Zealand prostitutes collective (NZPC); n.d. https://www.nswp.org/fr/node/3201.

[26] Robinson RS, Zayed H. Understanding the capacity of community-based groups to mobilise and engage in social action for health: results from Avahan. Glob Public Health 2021;16:1590—603. https://doi.org/10.1080/17441692.2020.1837912.

[27] Heywood M. South Africa's treatment action campaign: combining law and social mobilization to realize the right to health. J Hum Rights Pract 2009;1:14—36. https://doi.org/10.1093/jhuman/hun006.

[28] Medecins Sans Frontieres. Health care for the community by the community; n.d. https://www.msf.org/community-led-healthcare-internally-displaced-people.

Chapter 8

Democratization of health care in hospital settings—An integral part of public health practice

Ishwarpreet Kaur and Amarjeet Singh

Department of Community Medicine and School of Public Health, Postgraduate Institute of Medical Education and Research, Chandigarh, Punjab, India

Introduction

As a discipline, there has always been confusion and debate about the scope and potential of public health [1], e.g., 'Can we practice or apply public health principles in hospitals?' Any change in hospital practices which has the potential to benefit the masses through improvements in their quality of life essentially has a public health connotation. In this chapter, we propose to address this vital question by sharing some case studies from India and other countries.

The healthcare access process often leaves patients feeling dissatisfied due to endlessly bureaucratic, indifferent, and cold hospital services. Inadequate human resources and infrastructure, abrasive behavior of staff, long waiting hours and shorter consultation time result in poor quality of care, and lack of relief in symptoms [2–6].

Proper fulfillment of the patient's expectations of prompt care and the availability of a doctor listening to their health issues are impossible with a low doctor–population ratio. The duration of first-time doctor–patient consultation is less than 5 min for nearly half of the global population. The range varies from 48 s in Bangladesh to 22.5 min in Sweden. In India, it is reported as less than 2 min [7]. Even the best public sector tertiary care hospitals in India face difficulty delivering standard health care due to the huge patient load [8].

An unmanageable patient load is a major problem in most public sector hospitals, especially in low- and middle-income countries. Such high patient flow increases the workload of the doctors resulting in short consultation times

Principles and Application of Evidence-Based Public Health Practice
https://doi.org/10.1016/B978-0-323-95356-6.00012-4
127

and low quality of care. It can lead to medication errors due to inappropriate history taking before prescribing treatment and inadequate instructions while prescribing the drugs [5,9], all linked to low consultation time. Further, the doctor's advice mostly focuses on the medications rather than making the patient understand the disease and do self-care. The specialists often develop an unsympathetic and rude attitude due to questions related to the basics of disease and related self-care. It can be avoided if we stress conservative management of the health problems of the patients through decentralized counseling in outpatient departments and other settings. Adopting a health-promoting hospital approach can be useful in resolving this issue by enhancing patient satisfaction [10−12]. Otherwise, patient and/or caregiver education is one of the pillars of assessing and ensuring the quality of care. It has the potential to empower patients and their caregivers about self-care by building their knowledge, skills, and positive attitudes about health. It will also help the patients to adhere to the advice given by the doctors. Such democratization of health care has enabled people to make decisions that previously required highly trained professionals. They are encouraged to take a more active role and responsibility in managing their health. The engagement and active participation of patients or caregivers will improve the patient experience, reduce unnecessary hospital visits and medications and hence the cost, and ultimately result in better health outcomes [13,14].

Changes in the dynamics of doctor−patient interactions in hospitals

Over the last two decades, new concepts have emerged globally in hospital care, viz., patient and family-centered care, patient-as-person concept, non-pharmaceutical therapy, therapeutic education, social prescription, information therapy, and others [15−17]. There is a focus on active collaboration and shared decision-making between patients, families, and providers. The aim is to design and manage a customized and comprehensive care plan, emphasizing the patients' problems more than their diagnoses. Here, the focus is on respect for persons, individual right to self-determination, understanding, empathy, two-way communication, and eye-to-eye contact.

A pill-fixation mindset prevailing in our society forces people to expect a medicinal prescription for their maladies. Social prescription concept of wellness aims to encourage patients to factor in their needs and interests in the backdrop of their lifestyle and self-care besides the medicinal prescriptions [18−20].

With the advent of modern technology, 'information prescriptions' now support individuals to take greater control of their own health [21]. This concept was highlighted in 2006 as an IT-based model to prevent ill health through shared decision-making. Even otherwise, people do consult Google for their health-related issues. They even cross-check doctors' prescriptions before and after clinic visits.

Rising patient expectations have accelerated the shift toward more personalized and convenient care, i.e., consumerization of health care. Health tech developers are using data and a 'consumer-first' mindset to provide easily accessible and holistic care. Individualized therapy or personalized treatment plans are a must for a better quality of care. 'One-size-fits-all' formula cannot be used when there are individual differences. Individualized therapy involves tailor-made treatment plans keeping in mind the individuals' genetic, biological, physiological, environmental, social, and behavioral information/ aspects.

The human body's power of self-healing was strongly emphasized. The concept of salutogenesis was also popularized through an emphasis on the healing resources of a person [22,23]. This is related to social iatrogenesis, i.e., when medical bureaucracy creates ill health by lowering people's tolerance levels for discomfort or pain. The sphere of our lives which we control comes under the preview of doctors, i.e., the medicalization of life.

We lose internal locus when we ignore our salutogenic potential and 'self-efficacy.' This exposes us to iatrogenesis since we go to the doctors to get well, even for minor illnesses. More health damages are caused by the belief that we cannot cope with illness without modern medicine. Hence, medical interventions produce dependence. The system tends to ignore the individuals' coping ability and immunity and the supporting and healing aspects of the social and physical environments. Further, there is overprescription of medicine due to pressure from pharmaceutical companies.

Patients and their family caregivers are quite underutilized resources in healthcare systems. In fact, patient-engagement efforts have not been integrated into the fabric of everyday life. This necessitates the need to create a 'health culture' that seeks to change their behaviors. This approach incorporates devising strategies to foster patient as well as person-centered health care by emphasizing the creation of supportive social conditions in hospitals, care coordination, and wellness promotion, an effort to customize care to patients' needs and preferences.

Democratization

Self-care and empowerment are essentially linked to the concept of demedicalization, i.e., people gain control over the processes, which were until now under the preview of doctors/nurses. There is the active involvement of patients in managing their diseases. This is the essence of the concept of democratization of health care [24]. Democratization is one of the tools used to promote health among patients or populations. "Human rights, self-reliance, and joint responsibilities are the three major emphasizes of democracy [25]. In health care, democratization calls for changing the traditional doctor—patient relationship, which is more hierarchical, to more engaging and empowering relationships for shared decision-making and self-management of diseases.

It is the process of valuing the patients and their caregivers throughout the continuum of care, satisfying their needs irrespective of time and place [26]. The bulk of the routine issues in health care is set to be shifted to the public domain from the erstwhile, rather esoteric medical jargon-ridden paradigm. The paternalistic position of the doctors as the prescribing authority is slowly getting metamorphosized to an 'adult-adult' ego state dialogue initiator as per the transactional analysis conceptual framework [27].

Democratization of health care warrants the integration of physical, behavioral, and social service health for understanding the situation comprehensively and making the decision. Comprehension of health or disease situations is directly linked with health literacy. Hence, health literacy is one of the fundamental requirements and components of the democratization of health care. In addition to traditional tools, advanced technologies help to improve health literacy quickly and provide customized information for the current practices to make daily decisions [28]. Though democratization expects a shared decision-making, with equal power among stakeholders, different levels of health literacy of patients and their caregivers obviously will not lead to the same practically [25].

The Institute of Patient- and Family-Centered Care of the United States seeks to integrate education, consultation, and technical assistance; materials development and information dissemination; research; and strategic partnerships for changing the way health care is provided to the community. It seeks to build on the strengths of patients and families by enhancing their confidence and competence [29].

Democratization of health care, especially based on advanced technology, is envisioned as one of the futures of health by the survey participants (medical, engineering, and other professionals) from various countries [30]. The high burden of chronic and neurodegenerative diseases and increasing lifespan importantly emphasize the urgent need for democratization of health care which novel and advanced technologies can expedite.

Globally, there are many examples of adopting a patient-centered and holistic approach in clinics, e.g., 11th Street Family Health Services of Philadelphia which started in the 1990s. It serves the patients' physical, mental, spiritual, and social needs in the four public housing communities covered by Medicaid. The clinic uses an integrated service approach, e.g., for a family visit, a 5-year-old can get immunizations and a dental check-up. At the same time, art therapy is offered to her teen sibling as part of mental health care. It uses couples and family counseling. It organizes mindfulness training, cooking classes, housing, and food assistance [31,32].

Some case studies from India

As a matter of coincidence, parallel to the ideas of social prescription/patient-centered care and patient empowerment related to women's health was developed and practiced at the dawn of the present millennium by a group of researchers in India. It all began with ascertaining the impact of behavior therapy on urinary incontinence among adult women [33,34]. Women with urinary incontinence were randomized into two groups. The intervention group was taught behavior therapy (one-to-one) with the help of a self-management educational manual and the control group was given standard medical care without behavior therapy (Fig. 8.1). At the end of 8 months of follow-up, 52.5% of cases became continent without surgical intervention compared to 12.8% in the control group. Similar behavior therapy (one-to-one training to do Kegel exercise) for uterine prolapse was tested and found effective in outpatient settings (Fig. 8.1) [35].

Similarly, the effectiveness of exercise was compared with the use of hot water bottles in the management of primary dysmenorrhea among adolescent schoolgirls in Chandigarh [36,37].

To expand this health promotion and demedicalization concept, a separate multipurpose behavior therapy (MPBT) room was established in the Obstetrics and Gynecology (OG) OPDs of the study hospital. In addition to prolapse, urinary incontinence, and dysmenorrhea, this room also provided services to women with menopause, infertility (education and lifestyle modification for better physical, mental, and social health), ante- and post-natal cases (maternity preparedness-technology-based, couple counseling on congenital disorders, and postural and ergonomic health and fitness), common problems related to puberty and polycystic ovary disease (PCOD). In PCOD, the effectiveness of probiotic-based dietary and lifestyle regimes was tested,

FIGURE 8.1 Cover page of self-instructional manual on behavior therapy for (A) urinary incontinence and (B) pelvic organ prolapse among women.

which reported improvement in the irregular menstrual cycle, acne, and hirsutism. Even weight loss was witnessed.

Highlights of the MPBT room approach

This unique endeavor is the first of its kind in India. In MPBT, after a referral from OG OPD, counselors, in a separate room, give advice to women with minor problems related to pregnancy or those with dysmenorrhea, urinary incontinence, uterine prolapse, menopause, or infertility. It gives global exposure to the efforts of the researchers, as people from anywhere in the world can access the uploaded material, which has been successfully tested in OPDs. Readers may go through the uploaded material for their health-related issues. It will help in many of their related queries. Concerned doctors/departments may be contacted for further guidance or clarification.

Highlights of this strategy are family-centered, individualized, and group counseling sessions, adoption of a 'demedicalization' approach for tackling the routine OG problems of women + integrated training strategies used. The concept of democratization of health care was also ingrained in their approach [38–46].

- Couples are counseled together in the MPBT room for maternity care and in case of infertility. Thus, they can understand each other better and express their concerns without any hesitation.
- Patients are counseled along with their family members.
- Sufficient time (15–20 min) is devoted to each patient
- Efforts are made to improve adherence to dietary counseling and weight reduction regime
- Exercises are demonstrated individually in the room. Patients repeat them (Return demonstration). A special set of floor, as well as bed-based exercises, have been developed.
- Videos are shown regarding ante- and post-natal care
- Easy-to-understand IEC materials like charts, flip cards, and posters are displayed in the room
- Laptop-based IEC materials are shown
- Yoga and meditation-related counseling is also given
- Every patient is given a booklet on the relevant management aspect of her problem free of cost (Hindi/English/Punjabi)
- Patients are followed up on mobile phones to ensure compliance. Social media–based groups have been used for sharing information, interaction, and follow-up.
- Even adoption services have been linked with MPBT room for those who cannot conceive.
- Our results proved that they could be weaned off their dependence on a medicinal cure.

The MPBT room concept is based on multidisciplinary teamwork in OPDs, and the same can be replicated in similar settings elsewhere.

Lately, the above work on women's health has also harnessed the potential of the quaternary industrial revolution through action research projects. Mobile phone—based applications (mHealth) have been developed to improve patient engagement and remote monitoring. mHealth app included exercise videos or pictorial exercises pdf files, dietary instructions, possible complications and their sign and symptoms, and side effects of medicines customized to each patient. The developed applications were patient-friendly and prepared in such a way that patients could contact nurses, doctors, or other healthcare professionals whenever they have queries, doubts, and problems [47,48].

Democratization of healthcare beyond OG OPD and the study hospital

Knee osteoarthritis at Orthopedics OPD

Beyond the OG OPD, this approach was tried among patients with knee osteoarthritis (KOA). KOA management could be demedicalized to quite an extent by involving patients, their families, and paramedical personnel in treatment under the concept of "Knee school (KS)." Here, exercises significantly reduce knee pain, which significantly impacts their QOL [39,40,49]. In many patients, surgeries could be avoided.

Essentially, the approach involved active listening to the patients and giving them sufficient time. Their family caregivers are also involved in the therapy. Interdisciplinary interventions involving clinicians, nurses, physiotherapists, nutritionists, social scientists, and public health experts were provided. The main focus is on lifestyle changes. Weight loss, diet pattern modification, adequate fluid intake, physical exercises, yoga, meditation, and pain-coping skills form the intervention package. Due emphasis is given to adopting an individualized approach per the patients' physique, mindset, family situation, social class, education status, comprehension capacity, comorbidities and others. Further, the KS was converted to electronic KS during the COVID-19 pandemic integrating telemedicine components [50].

Care to patients who have undergone surgery at the hospital and home

Educational cum behavioral intervention was developed and tested to provide (a) nursing care of bedridden operated neurosurgery and orthopedic patients, (b) self-care after mastectomy among patients with breast cancer, and (c) self-care after therapy for cervical cancer and ovarian cancer.

Prevention and non-pharmacological management of low back pain in workplace setting

The research group simultaneously tested a prevention package (prevent the occurrence of low back pain [LBP]) and a treatment package (reduce the severity of LBP) among non-teaching women at a university in India. It was completely an individualized and social prescription-based approach where they taught the exercises (one-to-one/group) and corrected postural abnormalities among the study participants. Both prevention and treatment groups benefited from the intervention.

In India, women are not accustomed to exercising. Hence, asking them to do exercise involves significant behavior change through training. According to the BASNEF [51] model of behavior change, subjective norms and experiences do influence the compliance of a person with exercise regimes. Hence, having someone else to exercise with under supervision surely helps. In this study, the respondents were motivated to form groups of 2–3 and practice office stretching exercises whenever they found time during office hours. Often, people find it difficult to change their routine if left alone. The concept of 'group norm' was followed so that there was no apprehension of being singled out. It proved to be a successful strategy in this study.

Women were advised to maintain correct postures, e.g., while washing utensils or cooking to stand close to the sink and use a footrest by putting one foot on it, to be alternated every few minutes. Dietary management was advised as part of the intervention since women hardly took any fruits. They were also advised to drink adequate quantity of water everyday. Both prevention and treatment groups benefitted through this intervention.

The quality of therapist–patient interaction resulted in better patients' adherence to the prescribed regime. The faults in their posture were corrected. Workplace modifications were also advised, e.g., using a footrest while working on the desk or using a cushion on the chair to support the back, adjusting the monitor computer screen with respect to the eye level. The respondents were satisfied that their problems were heard with patience. This made them much more comfortable discussing their health problems with doctors. Measures like walking within the campus; taking the stairs instead of the lift; taking 10-min walk after lunch, etc., were also recommended. They were advised to follow the 20-min rule, i.e., to get up every 20 min from their seat, take a round in the room, and sit back for work again.

At follow-up visits, the respondents were asked to share their experiences with other respondents. This aroused a fellow feeling among them. Moreover, it improved self-management. Implementing the well-known concept of Back Schools can be effective in non-medicinal management and prevention of LBP.

Further, the researchers used a WhatsApp group that served as a platform for enhanced discussion and social support. Video and pictorial messages on exercises, correct posture, and dietary management were shared. Messages related to overall health were also posted in this group. This acted as a reminder for the respondents to do regular exercise and be conscious about their posture. Respondents were given the liberty to seek clarifications related to their overall health. The respondents felt a sense of personal touch through interactions with the researcher. This strengthened their faith in the investigator and the prescribed regime. The researchers found a definite benefit of this approach in bringing about patient behavior change.

Open access dissemination of educational content in the public forum

Further, all the tested educational content on self-care on various conditions has been uploaded to the website of the study institute for anyone to access (in **'Public Forum'** under **'Patient Empowerment'** drop-down menu). It could have helped hundreds of patients during the COVID-19 pandemic since the OPD was closed and routine health care was not easily accessible. Based on already developed and tested content, YouTube videos were created and disseminated as an introduction to accessing information from the institute's website regarding self-care at home for various conditions. These videos reflect the development of information, education and communication material for the general public to make them aware of the issues related to their health needs and assist them in self-care at home [52,53].

Democratization of health care through advanced technology

Do-It-Yourself automated insulin delivery system

The Do-It-Yourself automated insulin delivery system is one of the important examples of the use of advanced technology in democratizing healthcare. The evidence on its application included "intelligent computing, sharing information and privacy, security, and safety," the three pillars of democratization of health care using technology [54]. Intelligent computing accurately monitors the glucose continuously and automatically feeds (talks) into the software and builds an artificial pancreas system. Here, OpenAPS, AndroidAPS, and Loop software were used for intelligent computing. Sharing of information helped to overcome the technological barrier and to obtain technical (from healthcare providers), social, and emotional (from similar communities, peers, and families) support without compromising the security, safety, and privacy of the data.

Demonstrating the democratization of healthcare philosophy inherent in the use of behavior therapy/counseling and dealing the social health beyond physical health from the above examples empowered the respective patients or caregiver groups. This approach provides relief to them without resorting to any medication. This will also improve the socio-medical environment of 'pill for every ill' kind of medicalized life and mindset of citizens.

During the last 30—40 years, the patient empowerment movement has been boosted through the concept of quaternary prevention, i.e., actions taken to identify and protect patients from overmedicalization, the health activities to avoid the consequences of unnecessary intervention [55]. Public health professionals have opportunities to spread awareness about low-cost, easy treatments and their applications.

Conclusion

Democratization in health care is a recently recognized subfield of health promotion. Irrespective of the setting (hospital, community, workplace, or others), it is an integral part of the public health practice. Democratization calls for changing the traditional doctor—patient relationship, which is more hierarchical, to a more engaging and empowering relationship for shared decision-making and self-management of diseases. Adequate clinician time should be ensured for the patients, who should be integral to shared care planning and decision-making. In line with this, there is a need to change medical education and training across settings. The interventions like MPBT rooms may be tested and scaled in other settings. The Fourth Industrial Revolution—based healthcare innovations (mHealth, telehealth, and Artificial Intelligence) seek to bridge the gaps in the availability of services for all people. These advanced technologies must be used to empower the people at large, to make daily decisions regarding their health and to improve their health literacy.

References

[1] Singh AJ. What is (there) in a name. Indian J Community Med 2004;29(4):151—4.
[2] Rout SK, Sahu KS, Mahapatra S. Utilization of health care services in public and private healthcare in India: causes and determinants. Int J Healthc Manag 2021;14(2):509—16. https://doi.org/10.1080/20479700.2019.1665882.
[3] Cleary PD, McNeil BJ. Patient satisfaction as an indicator of quality care. Inquiry 1988;25(1):25—36.
[4] Kumar S, Haque A, Tehrani HY. High satisfaction rating by users of private-for-profit healthcare providers-evidence from a cross-sectional survey among inpatients of a private tertiary level hospital of north India. N Am J Med Sci 2012;4(9):405—10. https://doi.org/10.4103/1947-2714.100991.
[5] Kumari R, Idris M, Bhushan V, Khanna A, Agarwal M, Singh S. Study on patient satisfaction in the government allopathic health facilities of Lucknow district, India. Indian J Community Med 2009;34(1):35—42. https://doi.org/10.4103/0970-0218.45372.
[6] Farid M, Masood J, Hussain K. Determination of waiting time in patients visiting general male and female outdoor clinics. Pak Armed Forces Med J 2018;68(5):1368—71.
[7] Irving G, Neves AL, Dambha-Miller H, Oishi A, Tagashira H, Verho A, et al. International variations in primary care physician consultation time: a systematic review of 67 countries. BMJ Open 2017;7(10):e017902. https://doi.org/10.1136/bmjopen-2017-017902.

[8] Bajpai V. The challenges confronting public hospitals in India, their origins, and possible solutions. Adv Public Health 2014;2014:1−27. https://doi.org/10.1155/2014/898502.

[9] Patel RS, Bachu R, Adikey A, Malik M, Shah M. Factors related to physician burnout and its consequences: a review. Behav Sci 2018;8(11):98. https://doi.org/10.3390/bs8110098.

[10] Ahuja PK, Gupta AK, Jain B, Singh A, Bains P. Health promoting hospitals. In: Singh AJ, Goel S, Kathiresan J, editors. Health promotion: need for public health activism. Germany LAP LAMBERT Academic Publishing; 2013.

[11] Singh A, Anand A, Jaswal N, Kaur R. Editorial. Introducing health promoting orientation in hospitals: a desirable approach in the 21 century. Healthline 2019;10(2):3−4.

[12] Singh AJ, Goel S, Kathiresan J, editors. Health promotion: need for public health activism. Germany LAP LAMBERT Academic Publishing; 2013.

[13] Topol E. The patient will see you now: the future of medicine is in your hands. New York: Basic Books; 2015.

[14] Tang PC, Smith MD, Adler-Milstein J, Delbanco T, Downs SJ, Mallya GG, et al. The democratization of health care: a vital direction for health and health care. In: NAM perspect. Discussion paper. Washington, DC: National Academy of Medicine; 2016. https://doi.org/10.31478/201609s. vol. 6, Issue 9.

[15] Doyle C, Lennox L, Bell D. A systematic review of evidence on the links between patient experience and clinical safety and effectiveness. BMJ Open 2013;3(1):e001570. https://doi.org/10.1136/bmjopen-2012-001570.

[16] Webair HH, Ismail TAT, Ismail SB. Patient-centered infertility care from an Arab perspective: a review study. Middle East Fertil Soc J 2018;23(1):8−13. https://doi.org/10.1016/j.mefs.2017.10.003.

[17] McCormack B, Dewing J, McCance T. Developing person-centred care: addressing contextual challenges through practice development. Online J Issues Nurs 2011;16(2):3. https://doi.org/10.3912/ojin.vol16no02man03.

[18] Brandling J, House W. Social prescribing in general practice: adding meaning to medicine. Br J Gen Pract 2009;59(563):454−6. https://doi.org/10.3399/bjgp09X421085.

[19] Moffatt S, Steer M, Lawson S, Penn L, O'Brien N. Link worker social prescribing to improve health and well-being for people with long-term conditions: qualitative study of service user perceptions. BMJ Open 2017;7(7):e015203. https://doi.org/10.1136/bmjopen-2016-015203.

[20] Bickerdike L, Booth A, Wilson PM, Farley K, Wright K. Social prescribing: less rhetoric and more reality. A systematic review of the evidence. BMJ Open 2017;7(4):e013384. https://doi.org/10.1136/bmjopen-2016-013384.

[21] Sonika R, Sharma VL, Singh A. Information therapy: bridging the information gap between doctors and patients. SE Asia J Public Health 2015;4(2):47−50. https://doi.org/10.3329/seajph.v4i2.23695.

[22] Senjam S, Singh A. Study of sense of coherence health promoting behaviour in north Indian students. Indian J Med Res 2011;134(5):645−52. https://doi.org/10.4103/0971-5916.90989.

[23] Singh AJ, Goel S, Senjam SS, Goel S, editors. Mantras for healthy lifestyle—a salutogenic approach. New Delhi: Century Publications; 2012.

[24] Stanford Medicine. The democratization of health care. Stanford Medicine 2018 Health Trends Report; 2018. p. 1−21.

[25] Kincheloe M. Democratization in the therapeutic community. Perspect Psychiatr Care 1973;11(2):75−9. https://doi.org/10.1111/J.1744-6163.1973.TB00801.X.

[26] Barbazzeni B, Fritzsche H, Friebe M. Forecasting the future of healthcare democratization forecasting the future of healthcare democratization. Curr Dir Biomed Eng 2021;7(2):155—8. https://doi.org/10.1515/cdbme-2021-2040.

[27] Murray H. Transactional analysis—Eric Berne. Simply psychology. 2021. Available from: https://www.simplypsychology.org/transactional-analysis-eric-berne.html.

[28] Dzau VJ, McClellan MB, McGinnis JM, Burke SP, Coye MJ, Diaz A, et al. Vital directions for health and health care: priorities from a National Academy of Medicine Initiative. JAMA 2017;317(14):1461—70. https://doi.org/10.1001/JAMA.2017.1964.

[29] Johnson BH, Abraham MR. Partnering with patients, residents, and families: a resource for leaders of hospitals, ambulatory care settings, and long-term care communities. Bethesda, MD: Institute for Patient- and Family-Centered Care; 2012.

[30] Barbazzeni B, Haider S, Friebe M. Engaging through awareness: purpose-driven framework development to evaluate and develop future business strategies with exponential technologies toward healthcare democratization. Front Public Health 2022;10:851380. https://doi.org/10.3389/fpubh.2022.851380.

[31] Gloria D. Needs-based care for a community. Holist Nurs Pract 2009;23(3):127. https://doi.org/10.1097/01.HNP.0000351368.17093.0d.

[32] Gerrity P. And to think that it happened on 11th street: a nursing approach to community-based holistic care and health care reform. Altern Ther Health Med 2010;16(5):62—7.

[33] Kumari S, Singh AJ, Jain V, Mandal AK. Urinary incontinence—behaviour therapy for urine leakage. In: Singh AJ, Walia I, Dhaliwal LK, editors. Demedicalizing women's health. New Delhi: Gyan Publishing House; 2010.

[34] Kumari S, Jain V, Mandal AK, Singh A. Behavioral therapy for urinary incontinence in India. Int J Gynaecol Obstet 2008;103(2):125—30. https://doi.org/10.1016/j.ijgo.2008.06.019.

[35] Kashyap R, Jain V, Singh A. Comparative effect of 2 packages of pelvic floor muscle training on the clinical course of stage I—III pelvic organ prolapse. Int J Gynaecol Obstet 2013;121(1):69—73. https://doi.org/10.1016/j.ijgo.2012.11.012.

[36] Chaudhuri A, Singh A. How do school girls deal with dysmenorrhoea? J Indian Med Assoc 2012;110(5):287—91.

[37] Chaudhuri A, Singh A, Dhaliwal L. A randomised controlled trial of exercise and hot water bottle in the management of dysmenorrhoea in school girls of Chandigarh, India. Indian J Physiol Pharmacol 2013;57(2):114—22.

[38] Singhal N, Singh AJ. Promoting empowerment of adolescent girls by demedicalizing management of dysmenorrhea. Ind J Youth Adol Health 2015;2(4):1—3.

[39] Sharma M, Singh AJ, Dhillon MS. Spending time with patients significantly influences outcomes of non-operative treatment of knee osteoarthritis. J Postgrad Med Edu Res 2015;49(3):v—vi. https://doi.org/10.5005/jpmer-49-3-v.

[40] Sharma M, Singh A, Dhillon MS, Kaur S. Conservative therapy through adequate doctor patient interaction improves outcomes in patients suffering from mild and moderate knee osteoarthritis. Int J Healthc Edu Med Inform 2017;4(3&4):1—2. https://doi.org/10.24321/2455.9199.201712.

[41] Singh AJ, Suri V, Kaur S, editors. Health empowerment of women—a desirable strategy in 21st century hospitals. New Delhi: Kalpaz Publications; 2020.

[42] Kaur T, Dudeja P, Sharma M, Jain V, Singh A. Impact of running a behavior therapy room for various categories of urinary incontinence cases in different age groups in obstetrics-gynecology outpatient department of a tertiary care hospital. J Postgrad Med Edu Res 2014;48(4):159—63. https://doi.org/10.5005/jp-journals-10028-1124.

[43] Sharma R, Dhaliwal LK, Suri V, Kaushal P, Singhal N, Rana AK, et al. Implementing a life cycle approach through establishment of a multipurpose behavior therapy room in PGIMER, Chandigarh for enhancing the fitness level of women. J Postgrad Med Edu Res 2017;51(3):115−22. https://doi.org/10.5005/JPMER-51-3-115.

[44] Gupta R, et al. Managing common problems at puberty—a guide for adolescent girls. New Delhi: Century Publications; 2017.

[45] Kaur I, Singh A. Empowering adolescent girls to manage the impending epidemic of PCOS. Ind J Youth Adol Health 2016;3(2):1−3.

[46] Kaur I, Suri V, Rana SV, Singh A, Sachdeva N, Sahni N. Impact of lifestyle intervention for management of the modern life scourge of polycystic ovarian syndrome among girls—a case series. Ind J Youth Adol Health 2017;4(3):21−4.

[47] Development, validation and evaluation of the impact of implementation of E-counselling based macro—micronutrient and lifestyle intervention programme on PCOS related signs/symptoms in adolescent girls of Chandigarh (ICMR project-2020-contd).

[48] Sharma M, Dhillon MS, Singh A, Adhya B, Chalana A, Negi S, Verma N, Santosh P. Establishing 'e-knee school' for patients suffering from knee osteoarthritis during COVID 19. Chandigarh: Department of Community Medicine and School of Public Health, Orthopaedics and Physical Rehabilitation and Medicine, PGIMER; 2020 (CTRI/2020/03/024311).

[49] Sharma M, Singh A, Dhillon MS, Kaur S. Comparative impact of nonpharmacological interventions on pain of knee osteoarthritis patients reporting at a tertiary care institution: a randomized controlled trial. Indian J Palliat Care 2018;24(4):478−85. https://doi.org/10.4103/IJPC.IJPC_14_18.

[50] Bali S, Singh AJ. Mobile phone consultation for community health care in rural north India. J Telemed Telecare 2007;13(8):421−4. https://doi.org/10.1258/135763307783064421.

[51] Jeyashree K, Singh AJ. Theories in health promotion. In: Singh AJ, Goel S, Kathiresan J, editors. Health promotion: need for public health activism. vol. 1. Germany: LAP LAMBERT Academic Publishing; 2013.

[52] Kumari S, Singh A, Dhaliwal LK, Suri V, Jain V, Mandal AK. Self-management of urine leakage by women at home anytime! How can they do it themselves? India: Post Graduate Institute of Medical Education and Research, Chandigarh; 2020. Available from: https://youtu.be/AgeAPMd61l4.

[53] Managing the problem of acidity in pregnancy at home during COVID-19 lock down period. Chandigarh, India: Post Graduate Institute of Medical Education and Research; 2020. Available from: https://youtu.be/_mVDn7B2XQ0.

[54] Stanford Medicine. Health trends report. The democratization of health care. 2018. Available from: https://med.stanford.edu/content/dam/sm/school/documents/Health-Trends-Report/Stanford-Medicine-Health-Trends-Report-2018.pdf.

[55] Norman AH, Tesser CD. Quaternary prevention: a balanced approach to demedicalisation. Br J Gen Pract 2019;69(678):28−9. https://doi.org/10.3399/bjgp19X700517.

Chapter 9

Community participation for improving the coverage and quality of evidence-based public health practice

Hemant Deepak Shewade[1], Deepak H.J. Murthy[2] and Seetharam Mysore[2]

[1]*Division of Health Systems Research, Indian Council of Medical Research-National Institute of Epidemiology (ICMR-NIE), Chennai, Tamil Nadu, India;* [2]*Swami Vivekananda Youth Movement (SVYM), Saragur, Karnataka, India*

Introduction

In September 1978, the Declaration of Alma-Ata was adopted at the international conference on primary health care. This was the first international declaration that emphasized the importance of primary health care. The declaration called for urgent action by all (governments, health and development workers, world community) to protect and promote the health of all people. Community participation was at the heart of the Alma-Ata Declaration and is one of the pillars of primary health care. The fourth article of the declaration states, "*people have the right and duty to participate individually and collectively in the planning and implementation of their health care.*" The seventh article states that primary health care "*requires and promotes maximum community and individual self-reliance and participation in the planning, organization, operation, and control of primary health care.*"

Community participation is defined as "*a process by which people are enabled to become actively and genuinely involved in defining the issues of concern to them, in making decisions about factors that affect their lives, in formulating and implementing policies, in planning, developing and delivering services and in taking action to achieve change*" [1]. The World Health Organization in its world health report 2008 called for the involvement of communities in shaping their primary healthcare services [2]. Target 16.7 of

Principles and Application of Evidence-Based Public Health Practice
https://doi.org/10.1016/B978-0-323-95356-6.00013-6
141

the 2030 sustainable development goals (SDGs) reinforces responsive, inclusive, participatory decision-making at all levels [3]. Hence, community participation has a key role to play in advancing universal health coverage and meeting the targets of the 2030 SDGs. Globally, there are increasing calls for community participation to achieve universal health coverage.

There are other terminologies related to community participation and these are used synonymously in literature. These include, social participation, community monitoring and planning, community processes, communitization, and community action for health (CAH), among many others. Social participation can be defined as "*a person's involvement in activities providing interactions with others in community life and in important shared spaces, evolving according to available time and resources, and based on the societal context and what individuals want and is meaningful to them*" [4].

Public health practice and community participation

Globally, we are moving from expert-based medicine to evidence-based medicine for clinical management of individuals' disease. Similar to evidence-based medicine, we should move from expert-based to evidence-based public health practice. Data drive the decisions, data not just from routine monitoring mechanisms but also from systematic research (operational/implementation/health systems research) involving quantitative as well as qualitative data, facility-based as well as community-based.

In clinical practice, the patient (end user) visits the clinician with some felt need and listens to the prescriptions. However, in public health, the community (end users) being highly heterogeneous, there are multiple views regarding the problem as well as solutions. Hence, the public health interventions will not get successfully identified and implemented without community participation. Community participation will enable public health practitioners to understand the need or problems of the community in a better way and plan the interventions according to the need. Similar to patient-centered individualized clinical care involving the patient, community-centered public health interventions need to be adopted.

Ultimately, public health practitioners are concerned about two important outcomes following the implementation of any intervention, namely the coverage and quality of the service. For example, treatment for a disease is a service and successful treatment is the desired outcome of the service.

Coverage indirectly ensures the availability, acceptability, accessibility, affordability, and equity of the services and its delivery. Coverage of evidence-based public health practice may be ensured through a monitoring mechanism to track what proportion of eligible beneficiaries have received the service and/or the extent of desired outcome of the service. Community participation is the key to better coverage of public health interventions.

Quality has many dimensions and may be ensured through a set of standard guidelines and standard operating procedures in place. Though monitoring mechanisms have indicators to capture quality of services, advanced levels of community participation may potentially ensure better quality of care.

Community monitoring can help in identifying gaps in implementation from an emic (insider) perspective that routine monitoring, evaluation, and research may miss. Community planning can help in identifying and implementing local context-specific solutions. The need for community participation for improving coverage and quality of public health practice arises not only from the calls for *internal accountability for service providers (accountable to their administrative superiors)* but also *hybrid forms of accountability involving communities* [5].

Models of community participation

In 2020, *Júnior and Morais* summarized the four main models of community participation in international literature [6] (Table 9.1).

The first is the *Arnstein* "ladder" model that illustrates eight types of participation: with advanced participation corresponding to the highest rungs [7].

TABLE 9.1 Four commonly cited models of community participation in literature [6].

Model name	Description	
Arnstein "ladder" model	Three levels Citizen power	Eight types Citizen control (highest rung) Delegated power Partnership
	Tokenism	Placation Consultation Informing
	Nonparticipation	Therapy Manipulation (lowest rung)
Continuum of community participation	Mobilization, collaboration, and empowerment	
Theory of change	Informing, consultation, co-production, and community control	
Utilitarian and social justice	Utilitarian perspective—aims at improving program outcomes which may not impact living conditions of the population Social justice perspective—focus on community empowerment and development with social and structural changes at the center	

The eight types of participation are divided into three levels: non-participation (manipulation and therapy), tokenism (informing, consultation, and placation), and citizen power (partnership, delegated power, and citizen control). 'Continuum of community participation' is the second widely cited model. It is based on community mobilization, collaboration, and empowerment, the three ways by which communities participate in health action [8,9].

Popay developed the 'theory of change' model [10]. It consists of four approaches: informing, consultation, coproduction, and community control. The fourth model is based on a systematic review [11]. The authors suggested two meta-models of community participation in health systems: utilitarian and social justice. The utilitarian perspective aims at improving program outcomes and effectiveness, which may not impact living conditions of the population. The social justice perspective has the focus on community development and empowerment with structural components of social determinants at the center.

Scope of community participation in health

Community participation in health may be implemented horizontally or vertically (health/disease program specific). It may be implemented for primary clinical, preventive, and promotive care; or secondary care or even for tertiary level care.

Community participation is widely documented in health service delivery, their planning, implementation, monitoring and evaluation [12,13]; maternal and child health [14,15]; primary healthcare nurse's involvement in patient and community participation in the context of chronic diseases [16]; environment issues like water sanitation and hygiene (popularly known as WASH) [13,17]; marginalized and indigenous communities [14,18]; special conditions like disability [19]; and health systems research [9], among many others.

Success stories and challenges in community participation

Many find the principles of community participation that found a mention in the Alma-Ata Declaration as lofty principles that are difficult to implement in real-world programs, especially for poor people at the margins. On the 30-year anniversary (2008) of the Alma Ata Declaration, *Rifkin SB* opined that there are three reasons why integrating community participation into health programs was difficult [20]. First, the dominance of the biomedical paradigm that views community participation as merely an intervention to improve health outcome or program effectiveness. Second, the lack of in-depth analysis of the perceptions of community members regarding the use of community health workers. Third, theoretical frameworks of community participation are limited to what works. Despite these reasons, community participation has contributed toward local level improvements in health, especially in poor communities.

Hence, it will continue to be relevant to public health implementers, professionals, and researchers [20].

Across geographies, implementation of community participation has shown mixed results over a period of time [21]. It could be due to the profile of the community (health literacy, socioeconomic status, political system, and others) and the theory or methodology adopted. A lot of work has documented the limits of the implementation of processes aimed at strengthening community-based accountability across the world [22]. In the following paragraphs, we share some examples of successes and challenges.

In high and upper-middle income countries, community participation in non−disease-specific general health initiatives had a positive impact on health, particularly when substantiated by strong organizational and community processes [12]. A systematic review from Bangladesh, India, Malawi, and Nepal (examples of low resource settings) reported that where women's groups discussing mother and infant care involved at least 30% of local pregnant women, the newborn mortality rate fell by one third compared to control villages [23,24].

A systematic review from rural/remote areas of India, Nepal, Pakistan, and Bangladesh found that women's community participation interventions improved child births at a healthcare facility (facility births), but the evidence was of low quality. There was no impact on maternal mortality [15]. Despite few high-quality quantitative studies documenting the effect of community participation on improving skilled maternal and newborn care, the systematic review lacked information about why participation interventions succeeded or did not succeed. This gap can be filled through qualitative systematic enquiry and this calls for mixed-methods study design while evaluating community participation [25].

Coming to community participation in health systems research, there are many positive examples in the literature. Despite these, community participation in health systems interventions was variable, with few being truly community directed [9].

In Ireland, specific activities were undertaken as part of a community process in primary health care. *McEvoy et al.* opined that the likelihood of these becoming a routine way of working in Ireland was low due to lack of a clear understanding of the aims, objectives, and benefits among the stakeholders [26]. In rural South Africa, health committees were not operational as expected due to lack of clarity on roles, capacity, power, autonomy, and support. There was a lack of practical guidance from those in the health system which could be a reason for these challenges [13].

In a review by *Rifkin* in 2014, limited links were found between community participation and outcomes. Where links were found, they were situation-specific, unpredictable, and not generalizable. Hence, the argument that community participation should be better understood as a process [27]. This also brings us to the question of whether to implement community processes from a utilitarian or social justice framework. Further, the availability of skilled public health practitioners on social sciences, especially the traditional methods, is limited across settings which is needed to institutionalize community participation in routine practice.

Digital approaches to engage people and improve their participation is another contemporary area; its use was vastly practiced during the COVID-19 pandemic. Social media platforms (open or closed group), webpage forum/blogs, telephone calls, and mobile phone applications may be used for improving community participation [28]. The digital approaches will be useful since the participation can be ensured in anonymous manner in addition to providing flexibility.

Community participation in India

In the Indian context, the interventions to improve the community participation have persistently come up with mixed results [29]. The passage of Panchayati Raj Bill 73rd and 74th constitutional amendments (in 1995) provided the legal basis for the concept of community participation. This has resulted in devolution of powers to local government bodies. The early programs between 1970s and 1990s were largely community health worker-based programs. Later on, they were based on the accountability and rights-based approaches [21]. However, with the exception of well-documented example of Kerala [30] and Nagaland [31], most of these were *"symbolic events or merely to serve as means to predefined ends."* In other words, the actual large scale translation of the ideas of community participation by the government used more limited definitions of community participation [29,32]. Thus, smaller community-based projects with more involvement from non-governmental organizations witnessed the more empowering and rights-based approaches. The bureaucratic attempts invoked the limited utilitarian perspectives of community participation.

Recent efforts at the national level in India—under the India's National Health Mission (previously national rural health mission—NRHM)—have also met with mixed results, with potentially lower quality of accountability being achieved by the processes. In India, CAH (also referred to as community monitoring and planning) was introduced in 2005 as one of the pillars of NRHM to ensure community-based accountability of health services through community monitoring and planning. This involved setting up of village health sanitation and nutrition committees (VHSNCs) at revenue village level and health facility management committees at public health facility level. In the VHSNCs, one of the frontline workers is the member secretary and one of the members of the local government body is the chairperson.

The Ministry of Health and Family Welfare funded a central pilot project that was developed by a standing committee called as Advisory Group on Community Action. This pilot project was rolled out in 9 states and around

150 districts during 2008–09. The idea was to enable state government to familiarize themselves with this new idea and adopt it after necessary adaptations. In the postpilot stage (2010–14), only two out of nine states continued the implementation after piloting, and the other states either modified it significantly or completely discontinued it [33]. The two states that implemented the original model had a strong nongovernmental organization coalition which was able to build trust with key officials [33]. This was despite CAH finding a prominent place on the policy agenda. Barriers and challenges of implementing CAH in Maharashtra (in west India) and Tamil Nadu (in south India) have been documented by Shukla et al. and Gaitonde et al. [21,34].

Tamil Nadu was one of the two states where post-pilot CAH was implemented using the original model for two years and then abruptly stopped. There were dissonances in the diversity of perspectives regarding community-based accountability and its perceived role in the existing system [21]. This was similar to the lack of a clear understanding of the aims, objectives, and benefits among the stakeholders in Ireland [26]. There were disconnects in the form of lack of spaces and processes for sense-making in what authors called a *"largely hierarchically functioning system"* [21]. During the same time, in Maharashtra, health facility management committees ensured significant positive changes in the local health planning process [34].

Despite the presence of many iconic success stories involving people's participation using this mechanism, the field is more littered with 'failures' and 'unexpected effects' of these interventions and spaces than with successes [22,32,35]. In two rural districts in east India, VHSNCs performed few of their specified functions for decentralized planning and action (62% committees monitored community health workers, 7% checked subcenters, and 2% monitored drug availability with community health workers). Virtually none monitored data on malnutrition. Key challenges included irregular meetings, members' limited understanding of their roles and responsibilities, restrictions on planning and fund utilization, and weak linkages with the broader health system [36].

Due to varying understanding of the concept of participation, there is a major gap in the ways to utilize the processes and measure and link the outcomes. The rights-based approach is more visionary and works for more empowerment and redistribution of power in the system. The utilitarian approach is way short sighted compared to rights-based approach and sees participation more as a means to an end. While implementing a rights-based approach if most of the work was done by the staff from non-governmental organization (on behalf of VHSNCs or health facility management committees), it becomes impractical to replicate over long term. If community is passive, then there is limited institutional memory for long-term change.

Way forward

Notwithstanding the clash between the utilitarian and rights-based approach, community participation requires long-term investment as it takes time to mature and provide the desired results. It also generates many non-tangible benefits. Hence, the authors feel that expectations during initial phases of implementation should be realistic and process indicators based (tending toward a utilitarian bureaucratic approach), like number of monthly meetings conducted. Administrators should not expect that after an order was passed to formulate local health committees, they could start evaluating health facilities using a checklist (common mistake noted in field). It takes time for the local health committee members to get sensitized to health system, health services, concept of health especially the non-health determinants, levels of disease prevention and services provided at village level, primary healthcare level, and above. Hence, it is naturally expected that local health committees will take time to develop a healthy working relationship with the health system. Similarly, it takes time for the health systems to mature in the context of a functioning community participation processes and develop a healthy working relationship with the community.

The question that needs to be answered is how do we ensure that well performing and committed local health committees continue to remain so over long term. In the absence of feedback and motivation, it is idealistic to expect they will sustain their performance and motivation. To begin with, we should consider minimal realistic processes (monthly meeting self-reported data) that are systematically carried out by all local health committees (say in a district or a state). We may collect minimal quarterly data on health report card of their area (self-reported). We may track how many are presenting their annual reports to their local government bodies. Based on this, feedback can be provided as to where the local health committees stand in cumulative score and rank within the district. Cohort wise tracking of and awarding the best performing committees every year will help the well-performing local health committees remain motivated. Availability of mobile-based data capture that permits offline entry and synchronizing with cloud as and when data connectivity is available can assist this process.

This may be done for a period of one year and then we may expect some of them to contribute toward data for evidence-based public health practice. Few may proceed toward a more fruitful participation: moving toward the level of "citizen power" in the Arnstein ladder model, toward collaboration and empowerment in the 'continuum of community participation' model, toward community control in 'theory of change' model, and eventually toward 'social justice.' This requires mentorship that can be provided by non-governmental organizations, research institutions, and teaching hospitals with experience in public health and social and behavioral sciences. Mentorship for implementation strengthening and operational research for district and state level nodal officers should be explored.

Over time, depending on local context, local health organizations can work on non-health determinants like girl child education, waste management, and drinking water quality and work toward improving other socio-cultural—economic—political—ecological factors related to health of the community (social entrepreneurship, rural development, and strengthening of the local governance system).

The ongoing debate regarding definitions, models, and operational challenges to community participation [26] notwithstanding, it will require years of consistent commitment from the state for the community participation processes to mature for us to see its systematic, long-term, and large-scale impact on coverage and quality of public health practices. Despite policy-makers suggesting that community participation is good for rural communities, policy enactment must move beyond mandated tokenism for there to be a recognition that meaningful participation is neither easy nor linear [37]. The disconnects and dissonances have to be addressed through constant dialogue and discussion [21].

To ensure sustainability, vertical or disease-specific programs or projects (government funded or external funded) should not develop standalone community participation interventions. They should be mandated to build on the existing social capital. The untied funds required for implementation of decentralized decision-making should be released in a timely manner without needless audit objections. Communities are not going to go anywhere. Hence, we need consistence in policy concerning community participation and patience from the administrators to implement community participation holistically with realistic sustainable short-term process indicators, while the focus is on medium- and long-term goals.

References

[1] World Health Organization. Regional Office for Europe. Community participation in local health and sustainable development: approaches and techniques (No. EUR/ICP/POLC 06 03 05D (rev. 1)). Copenhagen, Denmark: World Health Organization (WHO); 2002.

[2] World Health Organization (WHO). The world health report 2008: primary health care, now more than ever. Geneva, Switzerland: World Health Organization (WHO); 2008.

[3] Word Health Organization (WHO). Health in 2015 from millennium development goals (MDG) to sustainable development goals (SDG). Geneva, Switzerland; 2015.

[4] Levasseur M, Lussier-Therrien M, Biron ML, Raymond É, Castonguay J, Naud D, et al. Scoping study of definitions of social participation: update and co-construction of an interdisciplinary consensual definition. Age Ageing 2022:51. https://doi.org/10.1093/ageing/afab215.

[5] Goetz AM, Jenkins R. Hybrid forms of accountability: citizen engagement in institutions of public-sector oversight in India. Publ Manag Rev 2001;3:363—83. https://doi.org/10.1080/14616670110051957.

[6] Júnior JPB, Morais MB. Community participation in the fight against COVID-19: between utilitarianism and social justice. Cad Saúde Pública 2020;36:e00151620. https://doi.org/10.1590/0102-311X00151620.

[7]   Arnstein SR. A ladder of citizen participation. J Am Inst Plann 1969;35:216−24. https:// doi.org/10.1080/01944366908977225.

[8]   Rifkin SB. Lessons from community participation in health programmes. Health Policy Plan 1986;1:240−9. https://doi.org/10.1093/HEAPOL/1.3.240.

[9]   George AS, Mehra V, Scott K, Sriram V. Community participation in health systems research: a systematic review assessing the state of research, the nature of interventions involved and the features of engagement with communities. PLoS One 2015;10:e0141091. https://doi.org/10.1371/journal.pone.0141091.

[10]   Popay J. Community empowerment and health improvement: the English experience. In: Morgan A, Davies M, Ziglio E, editors. Health assets in a global context: theory, methods and action. Ney York: Springer; 2010. p. 183−96.

[11]   Brunton G, Thomas J, O'Mara-Eves A, Jamal F, Oliver S, Kavanagh J. Narratives of community engagement: a systematic review-derived conceptual framework for public health interventions. BMC Publ Health 2017;17:944. https://doi.org/10.1186/s12889-017-4958-4.

[12]   Haldane V, Chuah FLH, Srivastava A, Singh SR, Koh GCH, Seng CK, et al. Community participation in health services development, implementation, and evaluation: a systematic review of empowerment, health, community, and process outcomes. PLoS One 2019;14:e0216112. https://doi.org/10.1371/journal.pone.0216112.

[13]   Hove J, D'Ambruoso L, Kahn K, Witter S, van der Merwe M, Mabetha D, et al. Lessons from community participation in primary health care and water resource governance in South Africa: a narrative review. Glob Health Action 2022;15:2004730. https://doi.org/ 10.1080/16549716.2021.2004730.

[14]   Smylie J, Kirst M, McShane K, Firestone M, Wolfe S, O'Campo P. Understanding the role of Indigenous community participation in Indigenous prenatal and infant-toddler health promotion programs in Canada: a realist review. Soc Sci Med 2016;150:128−43. https:// doi.org/10.1016/j.socscimed.2015.12.019.

[15]   Sharma BB, Jones L, Loxton DJ, Booth D, Smith R. Systematic review of community participation interventions to improve maternal health outcomes in rural South Asia. BMC Pregnancy Childbirth 2018;18:327. https://doi.org/10.1186/s12884-018-1964-1.

[16]   Heumann M, Röhnsch G, Hämel K. Primary healthcare nurses' involvement in patient and community participation in the context of chronic diseases: an integrative review. 2022. https://doi.org/10.1111/jan.14955.

[17]   Nelson S, Drabarek D, Jenkins A, Negin J, Abimbola S. How community participation in water and sanitation interventions impacts human health, WASH infrastructure and service longevity in low-income and middle-income countries: a realist review. BMJ Open 2021;11:e053320. https://doi.org/10.1136/bmjopen-2021-053320.

[18]   Snijder M, Shakeshaft A, Wagemakers A, Stephens A, Calabria B. A systematic review of studies evaluating Australian indigenous community development projects: the extent of community participation, their methodological quality and their outcomes Health behavior, health promotion and society. BMC Publ Health 2015;15:1154. https://doi.org/10.1186/ s12889-015-2514-7.

[19]   Chang FH, Coster WJ, Helfrich CA. Community participation measures for people with disabilities: a systematic review of content from an international classification of functioning, disability and health perspective. Arch Phys Med Rehabil 2013;94:771−81. https:// doi.org/10.1016/j.apmr.2012.10.031.

[20]   Rifkin SB. Lessons from community participation in health programmes: a review of the post Alma-Ata experience. Int Health 2009;1:31−6. https://doi.org/10.1016/j.inhe.2009.02. 001.

[21] Gaitonde R, San Sebastian M, Hurtig AK. Dissonances and disconnects: the life and times of community based accountability in the National Rural Health Mission in Tamil Nadu, India. BMC Health Serv Res 2020;20. https://doi.org/10.1186/s12913-020-4917-0.

[22] Cornwall A, Coelho VS, editors. Spaces for change? The politics of citizen participation in new democratic arenas. London and New York: Zed Books; 2006.

[23] Prost A, Colbourn T, Seward N, Azad K, Coomarasamy A, Copas A, et al. Women's groups practising participatory learning and action to improve maternal and newborn health in low-resource settings: a systematic review and meta-analysis. Lancet 2013;381:1736−46. https://doi.org/10.1016/S0140-6736(13)60685-6.

[24] Victora CG, Barros FC. Participatory women's groups: ready for prime time? Lancet 2013;381:1693−4. https://doi.org/10.1016/S0140-6736(13)61029-6.

[25] Marston C, Renedo A, McGowan CR, Portela A. Effects of community participation on improving uptake of skilled care for maternal and newborn health: a systematic review. PLoS One 2013;8:e55012. https://doi.org/10.1371/journal.pone.0055012.

[26] McEvoy R, Tierney E, MacFarlane A. "Participation is integral": understanding the levers and barriers to the implementation of community participation in primary healthcare: a qualitative study using normalisation process theory. BMC Health Serv Res 2019;19:515. https://doi.org/10.1186/s12913-019-4331-7.

[27] Rifkin SB. Examining the links between community participation and health outcomes: a review of the literature. Health Policy Plan 2014;29:ii98−106. https://doi.org/10.1093/heapol/czu076.

[28] Schroeer C, Voss S, Jung-Sievers C, Coenen M. Digital formats for community participation in health promotion and prevention activities: a scoping review. Front Public Health 2021;9:713159. https://doi.org/10.3389/fpubh.2021.713159.

[29] Murthy RK, Balasubramanian P, Bhavani K. Patient welfare societies, health committees and accountability to citizens on sexual and reproductive health: lessons from case studies from Tamil Nadu. Chengalpattu, India; 2009. https://doi.org/10.13140/RG.2.1.3623.9202.

[30] Isaac TMT, Heller P. Democracy and development: decentralized planning in Kerala. In: Fung A, Wright EO, editors. Deepening democracy. London and New York: Verso; 2003. p. 77−110.

[31] Department of Planning and Coordination, Government of Nagaland. Communitization and health—the Nagaland experience. Kohima, Nagaland; 2011.

[32] Coelho K, Kamath L, Vijayabaskar M, editors. Participolis: consent and contention in neoliberal urban India. New Delhi, India: Routledge; 2013.

[33] Gaitonde R, San Sebastian M, Muraleedharan VR, Hurtig AK. Community action for health in India's national rural health mission: one policy, many paths. Soc Sci Med 2017;188:82−90. https://doi.org/10.1016/J.SOCSCIMED.2017.06.043.

[34] Shukla A, Khanna R, Jadhav N. Using community-based evidence for decentralized health planning: insights from Maharashtra, India. Health Policy Plan 2018;33:e34−45. https://doi.org/10.1093/HEAPOL/CZU099.

[35] Manor J. User committees: a potentially damaging second wave of decentralisation? Eur J Dev Res 2004;16(1):192−213. https://doi.org/10.1080/09578810410001688806.

[36] Srivastava A, Gope R, Nair N, Rath S, Rath S, Sinha R, et al. Are village health sanitation and nutrition committees fulfilling their roles for decentralised health planning and action? A mixed methods study from rural eastern India. BMC Publ Health 2016;16. https://doi.org/10.1186/s12889-016-2699-4.

[37] Kenny A, Farmer J, Dickson-Swift V, Hyett N. Community participation for rural health: a review of challenges. Health Expect 2015;18:1906−17. https://doi.org/10.1111/hex.12314.

Chapter 10

Intersectoral coordination for concerted efforts to improve the population health using evidence-based public health practice

Seetharam Mysore[1], Deepak H.J. Murthy[1] and Hemant Deepak Shewade[2]

[1]*Swami Vivekananda Youth Movement (SVYM), Saragur, Karnataka, India;* [2]*Division of Health Systems Research, Indian Council of Medical Research-National Institute of Epidemiology (ICMR-NIE), Chennai, Tamil Nadu, India*

Introduction

Stable robust health for all members of the community is a basic prerequisite for the comprehensive growth and development of a society. Defined as the ability to be optimally productive, we see health as that essential 'state' of life, which bestows the ability to the individual and the collective to not only 'produce,' but also sustain and further enhance productivity in multiple dimensions to enrich the quality of life. Thus, health emerges as the basic building block for human development, both at the level of the individual as well as the collective.

As our understanding of sustainable 'Human Development' has deepened over the past decades, the need to be holistic and inclusive as well as the need to move away from excessive anthropocentricity has increasingly come to the fore. There is a greater acceptance of the interdependencies of different dimensions of human life—social, economic, cultural, geographic, ethnic, and educational—which influence health status. It hence follows that attainment of health is best achieved by collaborative and simultaneous progress in all these dimensions.

The determinants of health can be organized into broad categories like genetics, behavior, environment, healthcare systems, and social factors, all of which are interconnected. The Commission on Social Determinants of Health

Principles and Application of Evidence-Based Public Health Practice
https://doi.org/10.1016/B978-0-323-95356-6.00002-1

formed by the World Health Organization has emphasized that addressing social determinants of health is cardinal to achieving health equity. Global and regional cooperation is imperative in this mission [1].

Intersectoral coordination (ISC)

Definition of ISC

Intersectoral coordination is a formal relationship between a part or parts of the health sector with part or parts for another sector, which has been formed to achieve health outcomes in a way that is more effective, efficient, and/or sustainable than could be achieved by the health sector acting alone.

ISC is a tool that enables greater efficiency in health outcomes by filling in for potential gaps in knowledge, skills, and competencies in a given department of the health sector by leveraging the same from other departments/sectors. It can also be considered as an ongoing 'process' that when institutionalized leads to synergistic and coordinated progress toward the larger local, national, and global goals.

The purpose of coordination in health is to deliver better quality services to a larger number of people. Many models and frameworks of ISC have been tried out with varying success, each of which has lessons for replication and scaling in specific contexts.

Concepts and frameworks for ISC in health

The basic level of coordination is within the sector of health itself, with its complex levels of care and varied characteristics of providers. Such intra-health, and inter-departmental coordination can be envisaged across different

- Tiers of health care, with a seamless continuum of care from the primary care level to the tertiary/quaternary clinical care, as well as rehabilitative services.
- Programs and schemes, which address similar beneficiary groups
- Providers, like the government, private players, and non-governmental service organizations.

Going beyond the sector of health, it is essential to put health on the agenda of other related and contributing departments. Some platforms which provide for such collaborative initiatives are -

i **Health in All Policies:** A collaborative approach that identifies 'Health' as an outcome in the policies and programs of all departments (and not limited to health department) and at all levels [2].

ii **One Health:** A transdisciplinary strategy, spanning from local to global level, to achieve optimal health outcomes among humans, animals, plants, and the environment through understanding their interdependency and future sustainability [3].

iii **Sustainable Development Goals (SDGs) and the web of interconnectedness:** The SDGs consist of 17 goals and 169 targets which are interdependent and provide a blueprint to achieve economic prosperity, environmental protection, and safeguard the well-being of people around the world. The intricate web of interdependence of SDGs as seen in Fig 10.1, clearly depicts the inevitability of mutually supportive approaches, equally applicable to all SDGs, including SDG3 which pertains to the global goal for health [4].

These concepts and frameworks have been incorporated in contextually relevant ways in the national health policies and programs in different countries, for example, the National Health Mission and the National Health Policy of India.

Partnerships as enablers of intersectoral coordination

Partnership in health means bringing together all players involved in improving the health and well-being of people, namely sectors/departments, groups of providers, and social institutions (public or non-public) so that they work together for a common and shared goal, based on mutually agreed roles and principles.

Health services are being delivered by multiple service providers, including the government, the for-profit sector, private and corporate entities, as well as not-for-profit service organizations. All these agencies are involved in activities that continuously influence the determinants of health, thus influencing progress in health. Partnerships between service providers who work for a common objective irrespective of financial profit motives are being

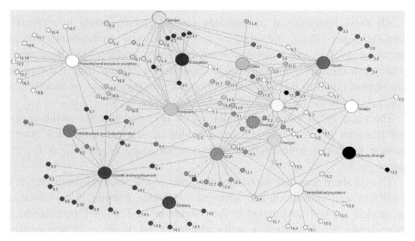

FIGURE 10.1   Web of interconnectedness of Sustainable Development Goals [4].

fostered as public—private partnerships to achieve faster and more holistic progress. Globally as well as within countries, formal frameworks for such coordination have been set forth, which define the institutional and financial mechanisms for the functioning of such partnerships. Despite bottlenecks and hurdles—like ambiguities in roles, lack of mutual trust, the tendency of the state to dominate, etc—such frameworks provide an excellent opportunity for sharing experiences, collectivizing the learnings, and generating evidence for rational policy advocacy.

There are a number of successful partnership models adopted across countries on different health priorities. Though the operational details of these models differ according to the setting or the health issue, they all provide important learnings about factors which make partnerships successful.

i **Global Polio Eradication Initiative (GPEI):** GPEI is a public—private partnership led by national governments with six global partners which has helped countries make huge progress in protecting populations from polio, which has declined by more than 99.9.%, saving an estimated 1.5 million lives and an estimated 16 million people from permanent paralysis [5].

ii **The Partnership for Maternal, Newborn and Child Health (PMNCH):** PMNCH is the world's largest global alliance working for the health of women, children, and adolescents. It has more than 1000 partner organizations representing 10 constituencies across 192 countries [6].

iii **The Universal Health Coverage (UHC) partnership:** The UHC partnership is promoting the member countries to adopt UHC through various strategic steps. Health experts in UHC partnership are actively facilitating policy dialogue on planning, governance, financing strategies, and development of effective cooperation in health across countries. The partnership has more than 115 member countries worldwide to date [7].

All these international partnerships set the global agenda in respective thematic areas, which are implemented by national governments. Implementation involves participation of many ministries, and multiple departments within each ministry. This pyramid of coordination extends from the national government down to the grassroots. Intersectoral coordination is an essential prerequisite for the implementation of these partnerships.

The Pulse Polio Initiative in India, which is considered one of the largest mobilizations of the workforce for a health cause, was successful mainly because personnel from multiple departments could be mobilized for the campaign.

Though there are some successful examples of intersectoral coordination, it is still not well understood and achieved in practice. Further, though most of the models are well articulated at the conceptual policy level, percolation to lower levels has remained a major challenge, as it requires well-defined institutional mechanisms to engage various stakeholders.

For example, the health department can undertake routine surveillance of water quality and observe the changing trends of a disease in a community, and use the data to predict any disease outbreak. However, the important functions related to the prevention of any outbreak are with other departments like public health engineering, water-sanitation department (to ensure safe and adequate water, and universal safe sanitation) and food safety department (to ensure safe and quality food). However, day-to-day coordination between health department and other departments to prevent such outbreaks is practically not happening, especially in most of low-resource settings. Further, the area of control and responsibilities of health and other departments during an outbreak are poorly defined.

The guiding principles for ISC

In a detailed study of factors influencing ISC under the CHORDIS PLUS initiative in Europe through a survey of 22 project partners in 14 countries and 2 workshops, an attempt was made to identify enablers and barriers for ISC [8]. There have been similar reviews and documentation of learnings of other experiments which provide insights into factors contributing to the success or failure of partnerships.

For effective intersectoral coordination, the implementing agencies involved should have a shared vision for the community that they are working for and understand the 'big picture' set out by the shared objective. Trust, a spirit of give-and-take, and willingness to work as a team are essential values that need to be nurtured by the leadership.

Based on the lessons learnt from the intersectoral initiatives, the authors feel that the factors impacting success of ISC positively or negatively can be categorized as potential Enablers and Disablers, as summarized as in Table 10.1.

The imperative for intersectoral approaches

With increasing challenges in addressing healthcare needs in developing countries, social protection platforms and intersectoral mechanisms are needed to address social determinants of health. These will have to be designed based on social, political, and economic contexts that are specific to population- or geographical-based interventions.

Even as we set out to achieve the idealistic goals set for the sector of health in the SDG framework, it is evident that these goals can be achieved only with the strong involvement of other key departments (Fig. 10.2).

It is imperative that the health as well as the other departments concerned, understand their mutual contributions and work for progress on SDG3. Some examples of such convergence are summarized in Table 10.2.

TABLE 10.1 Enablers and disablers of intersectoral coordination.

Enablers	Disablers
Shared vision, trust, and team-spirit for collective learning and generation of evidence	Lack of trust and understanding among the collaborators
Ensuring role clarity and shared ownership	Sense of fiefdom and reluctance to share information
Identification of potential areas of conflict and misunderstanding, and planning coping mechanisms	Fear of failure and blame
Building institutional mechanisms for implementation of partnerships	Fear of retribution
Ensuring joint monitoring processes	Tendency to cover up lacunae
Sharing and analysis of data, processes, and strategies	Tendency to falsify information to project nonexistent success
Dissemination on common platforms for evidence-based policy advocacy	Shirking responsibility and 'washing one's hands off'

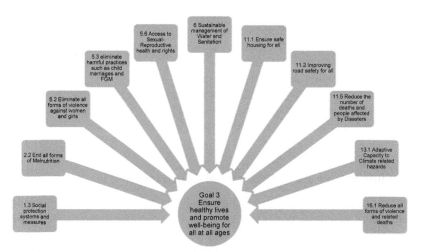

FIGURE 10.2 Sustainable Development Goal-3 and its essential direct linkages.

Advantages of ISC

ISC brings in the following advantages:

- Shared ownership of problems, and a shared sense of responsibility for solutions
- Transparency and accountability

TABLE 10.2 Imperative for intersectoral coordination for Sustainable Development Goal 3.

Goal statement	Limitations of direct interventions by health department	Enrichment by other departments for the overall achievement of the goal
SDG 3.5—strengthen the prevention and treatment of substance abuse, including narcotic drug abuse and harmful use of alcohol	The health department can set up treatment support and initiate behavior change communication	Success of the efforts will depend to a large extent on the efforts at prevention of first use among the youth and adolescents, a role best played by and in educational institutions. Similarly, enforcement of the legal provisions to reduce access to these destructive agents would be the purview of the police and enforcement authorities Public communication through multiple modalities will be essential for disseminating the correct information to community
SDG 3.6—by 2030, halve the number of global deaths and injuries from road traffic accidents.	The health department can at best look at building more hospitals for the care of accident victims and improve the quality of services provided.	Core responsibility of reducing the number of accidents itself would be with other departments like transport, roads, automobile safety, police and others.

- Equity and last-mile coverage
- Diverse viewpoints which add enrichment to the proposed solutions
- Broadening the perspectives and instilling a sense of holism
- Avoidance of duplication of services leading to greater cost-effectiveness and time-effectiveness
- Promotes sustainability

Use of evidence-based approaches in intersectoral coordination

Similar to evidence-based medicine, evidence-based public health practice is gaining importance globally. Jenicek et al. suggested a step-wise approach for incorporation of evidence in public health programs, and then going beyond implementation to evaluation and training. Evidence-based approaches need to be used along with intersectoral coordination to improve population health [9]. For effective use of such evidence-based approaches, it is essential that the framework be standardized and uniformly accepted by all the stakeholders.

It is often seen that in the administrative and implementational quagmire, data integrity and evidence application often lose out. It is hence important to trace the flow of data from its generation to its logical end of application as evidence in policy. Table 10.3, captures the flow of data, the often-observed pitfalls when multiple sectors are involved, and suggested strategies to pre-empt such failings.

Technology is emerging as a major game-changer, be it for improved quality of clinical care (e.g., use of point-of-care devices, and others.) or for prevention-promotion activities (e.g., more effective behavior change communication). Such advancements, especially in emerging fields like Artificial Intelligence, Information-Communication Technology, Data Analytics, etc. can contribute to greater success in initiatives involving intersectoral coordination.

A vision for the future

As we look toward the future, it would be important also to look back at the progress and the failings in the recent decades in the health sector and consolidate the learnings. While partnerships and coordination are essential, doubts remain about their efficacy and the ability of the stakeholders to stay steadfast in the principles of partnerships. The essential input and processes from different models of partnership need to be documented and reviewed. This will help the countries to form effective partnerships in future through adoption of proven best practices. However, it is important to acknowledge the limited availability of such evidence especially applicable to low resource settings. Despite the unprecedented damage caused to lives, lifestyles, and livelihoods, the COVID-19 pandemic had a silver lining in the form of the exemplary partnerships that emerged in different countries to manage the crisis. A systematic study of the stakeholders, their roles, and the mechanisms of management of these partnerships would provide valuable insights to expand the spirit of collaboration to non-COVID health situations as well. This would help us find some answers to the vexing questions: how do we succeed in breaking the barriers of vertical silos? How do we build trust and create the

TABLE 10.3 Data management framework for successful intersectoral coordination.

Stage of flow of data	Why is this important?	How to eliminate confusion and bring standardization?
Definitions for data	Definitions and characteristics of the parameter being measured are often different in different departments. E.g.: Water supply and safe sanitation coverage has a 'Safety Ladder' approach, and often different departments adopt variable levels of this ladder for their reports.	• Ensure that uniform and standard definitions for data are adopted and followed by all stakeholders • Clarify the definitions applicable for the coordination/partnership at the initial planning stage itself
	Baseline data of populations and specific beneficiaries eligible for services are often discordant in different departments. E.g.: Number of children with special needs as per education department is often different from the number shown in other departments like women and child welfare.	
Capacities to generate and handle data	Data often are represented in multiple ways—as in total numbers, ratios, proportions, targets, etc.	• Ongoing training of people involved are well trained in collection, inputting, reporting • Oversight by a set of experts at the district level
	Progress monitoring involves understanding of disaggregation in terms of geographies, social subgroups, vulnerable populations, etc.	
Data availability and usage	Data generated in specific departments often are not shared in the public domain. Critical information is thus not accessible to other related departments. E.g.: Information on ground water tables generated by departments involved in remote sensing is not always available for the departments planning water supply programs.	• Effective application of technology for information dissemination • Regular, real-time data should be available and accessible for all stakeholders • Project-specific dashboards to put data in the public domain

Continued

TABLE 10.3 Data management framework for successful intersectoral coordination.—cont'd

Stage of flow of data	Why is this important?	How to eliminate confusion and bring standardization?
	Activities and progress achieved by one department are not known to related departments working in the same sector, leading to duplication and inefficiencies E.g.: Distribution of aids and appliances for persons with disabilities	
	The generated data are not discussed and given due importance in the departmental and joint monitoring meetings.	
Data analysis for decision-making process	Emerging data are often overlooked in key decisions E.g.: Timely budget modifications in response to specific situations as in disease outbreaks and other crisis situations.	• Periodic macro- as well as micro-level reviews to ensure evidence feeds into planning
	Critical enabling and disabling factors for progress in a specific sector when shared across all related departments can help in collective learning and stratification. E.g.: Large-scale community phenomena like migration and natural phenomena like climate change	

spirit of WIN-WIN? How do we reassure all the stakeholders that their individuality would not be compromised in the name of synergy?

The ideal vision of health for the 21st century would be to seamlessly integrate all the social determinants of health with the components of healthcare system and provide a mechanism to link the various providers of these services [10]. This is essential to address the evident 'tip of the iceberg' burden of illnesses on one hand, and simultaneously address the root causes of poor health at the individual, family, and community level. The focus of health policies and programs needs to go beyond hospitals to assess and address the major determinants of health which can be achieved only through collective and coordinated efforts.

References

[1] World Health Organization. Commission on social determinants of health, 2005—2008. World report on social determinants of health equity; n.d. https://www.who.int/initiatives/action-on-the-social-determinants-of-health-for-advancing-equity/world-report-on-social-determinants-of-health-equity/commission-on-social-determinants-of-health. (Accessed 8 November 2022).

[2] Centers for Disease Control and Prevention. Health in all policies. AD for policy and strategy home. 2016. https://www.cdc.gov/policy/hiap/index.html. [Accessed 8 November 2022].

[3] World Health Organization. One health. Newsroom; 2017. https://www.who.int/news-room/questions-and-answers/item/one-health. [Accessed 8 November 2022].

[4] Morton S, Pencheon D, Squires N. Sustainable Development Goals (SDGs), and their implementation: a national global framework for health, development and equity needs a systems approach at every level. Br Med Bull 2017;124:81—90. https://doi.org/10.1093/bmb/ldx031.

[5] Groce NE, Banks LM, Stein MA. The Global Polio Eradication Initiative-polio eradication cannot be the only goal. Lancet Glob Health 2021;9:e1211. https://doi.org/10.1016/S2214-109X(21)00314-4.

[6] Pervaiz F, Shaikh BT, Mazhar A. Role of development partners in Maternal, Newborn and Child Health (MNCH) programming in post-reform times: a qualitative study from Pakistan. BMJ Open 2015;5:e008665. https://doi.org/10.1136/bmjopen-2015-008665.

[7] Bloom G, Katsuma Y, Rao KD, Makimoto S, Yin JDC, Leung GM. Next steps towards universal health coverage call for global leadership. BMJ 2019;365:l2107. https://doi.org/10.1136/bmj.l2107.

[8] van Dale D, Lemmens L, Hendriksen M, Savolainen N, Nagy P, Marosi E, et al. Recommendations for effective intersectoral collaboration in health promotion interventions: results from joint action CHRODIS-PLUS work package 5 activities. Int J Environ Res Public Health 2020;17. https://doi.org/10.3390/ijerph17186474.

[9] Jenicek M, Stachenko S. Evidence-based public health, community medicine, preventive care. Med Sci Monit 2003;9:SR1—7.

[10] World Health Organization. Intersectoral action for health: a cornerstone for health-for-all in the twenty-first century. In: International conference on intersectoral action for health. Halifax, Canada: World Health Organization; 1997. p. 1—50.

Chapter 11

Public health approaches to address substance use: An urgent need of multisectoral engagement

Cristina Rabadán-Diehl

Westat, Rockville, MD, United States

Introduction

The use of alcohol, tobacco, and other substances has been documented throughout history [1,2]. With the passage of time, society has categorized their use as socially acceptable (alcohol and tobacco), medically justified (prescription medications), or illegal (cocaine, heroin, and other stimulants). In the case of marijuana, some societies are in the midst of a reclassification, but whatever the historical reasons for these choices, the impact on public health has been enormous.

Regulatory agencies and health systems are responsible for the approval, purchase, prescription and dispensation of medications, and monitoring protocols allow for better control. Over decades we have accumulated evidence demonstrating alcohol's and tobacco's addictive properties and detrimental effects on health. However, while subjected to local laws and regulations, alcohol and tobacco products are still available for purchase and in most countries are considered legal. The mobilization of the public health community, the voices of scientific and advocacy groups, and the creation of international mechanisms like the "Framework Convention on Tobacco Control" [3] and the "Global Strategy to reduce the harmful use of alcohol" [4], to name a few, have significantly contributed to increasing our knowledge and helped build the evidence to develop effective public health interventions and policies. During the last 2 decades, tobacco use has decreased from 1.4 billion people in the year 2000 to about 60 million [5]; however, we have not been as successful on global alcohol intake. A recent study in the Lancet analyzed alcohol intake in almost 190 countries and concluded that the total volume of

Principles and Application of Evidence-Based Public Health Practice
https://doi.org/10.1016/B978-0-323-95356-6.00001-X

alcohol consumed globally has increased since 1990 [6]. While it is important to recognize that there is still much that needs to be done on tobacco and alcohol products, a connected global health community exists that is committed and works collaboratively to close the gap. Yet the same is not the case for another class substance: illicit drugs. Public health approaches are colored by our perception and judgment of the individuals who use them, the way they are obtained, and the punitive laws and actions that accompany them. People are discriminated against and marginalized based on the type of substance they use. Societal beliefs and media channels, like the entertainment industry, continue depicting a person with addiction as one who lacks character, chooses to live a life outside societal norms and expectations, and consequently deserves isolation, rejection, and punishment.

Regardless of the type of substance, we have come a long way in our understanding of addiction. During the last decades, much has been learned on its science, the struggles and suffering of the patients and families touched by the disease, and about preventive and treatment strategies that can help individuals enter into recovery and live fulfilling and productive lives. The current opioid epidemic in the United States (U.S.), the appearance of new and potent synthetic drugs, like fentanyl, and an increase in the drug supply chain facilitated by the dark web have triggered the alarm and a sense of urgency across the world. In this regard, globalization has a negative impact on the population health. Our public health successes, which resulted in longer life expectancy, better health indicators and improved quality of life, are now being threatened.

This chapter focuses on public health challenges, potential solutions, and opportunities for multisectoral collaborations to address substance use disorders (SUDs) in populations who primarily use internationally controlled substances (illicit drugs) as it is in this area where we see the most pressing need. Work relating to alcohol, tobacco, and prescription drugs is beyond the scope of this discussion though the use/possession of these substances not legally allowed can be classified as illicit use. As the U.S. is undergoing a public health emergency due to the opioid crisis, there are efforts and lessons learned that might be of benefit to public health professionals in other countries. Approaches highlighted here serve as potential examples and guidance and not an endorsement of best practices. While the term "substance use disorders" encompasses all substances, when used in this chapter it primarily refers to illicit drugs.

The magnitude of the problem and the science of addiction

SUDs have been identified as an important public health issue under the Sustainable Development Goals, Goal 3, Target 3.5: "Strengthen the prevention and treatment of substance abuse, including narcotic drugs abuse and harmful use of alcohol" [7].

While there are limitations in assessing the burden of drug use at the population level especially in low- and middle-income countries, newer methods like waste water−based surveillance or burden assessment in addition to modeling based on available data are getting popular. The United Nations Office on Drugs and Crime (UNODC) World Drug Report 2021 [8] estimates that around 275 million people used drugs worldwide in 2020 and that over 36.3 million people have a drug use disorder. Though cannabis is the most commonly used substance, opioids are responsible for the largest burden of the disease of addiction globally. Further, it is estimated that over 11 million people inject drugs globally and half of whom are living with hepatitis C. By 2030, the number of people using drugs is projected to rise in Africa by 40% and globally by 11% (Fig. 11.1). It is important to note that a proportion of disorders due to drug use are associated with the nonmedical use of pre-scription drugs for pain management (synthetic opioid analgesics), anxiety and sleep disorders or psychostimulants. The dark web has revolutionized illicit drug markets, and while currently it only accounts for a fraction of overall sales, there has been a fourfold increase in annual sales between mid-2017 to 2020.

The fact that many drugs have been labeled as "illegal" has prevented collective action from the public health perspective due to predominant linkage with the criminal justice system. While there are increasing discussions on proceeding with a public health approach rather with criminal justice system recently, and organizations like the World Health Organization (WHO), UNODC, and the National Institutes of Health (NIH) in the U.S. have done tremendous work in this space for a long time, it has only been in the past years, primarily driven by the opioid epidemic in the U.S., that a sense of urgency has arisen. The United States' crisis was initially driven by prescription medications like hydrocodone (Vicodin) and oxycodone (OxyContin, Percocet) in the 1990s and many countries argue that differences in health systems, economic models and societal dynamics will protect their populations from any similar situation. However, the sophistication of the drug supply, the appearance of fentanyl as a contaminant, and the interconnection between mental health and substance use pose a threat to any country. Drug abuse is now a global health problem and not just a problem of high-income countries [9].

The European Union (EU) Drug Market Report 2019 [10] shows that the market for synthetic opioids has been growing in Europe and that they account for the high morbidity and mortality rates associated with illicit drug con-sumption in the EU. Although the majority of illicit synthetic opioids come from outside the EU, medium and small-scale laboratories producing fentanyl inside the EU have been dismantled in the last few years (2015−18). Fentanyl has been identified as a main adulterant of other drugs in several European countries, and in others, like Estonia, it has displaced heroin as the main opioid being consumed. Cocaine is the second most commonly consumed illicit drug in the EU and other countries and digital technologies are enabling more effective supply chain modalities. Fentanyl is rapidly becoming the main

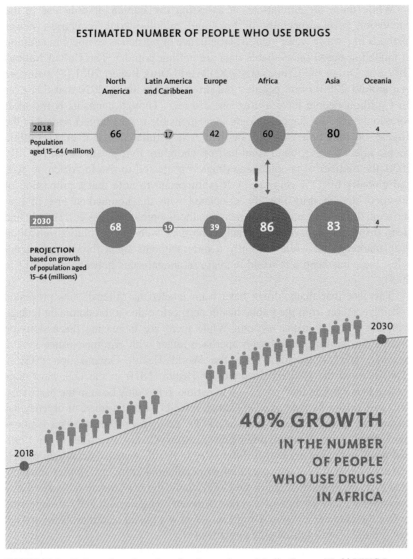

FIGURE 11.1 Estimated number of people who use drugs globally. *Source: World DRUG Report 2021. Booklet1—executive summary/policy implications (United Nations Publication, Sales No. E.21.XI.8).*

adulterant of cocaine and other drugs, and it contributes to cocaine and methamphetamines-related overdose deaths around the world. Fentanyl is an opioid 100 times more potent than morphine and 50 times more potent than heroin, extremely cheap to produce and very quickly becoming the cause of 90% of the deaths due to opioids in the U.S. (Fig. 11.2) [11] As a result, on

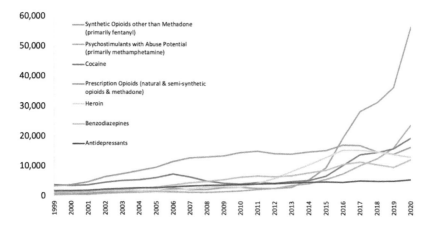

*Includes deaths with underlying causes of unintentional drug poisoning (X40–X44), suicide drug poisoning (X60–X64), homicide drug poisoning (X85), or drug poisoning of undetermined intent (Y10–Y14), as coded in the International Classification of Diseases, 10th Revision. Source: Centers for Disease Control and Prevention, National Center for Health Statistics. Multiple Cause of Death 1999-2020 on CDC WONDER Online Database, released 12/2021.

FIGURE 11.2   United States national drug-involved overdose deaths*. Number among all ages, 1999–2020. *Source: CDC WONDER online Database, released December 2021.*

October 26, 2017, the U.S. government declared the opioid crisis a national public health emergency. The COVID-19 pandemic that started in 2020 has exacerbated the use of substances around the world and in the U.S. has contributed to an unprecedented number of deaths [12]. In the period between December 2021 and December 2022, 109,680 lives were lost due to overdose, which translates into 300 lives a day (Fig. 11.3).

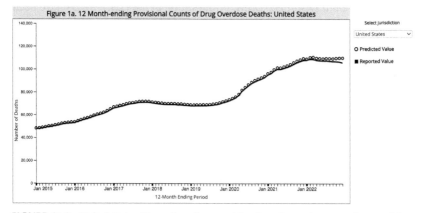

FIGURE 11.3   United States 12-month-ending provisional number and percent change of drug overdose deaths. December 2021-December 2022. *Source: CDC. National Center for Health Statistics. Based on data available for analysis on. May 7, 2023.*

Research has shown that drug addiction is not due to a lack of character or a weakness, but rather is a very complex disease. The potential gateway effect of tobacco or alcohol for other substance use in adulthood could not be proved [13]. The use of substances modifies biochemical and structural brain centers that result in behavioral changes in individuals. We are learning that some individuals seem more susceptible to develop addiction. Genetics, family history, trauma and violence, and drug exposures during childhood at times where the brain is undergoing critical changes increase vulnerability [14−18], but the existence of those risk factors does not mean that an individual will definitely develop the disease. Similarly, an individual without known risk factors can also develop addiction. We also know that the harmful consequences of drug abuse and addiction affect people of all ages and all groups. While currently we cannot predict who will develop an SUD, we know that we can prevent not only the initiation of the disease but also its progression. The medical and public health communities are actively exploring treatment modalities to help the individuals with SUDs recover and have productive lives, although there are still no cures. Relapses are frequent and, in many cases, bring devastating consequences. Therefore, public health interventions are critical.

As per the National Institute of Drug Abuse (NIDA) "*Addiction is defined as a chronic, relapsing brain disease that is characterized by compulsive drug seeking and use, despite harmful consequences. It is considered a brain disease because drugs change the brain-they change the structure and how it works. These brain changes can be long lasting, and can lead to the harmful behaviors seen in people who abuse drugs.*" [19] Currently, disorders due to substance use or addictive behavior have specific medical codes as established by the International Classification of Diseases (ICD-11) as well as by professional associations in several countries (The *Diagnostic and Statistical Manual of Mental Disorders*-DSM, Fifth Edition in the U.S.). SUDs have been embraced by the medical communities as a public health issue that deserves attention and urgent solutions.

It is not rare to find comorbidities among individuals who have an SUD [20]. Other conditions like mental or physical health disorders may be present at the same time or they may cooccur at different times in life. They can develop jointly, or one might increase the risk of the person developing the other. In addition to high risk for HIV and Hepatitis C Virus (HCV) infections, the individuals with SUD potentially have high exposure to NCD risk factors. Mental health disorders are among the 10 leading causes of burden of disease worldwide. Mental health disorders based on the 2019 global burden of disease data analysis [21] increased from 80.8 million DALYS (disability-adjusted life years) to 125.3 million between 1990 and 2019. In the same period, SUDs moved to 18th and 16th place in 2019 from 22nd place in causing DALYs among people aged 10−24 years and 25−45 years, respectively [22]. The COVID-19 pandemic has exacerbated depressive and anxiety disorders, with an additional 53.2 million cases of major depressive disorder globally due to the pandemic [23].

One of the main issues with comorbid substance use and mental disorders is that diagnosis and treatment can be very challenging. If treatments are not carried out in an integrated way, the patient's prognosis can be very poor, sometimes including death [24,25]. Screening and diagnostic instruments can help diagnose comorbidities; however they need to be administered by trained professionals with sufficient time and expertise. Therefore, an integrated health systems approach is needed where screening and management may be developed at outpatient, inpatient, and emergency settings levels, similar to tobacco and alcohol strategies. A comprehensive list of those instruments can be found in the UNODC publication on "Comorbidities in Drug Use Disorders" [20].

Global public health challenges and barriers

The Social Determinants of Health of substance use

Centuries of discrimination accompany the field of substance use. The Social Determinants of Health (SDoH) play a very important role not only on the onset and development of addiction but also on how we address the needs of communities that experience poverty, lack of education, homelessness, incarceration, and racial and ethnic discrimination. If the opioid epidemic in the U.S. has taught us anything it is that the disease of addiction itself does not discriminate. It can strike anyone and anywhere independently of one's social economic status, race or ethnicity, or level of education. However, there is no question that a social vulnerability to substance use exists and that the ecological model of public health applies at all its levels, i.e., individual, interpersonal, organizational, community, and public policy (Fig. 11.4), especially, when it comes to access to information, education, services, and treatment [26−29]. Socioeconomic factors at the macro, community, and individual levels have tremendous influence when it comes to drug use and access to treatment. Women, sexually diverse populations, indigenous and aboriginal peoples, ethnic groups and immigrants, and individuals living in rural areas are groups particularly impacted by socioeconomic disadvantages [30].

The SDoH have been the focus of dialogue when developing solutions to SUDs. The NIH HEAL (Helping to End Addiction Long-term) Initiative conducted a workshop in 2020 addressing the social determinants of opioid use, a recording of which is available online [31]. The workshop focused on establishing a research agenda to inform community- and system-level interventions. Similarly, a separate briefing by Bohler and colleagues [32] provides examples of programs targeting SDoH.

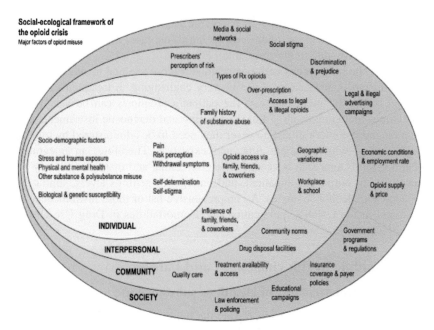

FIGURE 11.4 Social—ecological framework of substance use. *Source: Jalali MS, Botticelli M, Hwang RC, Koh HK, McHugh RK. The opioid crisis: a contextual, social-ecological framework. Health Res Policy Sys 2020;18:87. https://doi.org/10.1186/s12961-020-00596-8.*

The COVID-19 pandemic has challenged countries and public health systems, but it has also spurred innovations and expanded the use of telemedicine to deliver treatment services to individuals with SUDs. However, the economic and social impact of the pandemic has also interfered with treatment access and resulted in greater unemployment, poverty, inequities, and mental health conditions, especially in vulnerable populations at a higher risk of drug use. In addition, the rapid adoption of digital technologies during COVID-19 pandemic could potentially have a negative effect in increasing the access and use of illicit drugs.

Law enforcement and legal systems

Many diseases have devastating consequences to the individuals who suffer from them, but very few result in incarceration. Addiction is one of them. While not everyone with an SUD enters the criminal justice system, it is estimated that approximately 85% of inmates in the U.S. have either an active SUD (65%) or were under the influence of drugs or alcohol when the crime was committed (20%) [33].

The decriminalization of drugs is subject to an on-going dialogue and in some countries substances like marijuana are regulated but no longer illegal. A country worthy of study is Portugal, which enacted sweeping drug law reforms over 20 years ago. In 2001, Portugal decriminalized the possession and use of small quantities of drugs, although not drug trafficking. The law's primary purpose is to focus on supporting the individual. Persons found in possession of personal-use amounts are ordered to appear before a commission made up of legal, health, and social services officials who determine the degree to what this person might have a substance use problem. The commission can make referrals to voluntary treatment options or impose administrative sanctions. The law reform did not result in a change in the use of illicit drugs, but it had other significant positive impacts such as a reduction in use by adolescents and the number of people with drug dependencies. It also resulted in a reduced incidence of HIV/AIDS, higher numbers of people in drug treatment, reduced emergency visits, and drug-related deaths. Since it reduced the number of prisoners, the societal cost also reduced indirectly [34,35]. Similar to Portugal, Law Enforcement–Assisted Diversion program implemented in Seattle is an ongoing program since 2011, which allows the enforcement officers to sanction community-based services instead of incarceration. The results are positive on reducing the recidivism and cost in addition to access to basic needs to run a dignified life [36].

There are still many who believe that there is little use in offering treatment services until a patient has reached rock bottom. But there is no evidence that delaying the offer of treatment improves outcomes. To the contrary, two studies indicate that earlier interventions are associated with greater success, possibly because SUD is a progressive disease that may be increasingly harder to treat with time [37,38]. A decision not to intervene in early stages of the disease of addiction would be akin to denying treatment to prediabetic patients until they have overt diabetes or to wait for cancer patients to enter stage 4 before receiving appropriate treatment. The sooner we can intervene, the higher the likelihood of success and hence reduced morbidity and mortality.

Context and cultural competency

As with any other public health issue, it is important to recognize that one approach does not fit all, particularly given the divergent needs and realities in different communities. Cultural competencies as well as social, political, and health contexts are important considerations when we are developing prevention and treatment strategies for SUDs. Even where health care systems, societal structures, and cultural norms are similar, pilot programs in different communities need to be tested for their effectiveness using rigorous public health validation methods and being accompanied by careful process and outcome evaluations. Scalability, innovation, and sustainability are an integral part of the success of any public health program, but in this case the chronicity of the disease, the social barriers, and the rapid fluctuations in the drug market and supply chain call for dynamic and flexible approaches that can rapidly adapt to the changing environment. Tran et al. conducted a research mapping of SUDs and treatments over a 25-year period and provided a comprehensive narrative on the changes in treatment modalities over time [39]. The development of validated "tool kits" has been shown to be effective in providing resources to develop appropriate interventions; however, the creation of tool kits that take into account the social, political, and cultural environments of countries should be encouraged. Many local governments in the U.S. and elsewhere are developing their own tool kits. Although there are already a few examples [40–42], we need more multistakeholder joint efforts at the global level with input from countries and leadership from multilateral organizations.

Stigma: The biggest barrier of all

We have already discussed how addressing SUDs at the local, country, and global levels is a complex issue. Barriers at the micro- and macro level exist, which include political will, policy support, effective stakeholder coordination, and communication, adequate health infrastructures and resources, financing and community engagement. Although some successful policies and programs are highlighted in this chapter, it is important to recognize that in spite of the advancements made in our scientific knowledge of the disease of addiction, societal attitudes have not kept pace. Perception of SUDs has not significantly changed over time with the result that stigma remains one of the biggest barriers to progress [43]. Structural, political, community-level, and self-stigma persist and lead to discrimination. Stigma is a global health problem and one that needs to be addressed in every action we take at all levels in the public health framework.

The contribution of self-stigma is now recognized as a significant barrier to treatment seeking [44]. As with many other medical conditions, the individual needs to acknowledge and seek treatment for his or her disease. However, the profound changes in brain chemistry and the lens of discrimination used by society pose a significant impediment as they influence their own perception about the magnitude of the problem and how they identify with those that have the disease. Fear and shame also play a role, causing many to hide their addiction. Families are overtaken by fear as well. They may see changes in their loved ones' behavior but fail to understand the cause. Others may be in denial or not know to see help or support. Like patients, families may, out of fear and shame, hide their loved ones' disease. Addiction is a family disease yet; while other diseases inspire compassion and solidarity, this is a disease of isolation and silence. That silence is one of the biggest challenges we face as public health professionals.

One way we can break the silence is by changing the way we talk and using less stigmatizing language. Communication is the key to reduce and remove the stigma of SUD and link for treatment. On January 9, 2017, the Obama Administration released a document encouraging federal agencies to change the terminology when addressing substance use [45]. Subsequently, the National Institute on Drug Abuse launched an initiative called: "Words Matter-Terms to use and avoid when talking about Addiction" [46]. This is a short, comprehensive guide that provides terms to use, to avoid and an explanation of the reasoning behind it. For example, instead of using terms like "addict," "junkie," or "drunk," we should use "person with a substance use disorder," "person with opioid addiction," or a "person with alcohol disorder." By using those words, we are indicating that a person has a problem rather than the person is the problem. We acknowledge that it is a disease and not something that anyone chooses or the result of a lack of character. In addition, the Substance Abuse and Mental Health Administration (SAMHSA) developed a training resource entitled "Words Matter: How language choice can reduce stigma" [47]. This tool provides steps to ensure that we are using the right words in a positive and inclusive way. The National Academies of Sciences published in 2016 a book entitled *Ending Discrimination Against People with Mental and Substance Use Disorders: The Evidence for Stigma Change* [48]. The report provides six recommendations and, while they are directed to U.S. Federal Agencies, they can be applied to any government, as they revolve around education, advocacy, multistakeholder engagement (including the criminal justice system), and policy approaches.

While drug addiction is in some instances considered a subcategory of mental illness, there are many differences that support that stigma reduction should be tailored differently. Barry and colleagues conducted a survey of over 700 individuals to compare mental health and substance use attitudes about stigma, discrimination, treatment effectiveness, and policy support. Their findings indicate that the public opinion has significantly more negative views toward people with drug addiction compared to those with mental illness [49].

Developing the evidence base for combating stigma effectively is very much needed [50], but without a reliable measure of stigma the effectiveness of interventions is difficult to assess. Recently, Shatterproof and The Hartford, with the support of Indiana University, have developed the Shatterproof Addiction Stigma Index, a research tool that can be used to assess the degree of substance use—related stigma in a given population, and that can be incorporated into programs to determine the effectiveness of interventions [51].

As with any disease, advocacy, patient-based organizations, and humanizing narratives have significantly accelerated progress toward raising awareness and finding effective treatments that ultimately need to be delivered in primary healthcare settings and integrated with routine healthcare delivery. In the field of SUDs we are still lacking sufficient representation from patients and families affected by addiction. It is necessary that we engage people who use drugs [52], persons in recovery and families as critical stakeholders in dialogues and partnerships aiming to develop policies, public health programs, communications strategies, and international alliances. Lacking those important voices will likely result in a number of ineffective approaches and in some instances a waste of precious time.

Treatment and prevention

Treatment of SUDs is possible, and efforts are taking place around the globe to develop, validate, and implement evidence-based programs. However, effective treatment programs must take into consideration all the dimensions of drug addiction and therefore incorporate many of the elements we have highlighted above. The NIDA *Principles of Drug Addiction Treatment: a Research Based Guide* [53] is directed to different types of drugs, but the fundamental principles and critical elements apply to illicit drug abuse treatment as well (Fig. 11.5). Effective programs need to include different types of therapies and engage other services to tailor them to the needs of individuals. The programs must be adaptable and offer flexibility, to incorporate a variety of interventions, time frames, and monitoring. Because treatment modalities may need to be adjusted to the individual, one-time, short-term treatments are often not sufficient or effective, rather long-term investment in terms of trained staffs and resources is needed. As with other chronic conditions, relapses are likely but they are similar in frequency to those observed with hypertension, asthma, and diabetes. Perceptions that persons with SUDs are especially likely to relapse may stem from stigma.

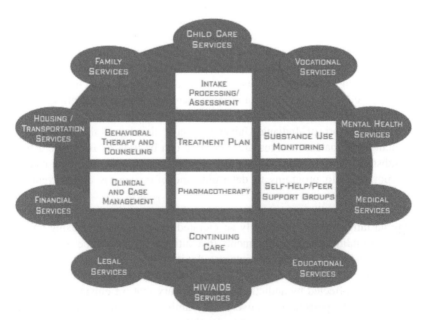

FIGURE 11.5 Components of comprehensive drug use treatment. *Source: National Institute on Drug Abuse; National Institutes of Health; U.S. Department of Health and Human Services.*

Treatments are most effective when designed around the needs of the individual (patient-centered approach) and can include a combination of not only behavioral but also pharmacological interventions, also referred to as medication-assisted treatment. For opioids use disorder, the term "Opioid Substitution Treatment" has been used, although proposed new terminology—"Opioid Agonist Treatment"—reflects more accurately the treatment approach and the mechanism of action [54]. Opioid agonist treatment includes the use of medications like methadone, buprenorphine, and naltrexone (including the long-acting formulations). Treatments using methadone or buprenorphine have been shown to be effective [55] and to reduce significantly the risk of death. The risk of death has been noticed to be highest during the initial 4 weeks of treatment; however, that risk was reduced by persistent engagement with treatment, and it increased if individuals discontinued the treatment during that time period [56].

Naloxone is not a treatment for opioid use disorder, but it can reverse the effects of an opioid overdose and save lives. In 2016, the WHO/UNODC launched the "Stop Overdose Safely (S—O—S)" Initiative [57]. This program provided training on how to recognize the risk of overdose and emergency care in the event of an overdose, including the use of naloxone. A multisite study on community management of opioid overdose was implemented in four countries (Kazakhstan, Kyrgyzstan, Tajikistan, and Ukraine) in 2019—20. About 40,000 kits of naloxone were distributed, and more than 14,000 people were trained. Evaluation of the project showed that it was widely accepted by stakeholders, ranging from people who use drugs through to health and law enforcement officials. However, fear of police intervention was identified as a barrier to carrying naloxone and intervening in an overdose situation [58].

The UNODC and WHO have created joint initiatives to support and facilitate knowledge exchange and collaboration. "Treatnet" was created in 2005 [59], an international network of drug dependence treatment and rehabilitation resource centers aiming to promote the use of effective addiction treatment practices. Treatnet is part of the UNODC-WHO joint program on drug dependence and treatment care [60]. Treatnet develops training tools, good practice documents, quality standards, and facilitates implementation of programs [61,62]. The "International Standards for the treatment of drug use disorders" (commonly known as *"The Standards"*) serve as a guidance to countries around the world on how to organize delivery of interventions. Led by the UNODC and the WHO, *the Standards* provide key principles, treatment systems, and potential interventions according to the different target populations such as pregnant women, children and adolescents, and people in contact with the criminal justice system [63].

Finding a suitable treatment program, whether inpatient or outpatient, that is a good fit for the individual's needs is often a challenge. Type of treatment, referral hesitancy, space availability at centers, cost, health insurance coverage, etc., are some of the barriers that patients and families encounter when identifying and making decisions. In the U.S., there are tools like the "SAMHSA Behavioral Health Treatment Services Locator" [64] and the "Shatterproof ATLAS (Addiction Treatment Locator, Assessment, and Standards platform)" [65] that provide very valuable information to persons seeking treatment options and facilities.

Prevention is one of the cornerstones of public health and it is critical to bending the curve in our fight against addiction. The international standards on drug use prevention provided by UNODC-WHO summarize the current evidence and the interventions or policies for prevention of substance use with a focus on young population [66]. Adolescence is a very important stage for preventing drug addiction due to brain vulnerability and life events, but prevention of substance use should follow a life course approach since addiction can develop at any age. Reduction of drug use and risky behaviors especially among children and the adolescent population has been shown to be effective by prevention strategies, but it requires involvement of a diverse group of stakeholders like schools, families, and communities [67]. Early childhood (prenatal period to 8 years of age) is often neglected in the prevention field, but it has been demonstrated that early years' experiences can set the stage for later substance use—related behaviors; therefore, it is a critical time for prevention [68].

There are different types of prevention programs: school-based and family-based programs, after-school programs, programs adapted to specific populations (e.g., high-risk families, individuals of certain cultures, race or ethnicity), and mass media programs. In searching for a gold standard, several reviews and metaanalysis studies have been conducted; however, the lack of methodological standardization related to design, assessment, and reporting has posed to be a challenge [69,70]. Mass media campaigns are frequently used to push health messages, but their effectiveness is controversial, including for preventing illicit drug use [71]. A good example of a prevention program for implementation in low- and middle-income countries at the family level is the "Family UNited skills program for prevention of negative social outcomes in LMICs" [72]. The program is directed to caregivers, children, and family members aiming to improve knowledge, communication skills, and child behavior with the ultimate goal of reducing violence, risky behaviors, and substance use and improving mental health.

Another angle of prevention is focusing on individuals who already use drugs. As already described, the use of naloxone is an effective measure to prevent deaths due to overdose. One of the limitations of the naloxone strategy is that in order for it to be effective people who are at risk of overdose should be near individuals who have been trained in recognizing the symptoms and on the use of naloxone. In addition, naloxone has to be available. Unfortunately, many of the deaths due to overdose, in particular opioid overdoses, occur when the person is alone. As a high percentage of those deaths are driven by fentanyl; rapid Fentanyl Test Strips (FTS) are an evidence-based effective way to prevent overdose deaths. Some studies have shown that individuals that use FTS are likely to change their own drug use behaviors and share the FTS with others who are at risk [73]. In an effort to reduce drug overdose, the U.S. federal government announced in April 2021 that federal funds could be used to purchase FTS. While there is some controversy about use and distribution of FTS, similar to that around needle exchange programs and wide condom distribution in the HIV field, there is no doubt that FTS saves lives and is a cost-effective harm reduction strategy [74].

Public policies

Countries around the world are aware that drug use is a global problem, but competing priorities like global health security, the COVID-19 pandemic, and the effects of climate change on health have diverted attention and resources. However, its importance has been recognized at the 2016 United Nations General Assembly special session on the world drug problem [75] as well as in other UN Member State resolutions and Ministerial Declarations [76]. The UN General Assembly has also declared June 26 as the "International Day Against Drug Abuse and Illicit Trafficking" and organizations around the world commemorate "International Overdose Awareness Day" on August 31 as a tribute to those who lost their lives to overdose, acknowledging the grief of families and friends left behind, and underscoring a commitment to end overdose deaths.

In spite of the damage to our societies, there is still a reluctance to work and develop strong public health policies around SUD treatment. As indicated in the drug abuse tool kit on "Investing in Drug Abuse Treatment: A Discussion Paper for Policy Makers" [77], there are three potential overarching reasons for a diminished role of SUD treatment in most public policies: (1) drug-related issues are still viewed as a criminal problem rather than a health one; (2) skepticism around the effectiveness of SUDs treatments; and (3) responsibility for the disease falls on the individual rather than society and therefore the societal benefit of treatment is not clear.

While some guidance principles and key elements have been developed to assist policy makers develop coherent drug policies [78], it is important that policy makers become active partners in the fight against addiction, in particular with the substance use research community to not only participate in the formulation and use of research, but also to promote the involvement of researchers in policy discussions [79]. Professional health associations are taking a stand in an effort to provide public policy recommendations. An example is the American College of Physicians' comprehensive position paper. It provides eight specific recommendations drawn from multistakeholder engagement, addressing among other things amendments of criminal justice laws, the implementation of harm reduction strategies, and expansion of the workforce of qualified professionals [80].

Final thoughts

The COVID-19 pandemic has highlighted more than any other health emergency the importance of global public health. It was the implementation of basic public health principles, especially before vaccines and some treatments were developed, that contributed significantly to prevent a much worse scenario. As we reflect about the global health crisis problem stemming from illicit drug use, let us remember some of the principles that drive many of us into the field of public health: compassion, commitment, and hope. Our compassion allows us to see the past stigma and embrace the millions of people who suffer from SUDs, their families, and the communities who struggle every day. Our commitment sustains us in the face of a confusing and complex health problem in the background of varying socio-economic and regulatory environments across the world that demands sustained effort and willingness to embrace innovative approaches, as we are part of the solution. And finally, there is hope. Through our work together, we will open a door to hope for individuals and families struggling with SUDs, allowing them to see that whatever other problems they may face, the larger community acknowledges their struggle and is working with them in the pursuit of peace and lives well lived [81]. Importantly, provision of all levels of preventions to combat the SUDs in primary and other levels of healthcare systems and prevention of young population from substance use through school and after-school based public health interventions may potentially facilitate the creation of a healthy society in the future.

References

[1]   Crocq MA. Historical and cultural aspects of man's relationship with addictive drugs. Dialogues Clin Neurosci December 2007;9(4):355–61. https://doi.org/10.31887/DCNS.2007. 9.4/macrocq.

[2]   Saah T. The evolutionary origins and significance of drug addiction. Harm Reduct J 2005;2:8. https://doi.org/10.1186/1477-7517-2-8.

[3]   World Health Organization. Framework convention on tobacco control. Geneva; 2005. Available from: https://fctc.who.int/.

[4]   World Health Organization. Global strategy to reduce the harmful use of alcohol. Geneva; 2010. Available from: https://www.who.int/publications/i/item/9789241599931.

[5]   World Health Organization. WHO global report on trends in prevalence of tobacco use 2000–2025. 3rd ed. Geneva; 2019 Available from: https://www.who.int/publications/i/item/ who-global-report-on-trends-in-prevalence-of-tobacco-use-2000-2025-third-edition.

[6]   Manthey J, Shield KD, Rylett M, Hasan OS, Probst C. Global alcohol exposure between 1990 and 2017 and forecast until 2030: a modelling study. Lancet June 22, 2019;393(10190):2493–502. https://doi.org/10.1016/S0140-6736(18)32744-2.

[7]   United Nations Department of Economic and Social Affairs. The 2030 agenda for sustainable development. 2015. Available from: https://sdgs.un.org/goals.

[8]   United Nations Office of Drugs and Crime. World drug report 2021. 2021. Available from: https://www.unodc.org/unodc/en/data-and-analysis/wdr2021.html.

[9] World Health Organization. The public health dimension of the world drug problem: how WHO works to prevent drug misuse, reduce harm and improve safe access to medicine. Geneva: World Health Organization; 2019 (WHO/MVP/EMP/2019.02). License: CC BY-NC-SA 3.0 IGO. Available from: https://www.who.int/publications/i/item/WHO-MVP-EMP-2019.02.

[10] Europeans Monitoring Centre for Drugs and Drug Addiction and Europol. EU drug markets report 2019. Luxemburg: Publications Office of the Europe Union; 2019. Available from: https://www.emcdda.europa.eu/publications/joint-publications/eu-drug-markets-report-2019_en.

[11] National Institute on Drug Abuse. Overdose death rates. Trends and statistics. 2022. Available from: https://nida.nih.gov/research-topics/trends-statistics/overdose-death-rates. [Accessed 14 September 2022].

[12] Wainwright JJ, Mikre M, Whitley P, Dawson E, Huskey A, Lukowiak A, et al. Analysis of drug test results before and after the US declaration of a national emergency concerning the COVID-19 outbreak. JAMA October 27, 2020;324(16):1674−7. https://doi.org/10.1001/jama.2020.17694.

[13] Nkansah-Amankra S, Minelli M. "Gateway hypothesis" and early drug use: additional findings from tracking a population-based sample of adolescents to adulthood. Prev Med Rep December 2016;4:134−41. https://doi.org/10.1016/j.pmedr.2016.05.003.

[14] Kilpatrick DG, Acierno R, Saunders B, Resnick HS, Best CL, Schnurr PP. Risk factors for adolescents substance abuse and dependence: data from a national sample. J Consult Clin Psychol 2000;68(1):19−30. https://doi.org/10.1037/0022-006X.68.1.19.

[15] Cox RG, Zhang L, Johnson WD, Bender DR. Academic performance and substance use: findings from a state survey of public school students. J Sch Health 2007;77:109−15. https://doi.org/10.1111/j.1746-1561.2007.00179.x.

[16] MacDonald A, Danielson CK, Resnick HS, Saunders BE, Kilpatrick DG. PTSD and co-morbid disorders in a representative sample of adolescents: the risk associated with multiple exposures to potentially traumatic events. Child Abuse Negl October 2010;34(10):773−83. https://doi.org/10.1016/j.chiabu.2010.03.006.

[17] Kurti AN, Keith DR, Noble A, Priest JS, Sprague B, Higgins ST. Characterizing the intersection of co-occurring risk factors for illicit drug abuse and dependence in a U.S. nationally representative sample. Prev Med 2016;92:118−25. https://doi.org/10.1016/j.ypmed.2016.09.030.

[18] Zubak Z, Zeniic N, Ostojic L, Zubak I, Pojskic H. A prospective study on the influence of scholastic factors on the prevalence and initiation of illicit drug misuse in adolescence. Int J Environ Res Public Health 2018;15(5):847. https://doi.org/10.3390/ijerph15050874.

[19] National Institute on Drug Abuse. The science of drug use and addiction: the basics. 2018. Available from: https://archives.drugabuse.gov/publications/media-guide/science-drug-use-addiction-basics#:∼:text=Addiction%20is%20defined%20as%20a,disorder%20and%20a%20mental%20illness.

[20] United Nations Office on Drugs and Crime. Comorbidities in drug use disorders: discussion paper pre-publication draft. Vienna; 2022. Available from: https://www.unodc.org/documents/commissions/CND/CND_Sessions/CND_65/CRPs/ECN72022_CRP12_V2201355.pdf.

[21] GBD 2019 Mental Disorders Collaborators. Global, regional, and national burden of 12 mental disorders in 204 countries and territories, 1990−2019: a systematic analysis for the Global Burden of Disease Study 2019. Lancet Psychiatry 2022;9:137−50. https://doi.org/10.1016/S2215-0366(21)00395-3.

[22] GBD 2019 Diseases and Injuries Collaborators. Global burden of 369 diseases and injuries in 204 countries and territories, 1990−2019: a systematic analysis for the Global Burden of Disease Study 2019. Lancet 2020;396(10258):1204−22. https://doi.org/10.1016/S0140-6736(20)30925-9.

[23] COVID-19 Mental Disorders Collaborators. Global prevalence and burden of depressive and anxiety disorders in 204 countries and territories in 2020 due to the COVID-19 pandemic. Lancet 2021;398:1700−12. https://doi.org/10.1016/S0140-6736(21)02143-7.

[24] Flynn PM, Brown BS. Co-occurring disorders in substance abuse treatment issues and prospects. J Subst Abuse Treat 2008;34(1):36−47. https://doi.org/10.1016/j.jsat.2006.11.013.

[25] Wüsthoff LE, Waal H, Gråwe RW. The effectiveness of integrated treatment in patients with substance use disorders co-occurring with anxiety and/or depression—a group randomized trial. BMC Psychiatry 2014;14(1):1−12. https://doi.org/10.1186/1471-244X-14-67.

[26] Amaro H, Sanchez M, Bautista T, Cox R. Social vulnerabilities for substance use: stressors, socially toxic environments, and discrimination and racism. Neuropharmacology 2021;188:108518. https://doi.org/10.1016/j.neuropharm.2021.108518.

[27] Sulley S, Ndanga M. Inpatient opioid use disorder and social determinants of health: a nationwide analysis of the National inpatient sample (2012−2014 and 2016−2017). Cureus 2020;12(11):e11311. https://doi:10.7759/cureus.11311.

[28] Williams N, Bossert N, Chen Y, Jaanimägi U, Markatou. Influence of social determinants of health and substance use characteristics on persons who use drugs pursuit of care for hepatitis C virus infection. J Subst Abuse Treat 2019;102:33−9. https://doi.org/10.1016/j.jsat.2019.04.009.

[29] Snijder M, Lees B, Stearne A, Ward J, Bock SG, Newton N, et al. An ecological model of drug and alcohol use and related harms among Aboriginals and Torres Strait Islander Australians: a systematic review of the literature. Prev Med Rep 2021;21:101277. https://doi.org/10.1016/j.pmedr.2020.101277.

[30] United Nations Office on Drugs and Crime. World drug report. Socioeconomic characteristics and drug use disorders 2020. (United Nations Publication, Sales No. E.20.XI.6). Available from: https://wdr.unodc.org/wdr2020/en/socioeconomic.html.

[31] National Institutes of Health. Helping to end addiction long-term. In: Social determinants of opioid use: establishing a research agenda to inform community and system level interventions; 2020. Available from: https://apps1.seiservices.com/SocialDeterminants/Default.aspx.

[32] Bohler R, Thomas CP, Clark TW, Horgan CM. Addressing the opioid crisis through social determinants of health: what are communities doing? Opioid Policy Research Collaborative at Brandeis University; 2021. Available from: https://opioid-resource-connector.org/sites/default/files/2021-02/Issue%20Brief%20-%20Final.pdf.

[33] National Institute on Drug Abuse. Criminal justice DrugFacts. 2020. Available from: https://nida.nih.gov/publications/drugfacts/criminal-justice#:~:text=While%20the%20exact%20rates%20of,population%20has%20an%20active%20SUD.

[34] Hughes CE, Stevens A. What can we learn from the Portuguese decriminalization of illicit drugs? Br J Criminol 2010;50(6):999−1022. https://doi.org/10.1093/bjc/azq038.

[35] Drug Policy Alliance. Drug decriminalization in Portugal. Learning from health and human-centered approach. New York; 2019. Available from: https://drugpolicy.org/sites/default/files/dpa-drug-decriminalization-portugal-health-human-centered-approach_0.pdf.

[36] King County. Law enforcement assisted diversion (LEAD). 2022. Available from: https://leadkingcounty.org/. [Accessed 14 September 2022].

[37] Chandler RK, Fletcher BW, Volkow ND. Treating drug abuse and addiction in the criminal justice system: improving public health and safety. JAMA January 14, 2009;301(2):183−90. https://doi:10.1001/jama.2008.976.

[38] United Nations Office on Drugs and Crime. Treatment and care for people with drug use disorders in contact with the criminal justice system. Alternatives to conviction or punishment. 2019. Available from: https://www.unodc.org/documents/UNODC_WHO_Alternatives_to_conviction_or_punishment_ENG.pdf.

[39] Tran BX, Moir M, Latkin CA, Hall BJ, Nguyen CT, Ha GH, et al. Global research mapping of substance use disorder and treatment 1971−2017: implications for priority setting. Subst Abuse Treat Prev Policy 2019;14:21. https://doi.org/10.1186/s13011-019-0204-7.

[40] United Nations Office on Drugs and Crime. Substance abuse treatment and care for women: case studies and lessons learned. Viena; 2004. Available from: https://www.unodc.org/documents/drug-prevention-and-treatment/UNODC_Women_Treatment_Case_Studies_E.pdf.

[41] United Nations. United Nations toolkit on synthetic drugs. 2020. Available from: https://syntheticdrugs.unodc.org/syntheticdrugs/toolkit.html.

[42] Substance Abuse and Mental Health Services Administration. SAMHSA opioid overdose prevention toolkit. HHS publication no. (SMA) 18-4742. Rockville, MD: Substance Abuse and Mental Health Services Administration; 2018. Available from: https://store.samhsa.gov/sites/default/files/d7/priv/sma18-4742.pdf.

[43] Volkow ND. Stigma and the toll of addiction. N Engl J Med April 2020;2382(14):1289−90. https://doi:10.1056/NEJMp1917360.

[44] Matthews S, Dwyer R, Snoek A. Stigma and self-stigma in addiction. Bioeth Inq 2017;14:275−86. https://doi.10.1007/s11673-017-9784-y.

[45] Executive Office of the President. Memorandum to heads of executive departments and agencies. 2017. Available from: https://obamawhitehouse.archives.gov/sites/whitehouse.gov/files/images/Memo%20-%20Changing%20Federal%20Terminology%20Regarding%20Substance%20Use%20and%20Substance%20Use%20Disorders.pdf.

[46] National Institute on Drug Abuse. Words matter: preferred language for talking about addiction. 2021. Available from: https://nida.nih.gov/drug-topics/addiction-science/words-matter-preferred-language-talking-about-addiction.

[47] Substance Abuse and Mental Health Services Administration. SAMHSA's Center for the Application of Prevention Technologies. Words matter: how language choice can reduce stigma. 2017. Available from: https://facesandvoicesofrecovery.org/wp-content/uploads/2019/06/Words-Matter-How-Language-Choice-Can-Reduce-Stigma.pdf.

[48] National Academies of Sciences, Engineering, and Medicine. Ending discrimination against people with mental and substance use disorders: the evidence for stigma change. Washington, D.C.: The National Academies Press; 2016. https://doi.org/10.17226/23442.

[49] Barry CL, McGinty EE, Pescosolido B, Goldman HH. Stigma discrimination, treatment effectiveness and policy support: comparing public views about drug addiction with mental illness. Psychiatr Serv October 2014;65(10):1269−72. https://doi:10.1176/appi.ps.201400140.

[50] McGinty EE, Barry CL. Stigma reduction to combat the addiction crisis—developing an evidence base. N Engl J Med April 2, 2020;382:1291−2. https://doi:10.1056/NEJMp2000227.

[51] Shatterproof. Shatterproof addiction stigma Index. October 2021. Available from: https://www.shatterproof.org/our-work/ending-addiction-stigma/shatterproof-addiction-stigma-Index.

[52] Ti L, Tzemis D, Buxton JA. Engaging people who use drugs in policy and program development: a review of the literature. Subst Abuse Treat Prev Policy 2012;7:47. https://doi.org/10.1186/1747-597X-7-47.

[53] National Institutes of Health, National Institute on Drug Abuse. Principles of drug addiction treatment: a research-based guide. 3rd ed. 2018 Available from: https://www.drugabuse.gov/publications/principles-drug-addiction-treatment-research-based-guide-third-edition/preface.

[54] Samet JH, Fiellin DA. Opioid substitution therapy-time to replace the term. Lancet 2015;357:1508−9. https://doi:10.1016/S0140-6736(15)60750-4.

[55] Wakeman SE, Larochelle MR, Ameli O, Chaisson CE, McPheeters JT, Crown WH, et al. Comparative effectiveness of different treatment pathways for opioid use disorder. JAMA Netw Open 2020;3(2):e1920622. https://doi:10.1001/jamanetworkopen.2019.20622.

[56] Sordo L, Barrio G, Bravo MJ, et al. Mortality risk during and after opioid substitution treatment: systematic review and meta-analysis of cohort studies. BMJ 2017;357:j1550. https://doi.org/10.1136/bmj.j1550.

[57] World health Organization and United Nations Office on Drugs and Crime. UNODC-WHO Stop-Overdose-Safely (S-O-S) project implementation in Kazakhstan, Kyrgyzstan, Tajikistan and Ukraine: summary report. Geneva: World Health Organization and United Nations Office on Drugs and Crime; 2021. License: CC BY-NC-SA 3.0 IGO. Available from: https://apps.who.int/iris/bitstream/handle/10665/340497/9789240022454-eng.pdf?sequence=1&isAllowed=y.

[58] Walker S, Dietze P, Poznyak V, Campello G, Kashino W, Dzhonbekkov D, et al. More than saving lives: qualitative findings of the UNODC/WHO stop overdose safely (S-O-S) project. Int J Drug Policy 2022;100:103482. https://doi.org/10.1016/j.drugpo.2021.103482.

[59] Tomás-Rosselló J, Rawson RA, Zarza MJ, Bellows A, Busse A, Saenz E, et al. United Nations Office on Drugs and Crime international network of drug dependence treatment and rehabilitation resource centres: Treatnet. Subst Abuse 2010;31(4):251−63. https://doi.org/10.1080/08897077.2010.514243.

[60] United Nations Office on Drugs and Crime. UNODC-WHO Joint Programme on drug dependence treatment and care. Vienna; 2009. Available from: https://www.unodc.org/docs/treatment/unodc_who_programme_brochure_english.pdf?msclkid=2d75b2dab43411ec9e548f2d77f9a967.

[61] United Nations Office on Drugs and Crime. Treatnet training package. 2010. Available from: https://www.unodc.org/unodc/en/treatment-and-care/treatnet-training-package.html.

[62] Busse A, Kashiro W, Suhartono S, Narotama N, campello G, Irwanto DP, et al. Fidelity Assessment of the Treatnet family (TF): a family-based intervention for adolescents with substance use disorders and their families. Addict Behav Rep June 11, 2021:14. https://doi.org/10.1016/j.abrep.2021.100363.

[63] World Health Organization and United Nations Office on Drugs and Crime. International standards for the treatment of drug use disorders: revised edition incorporating results of field-testing. Geneva: World Health Organization and United Nations Office on Drugs and Crime; 2020. License: CC BY-NC-SA 3.0 IGO. Available from: https://www.unodc.org/documents/drug-prevention-and-treatment/UNODC-WHO_International_Standards_Treatment_Drug_Use_Disorders_April_2020.pdf.

[64] Substance Abuse and Mental Health Services Administration. SAMHSA treatment finder. 2022. Available from: https://findtreatment.samhsa.gov/.

[65] Shatterproof. Shatterproof Atlas. 2022. Available from: https://www.treatmentatlas.org/?utm_source=google&utm_medium=cpc&utm_campaign=fdgsatlasbrandgeneral&utm_content=atlas&gclid=EAIaIQobChMI0633-ZD79gIVvebjBx2Cyw8eEAAYASAAEgK3CPD_BwE.

[66] United Nations Office on Drugs and Crime, World Health Organization. International standards on drug use prevention. 2nd ed. Vienna: United Nations; 2018. p. 60. Second updated edition.

[67] National Institute on Drug Abuse. Preventing drug use among children and adolescents. A research-based guide for parents, educators and community leaders. 2nd ed. 2003 Available from: https://nida.nih.gov/sites/default/files/preventingdruguse_2_1.pdf.

[68] National Institute on Drug Abuse. Principles of substance use prevention for early childhood. A research-based guide. 2016. Available from: https://nida.nih.gov/publications/principles-substance-abuse-prevention-early-childhood/table-contents.

[69] Valero de Vicente M, Ballester Brage L, Orte Socías, Amer Fernandez JA. Meta-analysis of family-based selective prevention programs for drug consumption in adolescence. Psicothema 2017;29(3):299−305. https://doi:10.7334/psicothema2016.275.

[70] Department of Justice. Office of Juvenile Justice and Delinquency Prevention. Substance use prevention programs. Literature review: a product of the model programs guide. 2022. Available from: https://ojjdp.ojp.gov/model-programs-guide/literature-reviews/substance-use-prevention-programs.

[71] Ferri M, Allara E, Bo A, Gasparrini A, Faggiano F. Media campaigns for the prevention of illicit drug use in young people. Cochrane Database Syst Rev June 5, 2013;6:CD009287. https://doi:10.1002/14651858.CD009287.pub2.

[72] United Nations Office on Drugs and Crime. Family UNited skills program for prevention of negative social outcomes in LMICs. 2020. Available from: https://www.unodc.org/res/listen-first/parenting-under-covid-19_html/Family-UNited-leaflet-20200218.pdf.

[73] Goldman JE, Wayne KM, Pereira KA, Krieger MS, Yedinak JL, Marshall BD. Perspectives on rapid fentanyl test strips as a harm reduction practice among young adults who use drugs: a qualitative study. Harm Reduct J 2019;16:3. https://doi.org/10.1186/s12954-018-0276-0.

[74] Johns Hopkins Bloomberg School of Public Health. Detecting fentanyl. Saving lives. 2018. Available from: https://americanhealth.jhu.edu/fentanyl.

[75] UNGASS. Special session of the United Nations general assembly on the world drug problem. 2016. Available from: https://www.unodc.org/ungass2016/.

[76] United Nations Office on Drugs and Crime. Implementation of all international drug policy commitments. Vienna; 2019. Available from: https://www.unodc.org/documents/hlr/19-V1905795_E_ebook.pdf.

[77] United Nations Office on Drugs and Crime. Investing in drug abuse treatment: a discussions paper for policy makers. Vienna; 2003. Available from: https://www.unodc.org/documents/drug-prevention-and-treatment/UNODC_Investing_in_drug_abuse_treatment_2003.pdf.

[78] Pompidou Group. Policy paper providing guidance to policy makers for developing coherent policies for licit and illicit drugs. Council of Europe; 2011. Available from: https://rm.coe.int/CoERMPublicCommonSearchServices/DisplayDCTMContent?documentId=09000016806f40ab.

[79] Meisel ZF, Mitchell J, Polsky D, Boualam N, McGeoch E, Weiner J, et al. Strengthening partnerships between substance use researchers and policy makers to take advantage of a window of opportunity. Subst Abuse Treat Prev Policy 2019;14:12. https://doi.org/10.1186/s13011-019-0199-0.

[80] Crowley BSJ, Kirschner N, Dunn AS, Bornstein SS, Abraham G, Bush JF, et al. Health and public policy to facilitate effective prevention and treatment of substance use disorders involving illicit and prescription drugs: an American College of Physicians Position Paper. Ann Intern Med 2017;166(10):733−6. https://doi.org/10.7326/M16-2953.

[81] World Health Organization. Drug overdose: a film about life. 2020. Available from: https://www.who.int/publications/i/item/WHO-MVP-EMP-2019.02.

Chapter 12

Healthy aging and quality of life of the elderly

Arun Chockalingam[1,2,3], Amarjeet Singh[4] and Soundappan Kathirvel[4]

[1]Medicine and Global Health, Faculty of Medicine, University of Toronto, Toronto, ON, Canada; [2]Health Sciences, York University, Toronto, ON, Canada; [3]Global Health, National Heart, Lung and Blood Institute at the National Institute of Health, Bethesda, MD, United States; [4]Department of Community Medicine and School of Public Health, Postgraduate Institute of Medical Education and Research, Chandigarh, Punjab, India

Introduction

Global demographic changes

Most regions of the world and countries are experiencing unprecedentedly rapid demographic change. The global population grew by 4 billion since 1950. The population of people over the age of 65 is currently 713 million, which is projected to double to 1.5 billion in 2050. Demographic processes are also undergoing extraordinary change: fertility has dropped rapidly, and life expectancy has risen to new highs due to epidemiological transition of the diseases from acute and fatal communicable diseases to chronic non-communicable disease (NCD). Due to reduced mortality in early age, there is an increasing population over the age of 65, practically in all parts of the world [1]. The 2020 World Population Data Sheet indicates that world population is projected to increase from 7.8 billion in 2020 to 9.9 billion by 2050 [2].

The age-dependency ratio

A changing age distribution has significant consequences for any government in its social and economic responsibility, e.g., resource allocation of education, health care, and social security to the young and old. The age-dependency ratio (ADR) is defined as the ratio of population aged below 15 and over 65 to the population of age 15−64. This ratio aims to measure how many 'dependents' there are for each person in the 'productive' age group. Although ADR rises during the initial stages of demographic transition, it falls sharply due to decline in fertility to reduce the proportion of the population under age 15.

Principles and Application of Evidence-Based Public Health Practice
https://doi.org/10.1016/B978-0-323-95356-6.00007-0

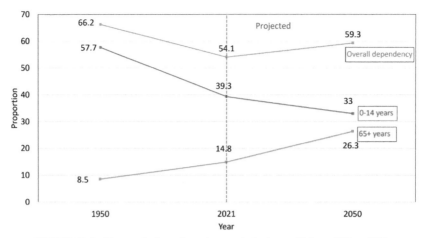

FIGURE 12.1   The trend of age-dependency ratio in the world from 1950 to 2050.

This decline creates a so-called 'demographic dividend,' which boosts economic growth by increasing the size of the labor force relative to dependents and by stimulating savings [3]. Finally, at the end of the transition, the ADR increases again as the proportion of the population over age 65 rises and their working status decreases. See Fig. 12.1 [2].

World of the elderly

Globally, the share of the population aged 65 years or over increased from 6% in 1990 to 9% in 2019 [4]. What is more, the proportion of adult life spent beyond age 65 increased from less than 20% in the 1960s to more than a quarter in most developed countries today. According to the Population Division of the United Nations Department of Economic and Social Affairs, it is in 17 developed countries and its projections to the end of the century indicate that 61% of the world population will be over 65 years in 2100 for 155 countries.

Population aging has been fastest in Eastern and South-Eastern Asia, Latin America, and the Caribbean and slow growth in North America and Europe which have already attained the net reproduction rate below the replacement level; see Table 12.1 [4].

However, not all older people are dependents. In many countries, people over the age of 65 are still active. This changes the perception significantly, in terms of the mortality risks, health status, type and level of activity, productivity, and other socioeconomic characteristics of older persons, particularly in the last few decades. This has led to a revolutionized thinking of what aging means and to provide context-specific structure, in terms of the living conditions and living arrangements of older persons, their productive and other

TABLE 12.1 Number of persons aged ≥65 years in SDG region, 2019 and 2050.

Region	Number of persons aged ≥65 in 2019 (millions)	Number of persons aged ≥65 in 2050 (millions)	Percentage change between 2019 and 2050
World	702.9	1548.9	120
Sub-Saharan Africa	31.9	101.4	218
Northern Africa and Western Asia	29.4	95.4	226
Central and Southern Asia	119.0	328.1	176
Eastern and South-Eastern Asia	260.6	572.5	120
Latin America and the Caribbean	56.4	144.6	156
Australia and New Zealand	4.8	8.8	84
Oceania, excluding Australia and New Zealand	0.5	1.5	190
Europe and Northern America	200.4	296.2	48

Source: United Nations. World population prospects 2019. Department of Economic and Social Affairs, Population Division; 2019.

contributions to society, and their need for social protection and health care. Older persons must be recognized as the active agents of societal development to achieve truly transformative, inclusive, and sustainable development outcomes. Such transformation could happen through effective public health policies. These approaches to understanding and measuring aging also carry important implications for internationally agreed development goals as cited by the 2030 Agenda for Sustainable Development [5].

Ageism and abuse of older people

A persistent challenge throughout the globe is ageism, meaning prejudice and discrimination toward older persons at individual and institutional levels that undermines older persons' status as rights holders including their right to autonomy, participation, access to education and training, health and social care, security, and decent employment. Moreover, old age aggravates existing disadvantages that individuals struggle with throughout their lives based on gender, race, ethnicity, disability, religion, or other factors. Abuse of older women and men can be of three types—physical, emotional, or financial. The United Nations (UN) and the World Health Organization (WHO) acknowledged the abuse of older people as a growing concern for all countries, regardless of their level of development [6,7].

Role of public health in aging process

With the population aging and a dramatic increase in the number of senior citizens, public health systems will be increasingly burdened with the need to deal with the care and treatment of individuals with multiple chronic diseases including dementia. The WHO together with the UN declared 2021−30, as the Decade of Healthy Ageing. The Decade of Healthy Ageing is a global collaboration bringing together governments, civil society, international agencies, professionals, academia, the media, and the private sector for 10 years of concerted, catalytic, and collaborative action led by WHO to foster longer and healthier lives [8]. The Decade of Healthy Ageing (2021−30) seeks to reduce health inequities and improve the lives of older people, their families and communities through collective action in four areas: changing how we think, feel, and act toward age and ageism; developing communities in ways that foster the abilities of older people; delivering person-centered integrated care and primary health services responsive to older people; and providing older people who need it with access to quality long-term care (LTC). The WHO identified 10 facts on aging and health as listed in Table 12.2 [9].

Effect of aging

People's physiological features, health of internal organs, as well as mental conditions change as they grow older and reach an old age. Biological bodily changes of aging have a cumulative effect on the health and fitness. Many elderlies become anxious and depressed due to their body image issues [10−13]. People commonly associate old age with disability and diseases.

External physical appearance [13]

All elderly eventually lose height and assume a stooped, forward bent posture; their hips and knees are flexed with arms bent at elbows. They tend to tilt their head backward to maintain eye contact—see Fig. 12.2A [13a]. **Skin** becomes dry, thin, rough, lax, furrowed, and itchy. Purpuric patches appear on the trunk and upper extremities. **Hair** loss occurs as seen in Fig. 12.2B [13b]. **Nails** lose shape and color, become brittle, hard, and difficult to clip. Subcutaneous fat

TABLE 12.2 WHO's 10 facts about aging and health.

The world's population is rapidly aging	People over the age of 60 years will rise from 900 million to 2 billion between 2015 and 2050 (increase 12%–22% of the total global population). Population aging is happening more quickly than in the past, e.g., France—10%–20%; Brazil, China, and India will have 20+ years of increased life expectancy.
There is little evidence that older people today are in better health than their parents	The proportion of older people in high-income countries needing help from another person to carry out basic activities such as eating and washing may have declined slightly over the past 30 years. However, there has been little change in the prevalence of less severe limitations in functioning.
The most common health conditions in older age are noncommunicable diseases	Older people in LMICs carry a greater disease burden than the rich world. Regardless of where they live, the biggest killers of older people are heart disease, stroke, and chronic lung disease. The greatest causes of disability are sensory impairments (particularly in LMIC), back and neck pain, COPD, depressive disorders, falls, diabetes, dementia, and osteoarthritis.
When it comes to health, there is no 'typical' older person	Biological aging is only loosely associated with person age in years. Some 80-year-olds have physical and mental capacities similar to many 20-year-olds. Other people experience declines in physical and mental capacities at much younger ages.
Health in older age is not random	Although older people's health reflect their genetic inheritance, most is due to their physical and social environments, and the influence of these environments on their opportunities and health behavior.
Ageism may now be more pervasive than sexism or racism	Ageism—discrimination against a person on the basis of their age—has serious consequences for older people and societies at large, e.g., prejudicial attitudes, discriminatory practices, or policies that perpetuate ageist beliefs. It can obstruct sound policy development and significantly undermine the quality of health and social care that older people receive.

Continued

TABLE 12.2 WHO's 10 facts about aging and health.—cont'd

Comprehensive PH action will require fundamental shifts in how we think about aging and health	Health is not the absence of disease. Healthy aging is achievable by every older person. It is a process that enables older people to continue to do the things that are important to them. While health and social care expenditures for older people are often viewed as costs to society, these should be understood as investments in realizing opportunities and enabling older people to continue to make their many positive contributions.
Health systems need to be realigned to the needs of older populations	Most health systems around the world are ill-prepared to address the needs of older people, who often have multiple chronic conditions or geriatric syndromes. Systems must be capable of providing older person-centered and integrated care and focus on maintaining capacities as people age.
In the 21st century, all countries need an integrated system of long-term care	In some countries, this means building a system from almost nothing. In others, it means rethinking long-term care: from a basic safety net for the most vulnerable, toward a broader system that maximizes older people's functional ability and upholds their autonomy and dignity. The number of older people who need support for activities of daily living in developing countries is forecast to quadruple by 2050.
Healthy aging involves all levels and sectors of government	Examples of intersectoral action include establishing policies and programs that expand housing options; making buildings and transport accessible; promoting age-diversity in working environments; and protecting older people from poverty through social protection schemes. Making progress on healthy aging will also require a far better understanding of age-related issues and trends.

diminishes. The elderly become more vulnerable to heat/cold. Light touch, vibration, corneal sensitivity, two-point discrimination, and spatial acuity assessment diminish. Antigravity lymphatic and venous flow in the lower limbs get compromised due to incompetence of vessels in old age. There may be excessive hair growth in the eyebrows, nares, and ears; elderly females acquire male pattern of hair growth. **Sarcopenia** or loss of muscle tissue is a natural part of the aging process. Movements get restricted due to skeletal

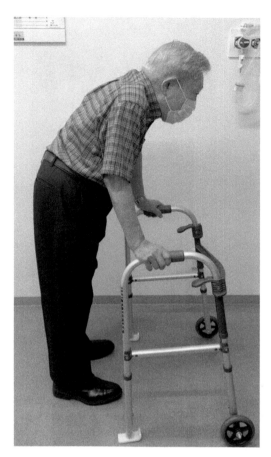

FIGURE   12.2A The physical appearance of elderly people. *Inanami H, Iwai H, Kato S, Takano Y, Yuzawa Y, Yanagisawa K, et al. Partial resection of spinous process for the elderly patients with thoracolumbar kyphosis: technical report. Medicina   (Kaunas,   Lithuania) 2021;57(2). https://doi.org/10.3390/medicina57020087.*

muscles atrophy and physical inactivity. There is decreased intervertebral distance. The elderly tend to have stooped posture, and at one stage, walking by themselves becomes difficult and resort to the help of walking stick or a walker. Resorption of bone leads to painful sensation in bones. Joints become rigid. Strength and stamina decrease. It leads to **proprioceptive problems** meaning there may be a regular feeling of unsteadiness in the older people due to various reasons, e.g., sensory loss in the feet, diabetes, cervical spondylosis, or peripheral neuropathy.

Sensory changes [13]

Elderly people experience loss of all sensory functions, gradually, one at a time—vision, hearing, smell, touch sensation, and taste. They find it difficult to communicate. In old age, voice becomes slow, deep, and quiet. Speech tremors

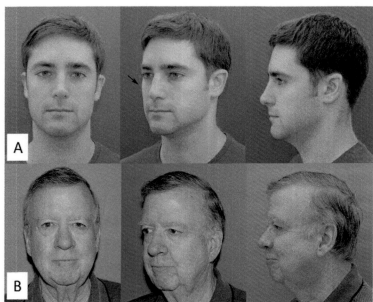

A-Younger; B-Older

FIGURE 12.2B The changes in face from youth to elderly. *Ko AC, Korn BS, Kikkawa DO. The aging face. Survey Ophthalmol 2017;62(2), 190−202. https://doi.org/10.1016/j.survophthal.2016. 09.002.*

may develop. Articulation becomes less precise. Fewer syllables can be uttered per breath. Elderly people may use more time and words to convey the same idea and have difficulty recalling a name. Hearing loss further complicates the issues. It causes frustration. The pinna loses flexibility and becomes longer. The aging nasal cartilage ossifies to produce a droopy **nose**. The resistance to infection and the smelling power decrease. There may be postnasal dripping and nasal drain. The chronic rhinitis and nasal stuffiness are common. Swallowing problems is also there. **Facial changes** occur. Loss of bone mass of lower jaw accentuates the size of upper mouth, nose, and forehead. Loss of lip appearance occurs because of tooth loss. Eyelids appear swollen because of redistribution of fat deposits. **Taste buds'** atrophy and decline in numbers and lose their capacity to relay flavor. **Vision:** all older persons eventually have some decline in the visual capacity. There is decreased lubrication of the eye. The pupil size is reduced. They feel difficulty with bright and changing light levels, reduced color vision; decreased contrast perception; and mobility difficulties related to loss of depth and contrast cues. Cataract may eventually develop. Some elders may experience distorted vision and reduced contrast perception. Eyelids become flaccid; skin around the eyes droop and sag. The lens loses elasticity with inability to focus clearly on near objects. Spots and specks appear across field of vision. **Mouth and teeth:** Bad breath and dry mouth are common among the elderly. Oral mucosa becomes thin, pale, and dry. Saliva is thick, foamy, and ropey. The teeth become sensitive to hot or cold stimuli with wearing off the enamel. Cracked teeth or tooth loss are common. Gums recede and eventually become edentulous (lack of teeth).

Aging and internal organs [14]

With aging, body functions and all system begin to slow. Such slowing includes circulatory, respiratory, gastrointestinal, reproductive, neurological, musculoskeletal, and immunological systems, and the functioning of major organs such as heart, lungs, kidneys, liver, pancreas, and brain.

Cardiovascular changes: The deposition of fat, collagen, and elastin fibers results in decreased number of pacemakers. It decreases the conductivity of heart. Arteriosclerosis increases the risk for hypertension, congestive heart failure, and other heart and circulatory diseases in the elderly. **Respiratory changes:** There is reduced efficiency of the **lungs,** in their ventilation with a decline in exercise tolerance. The chest wall becomes stiffer; the respiratory muscles lose strength. The oxygen saturation of blood decreases. Nose elongates downwards, leading to mouth breathing during sleep leading to dry mouth. The larynx and trachea calcify. Recurrent respiratory infections are common. Downward slant of the ribs limits the expansion of the chest. Alveoli have remained enlarged at the end of inspiration. There is trapping of air in alveoli.

Urinary system changes: Decreased blood flow and reduction in filtering units (nephrons) in **kidneys** affect their functioning. Bladder incontinence is seen.

Gastrointestinal muscle strength and mobility decrease. It slows peristalsis causing digestive discomfort. Loss of smooth gastric muscles delays emptying time. It exposes the epithelial lining to extended contact with gastric contents. **The liver** reduces in volume and weight. This results in decreased blood flow. Liver loses the ability to breakdown the protein. There is accumulation of abnormal proteins. Decline in the absorption of nutrients (fatty acids and cholesterol) is seen. **Pancreas** becomes fibrotic. Its enzymes of food digestion get reduced. Tissue sensitivity to insulin reduces. Elderly people become highly prone to diabetes mellitus. Diminished hormone levels lead to atrophy of ovaries, uterus, and vaginal tissue of aged women. Men develop firmer testes with prostatic hypertrophy.

Reproductive System: Due to change in hormonal secretion (decreased testosterone among men and estrogen in women after menopause), the sexual activity and interest may be affected. Menopausal symptoms lead to physical and emotional disturbances among elderly women [12]. **Immunological functions**: A decline in general immunity and the elderly become prone to many diseases. Aging is linked with **sleep disorders**. They are more easily aroused from sleep by auditory stimuli than young people.

Aging is a high-risk factor for practically all chronic NCDs, including cancer, heart disease, stroke, chronic kidney disease (CKD), diabetes, chronic respiratory diseases, and more. As per WHO (IARC), more than 70% of the mortality associated with many cancers including lung, prostate, colorectum, breast, stomach, liver, esophagus, pancreas, cervix and ovary, and bladder occur in patients over the age of 65 [13].

Aging and neurological changes

Brain changes in the elderly [14a,15]

Brain weight decreases and cerebral ventricles enlarge in size. Amyloid deposition plaques and tangles are formed in the nerve fibers. The conduction time of reflexes in the peripheral nerves decreases. Impaired memory, rigidity of outlook, and dislike of change are some of the commonest mental changes in the aged. Dizziness in the elderly is common. Overall, the elderly may have many health and social problems as listed in Table 12.3.

Loneliness and depression

Depression is a common complication of dementia. There is social isolation, loneliness, social withdrawal, and hopelessness. They become introspective and reflective. The fear of losing power, independence, and the ability to

TABLE 12.3 Common health and social problems faced by the elderly.

Mood—tearfulness, anxiety, worry, loss of pleasure, depression
Behavior—irritability, uncooperativeness, loss of interest in hobbies
Appetite—eating less or eating more
Thoughts—low self-esteem, fearfulness, guilt
Sleep—difficulty falling asleep, awakening earlier than usual
Energy—loss of energy, apathy, withdrawal
Anxiety disorders, panic attacks, agoraphobia, somatization syndrome, and malingering.
Have problems talking properly (language problems)
Not know time or day of the week
Have significant memory loss particularly for things that have just happened, recent events, and people's names
Become lost in familiar places
Become inactive and unmotivated
Have difficulty in making decisions
React unusually angrily or aggressively on occasions
Difficulty in hearing
One or the other medical diseases

make choices may be the greatest fear of getting old. For the dependent elderly, the vulnerability for physical and psychological exploitation is an additional problem. There is loss of drive, apathy, indecisiveness, sexual disinhibition, agitation, pacing, and purposeless behaviors (e.g., packing and unpacking).

Stress and anxiety

The elderly face lowering of standard of living, social disempowerment, social discard, and worthlessness. Economic insecurity is common with reduced income and low self-earning with mental and social consequences. After retirement, bulk of the time of the elderly is free with nothing to do. They have a lot of spare time to worry about how to pass it.

Dementia

Dementia is a brain disease characterized by loss of memory, speech, reasoning, and other cognitive functions; Alzheimer's disease is the most common form likely accounting for 70% of all cases. **Alzheimer's disease** begins gradually and gets worse over a period of few years (amnesia, aphasia, apraxia, and agnosia).

Dementia in adults (aged 40 years and older) is expected to nearly triple globally, from an estimated 57 million in 2019 to 153 million in 2050, due primarily to population growth and population aging [16]. The onset of dementia is devastating for individuals and their families. It profoundly affects the quality of life and takes a severe economic toll. As the disease progresses, limitations become clearer and more restricting. The elderly have difficulty with day-to-day living and face conditions listed in Table 12.4. In advanced stages, the person may become extremely dependent on their family members and caregivers for many things as listed in Table 12.5 [13]. There has been a recent increasing interest on tailoring the assessments and interventions among those living with dementia and their families to enhance outcomes and reduce distress by taking into account the impact of culture, beliefs, values, and perspectives [17]. Cultural competence within healthcare professionals is important for improving patient and family satisfaction with services.

Global burden of diseases

Changing pattern of diseases and growing NCD epidemic

The International Classification of Diseases (ICD) from the WHO is the world's standard tool to capture mortality and morbidity data. In the 1900s, most prevalent cause of death and disability was due to infectious diseases (IDs) or communicable diseases. In the late 1960s, as the scientific advancements found cure for infectious viruses through vaccines, many of the major IDs were eliminated in most part of the world. This led to increased longevity

TABLE 12.4 Conditions faced by the elderly as dementia progresses.

Has increased difficulty with speech
Becomes lost at home as well as outside
Shows problems with wandering and other behavior problems such as repeated questioning and calling out, clinging, and disturbed sleep
Put on clothing backwards
May loose urinary and bowel continence
Is unable to do activities of daily living
Rummages through wardrobes and drawers (own as well as others)
No longer manages to live alone without problems
May hallucinate

TABLE 12.5 Extreme dependency on family members and caregivers in advanced stages of dementia.

Have difficulty eating
Be incapable of communicating
Not recognize relatives, friends, and familiar objects (e.g., standing in front of a refrigerator looking at the milk not recognizing what it is!)
Have difficulty understanding what is going on around them
Have difficulty walking
Be unable to find their way around at home
Have bowel and bladder incontinence
Have altered eating habits leading to loss of weight
Display inappropriate behavior in public

of the population in every part of the world, which changed the pattern of mortality and morbidity to degenerative atherosclerotic diseases and neoplasm as well as mental health issues [18].

The global burden of diseases (GBDs) categorized all deaths into three groups, as follows: **Group 1** diseases—the communicable (or IDs), maternal, perinatal, and nutritional conditions; **Group 2** diseases—all NCDs that are noncontagious from person-to-person, such as heart disease, including high blood pressure, stroke; diabetes mellitus; all forms of neoplasm (or cancer); lung diseases; and mental disorders; and **Group 3** diseases—injuries including traffic-related, family violence, and war-related deaths [19]. The GBD provides a tool to quantify health loss from hundreds of diseases, injuries, and risk factors, so that health systems can be improved, and disparities can be eliminated. The GBD is a critical resource for informed policy making. The GBD study is a comprehensive global, regional, and national research program of disease burden that assesses mortality and disability from major diseases, injuries, and risk factors [20].

Until 2019, the GBD estimated Group 2 diseases accounting for two thirds of global mortality, one-fourth due to Group 1, and less than 10% to Group 3 diseases. Fig. 12.3 shows the changing pattern of diseases between 1990 and 2020 indicating the rapid rise in NCD [18]. However, things have changed dramatically since 2020 when COVID-19 took the world by surprise causing millions of deaths and disabled more [21].

Management of NCD in the elderly

Most aged people have one or more risk factors for NCD and or having one or more NCDs. Multimorbidity is common among the elderly population. Studies from different parts of the world confirmed multimorbidity NCDs among the elderly [22–25]. The United States Centers for Disease Control reported that NCDs affect older adults disproportionately, contributing to disability, diminished quality of life, and increased health- and long-term–care costs. While increased life expectancy reflects, at least in part, the success of public health interventions, public health programs must now respond to the challenges of growing burden of chronic illnesses, injuries, and disabilities and increasing concerns about future caregiving and healthcare costs [22].

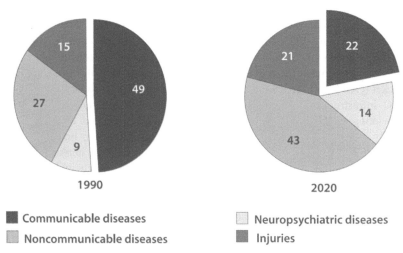

FIGURE 12.3 The changing pattern of global burden of disease between 1990 and 2020. *Chockalingam A, Balaguer-Vintro I. (Eds). Impending global pandemic of cardiovascular diseases. World Heart Federation. Barcelona, Spain, Philadelphia, USA: Prous Science; 1999.*

Equitability in health care for elderly population

Age is in fact estimated to be the most important determinant of health. Yet while older persons on average have greater healthcare needs than younger age groups, they also face distinct disadvantages in accessing appropriate, affordable, and quality care [26]. Despite their increased health risks, a large number of older persons across countries lack access to adequate levels and quality of health care [27]. Accessibility is another significant barrier to health care, particularly for those older persons with limited mobility and in rural areas with poor transportation infrastructure and where long distances must be traveled to reach health facilities [28]. A study of 12 European countries observed disparity by education level among people aged 50 and over in visits to medical specialists and dentists [29]. In the United Kingdom, lower socio-economic groups of older persons are disadvantaged in terms of access to and utilization of several health services, including mammography screening, vaccinations, eye and dental exams, and heart surgery [30,31].

Healthcare expenditure with aging

The increased number of persons aged $\geq$65 years will potentially lead to increased healthcare costs. The healthcare cost per capita for persons aged $\geq$65 years in the United States and other developed countries is three to five times greater than the cost for persons aged <65 years, and the rapid growth in the number of older persons, coupled with continued advances in medical technology, is expected to create upward pressure on health- and long-term—care spending [32]. In 1997, the United States had the highest health care spending per person aged $\geq$65 years ($12,100), but other developed countries also spent substantial amounts per person aged $\geq$65 years, ranging from approximately $3600 in the United Kingdom to approximately $6800 in Canada [33]. However, the extent of spending will depend on other factors in addition to aging [32]. In many of the low- and middle-income countries (LMICs), where there is no universal health coverage system in place, the elderly have to meet the cost of chronic disease management, through out-of-pocket (OOP) [34]. To shoulder the burden, resources need to be funneled into LTC infrastructure, relieving financial strain for patients and families and alleviating the burden on informal caregiving, often carried out by female family members [34]. While OOP in the Organisation for Economic Co-operation and Development countries average to 19.0% of healthcare costs (7.5% in France to 36.4% in Russia) [35], the people living in LMIC bear almost the entire of health care costs. A study from India reported that the monthly per capita health spending of elderly households is 3.8 times higher than that of non-elderly households [33].

Healthy aging and healthy living

Among the older adults aged 65 years and older, the percentage of people aged 80 years and older is expected to increase most rapidly in Taiwan, from approximately 24.4% in 2010 to 44% in 2060 [36]. A person-centered approach among older adults in Taiwan examined the factors influencing the healthy aging through a 14-year longitudinal survey, at five waves (carried out in 1993, 1996, 1999, 2003, and 2007) [37]. This study showed that with increasing age, the combined effects of the physical functioning, cognitive and emotional health, and comorbidities of older adults significantly impact their health changes. Apart from health deteriorating with age and sex disparities, educational and economic status, health behaviors, and social participation at the individual level were found to be the robust factors in predicting healthy aging.

Determinants of healthy aging

The COVID-19 pandemic has disproportionately affected people over 65 years of age, who had previously been in good health [38]. A systematic review identified, in light of global impact of COVID-19, **10 determinants of healthy aging** that can be applicable across different communities and countries to build their path to better health as follows: include physical activity, diet, self-awareness, outlook/attitude, life-long learning, faith, social support, financial security, community engagement, and independence [39]. Lu and colleagues identified almost similar determinants to assess healthy aging, by listing (a) physical capabilities, (b) cognitive functions, (c) metabolic and physiological health, (d) psychological well-being, and (e) social well-being as essential domains for future epidemiological research [40].

Substantial increases in the relative and absolute number of older persons in our society pose a challenge for biology, social and behavioral science, and medicine. As early as 1997, Rowe and Kahn defined **successful aging** according to three standards: 1. low probability of disease and disease-related disability and related risk factors; 2. high cognitive and physical functional capacity; and 3. active engagement with life [41]. Social networks and engagement and social capital are part of the third important standard for successful aging. In 2015, Stowe and Cooney refined Rowe and Kahn's successful aging model, as it emphasized only on personal control over one's later-life outcomes, and neglect of historical and cultural context, social relationships, and structural forces in influencing later-life functioning. Stowe and Cooney's revised model viewed development as a dynamic lifelong process, embedded in historical time and place, influenced by the web of relationships individuals are linked to, as well as more distal social structural factors showing how successful aging can better align with micro- and macrolevel issues through utilization of a life course perspective [42].

"Healthy aging" is often used interchangeably with terms such as "active", "successful," or "productive aging." The WHO reported how the broad determinants of health affect the process of aging. These determinants apply to the health of all age groups, with emphasis on the health and quality of life of older persons. Although, it is not possible to attribute direct causation to any one determinant, substantial body of evidence on what determines health suggests that all of these factors (and the interplay between them) are good predictors of how well both individuals and populations age. Fig. 12.4 illustrates how the following six determinants contribute to *active aging: economic, social, physical, personal and behavioral determinants, as well as health and social services* [43]. Culture is a cross-cutting determinant within the framework for understanding active aging. To promote active aging, health systems need to take a **life course perspective** providing continuum of care that focuses on health promotion, disease prevention, and equitable access and gender equity to quality primary health care and LTC [43].

Wellness and healthy aging

The Global Wellness Institute defines wellness as the active pursuit of activities, choices, and lifestyles that lead to a state of holistic health [44]. Wellness is not a passive or static state but rather an "active pursuit" associated with intentions, choices, and actions as we work toward an optimal state of health and well-being. Wellness is also linked to holistic health.

The term "Wellness" is perceived more holistically by today's consumers, and marketers are trying to target at least some of the elements through their product launches and marketing. Wellness has three main components: physical, mental, and social. Physical wellness includes avoiding sickness/allergies, fitness, and look. Mental wellness includes enjoyment of life, less

FIGURE 12.4 Determinants of active aging. *World Health Organization Noncommunicable Disease Prevention and Health Promotion Department. Active Ageing. A Policy Framework; 2002.*

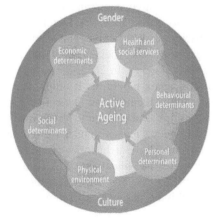

stress, maintaining energy levels, and work—life balance. Social wellness deals with social participation, environments, responsibilities, community, and success and sense of accomplishment. Healthy aging is about developing and maintaining functional abilities that enable well-being in old age.

Healthy living

Based on the determinants of healthy aging, seen above, the following five tips are helpful for seniors [45]:

1. **Prevention Is Key.** It does not take a lot of time or money to keep our muscles, bones, immune system, and mind healthy. Preventative care should include regular check-ups and screenings including disability assessments; proper medication management; and keeping up with vaccines and maintaining a proper diet and exercising often. These include heart diseases, obesity, high blood pressure, and more. In addition, the seniors must focus on fall prevention. Use appropriate and affordable technology-based devices for improving the activities of daily living, adherence to medications (as a reminder), and prevention of injuries.

2. **Eat Healthy Foods.** With a few simple changes in dietary, you can boost your immune system and reduce the risk of serious illnesses, lethargy, and depression. Here are some basic dietary recommendations that will fit the needs of people with diabetes, obesity, high blood pressure, high cholesterol, and hyperthyroidism. 1. Eat complex carbohydrates (fruits, vegetables, whole grains), 2. Reduce or eliminate simple carbohydrates (candy, soda, cake, cookies), 3. Drink plenty of water and limit or quit your alcohol intake, 4. Reduce intake of saturated fat and cholesterol (meats, oils, and high-fat dairy products), and 5. Try to avoid excess salt, eggs, and dairy products.

3. **Commit to Staying Active.** Exercise keeps our heart, lungs, and bones healthy. It increases lung capacity, endurance, balance, and metabolism, too. Some safe-for-senior exercises include walking, swimming, yoga, playing with the grandkids, Tai Chi, gardening, (light) house cleaning, and grocery shopping. For the more adventurous, you can also take up dance classes, water aerobics and others.

4. **Mental Health Is Just as Important as Physical Health.** It's not unusual for people to focus more on their physical health than their mental health. Besides regular physical exercise, some of the following activities that help with cognition and memory highly recommended are playing games (e.g., Scrabble, sudoku, crosswords, and puzzles), singing songs and playing instrument—Music can help strengthen memory and restart a decaying brain, sign up for a class—learn a new language, cooking, group fitness, arts and crafts, or a book club. Meditation is another great exercise for the brain. Seniors should also incorporate more brain foods into their diet to

help preserve memory and promote healthy cognitive functions. Some of the memory food include dark, leafy greens (like kale, spinach, and broccoli), dark fruits (like blueberries, blackberries, plums, and cherries), good fats (like **monounsaturated** fats found in nuts, cold-press olive oil, and avocado, and **polyunsaturated fats** such as omega 3 are found in oily fish like salmon, flaxseeds, and chia seeds) and **anti-inflammatories** (like coffee, dark chocolate, turmeric, cinnamon, tomatoes, and beats).

5. **Seniors Should Spend Time Doing What They Love**—When we do what we love, we feel better, and experience increased energy, better sleep, and overall mood improvement. Seniors who have hobbies should stick with them! Additionally, family and friends are great helpers when they come to creating healthy goals.

Encouraging healthy lifestyles can prevent, minimize, or even reverse poor health in old age, resulting in savings to healthcare system and better quality of life for seniors. Fig. 12.5 illustrates the effect of various factors influencing healthy aging. **Environmental influences** (adequate housing, family composition, education level/economical level of parents, ID exposure, toxic substance exposure, radiation exposure, violence exposure) account for 21%; **Health Care influences** (availability, accessibility, affordability) account for 16%, **Lifestyle influences** (tobacco use, unhealthy diet, inadequate activity, alcohol abuse, risky behaviors leading to injuries) account for 53% and **Unmodifiable risks influences** (race, sex, genetics) account for 10%.

Government support and care for the elderly

Majority of the seniors depend on their lifelong savings and government pensions. High-income countries provide Old Age Security Assistance in addition to the pension. While the government assistance alone is not sufficient, many of the high-income countries provide seniors' housing, subsidized

FIGURE 12.5 Influencers of healthy aging.

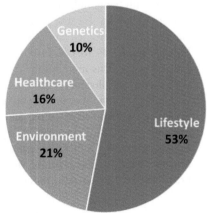

or free healthcare including drug plans, social assistance including LTC homes, and finally palliative care. Even in high-income countries many middle-income seniors will have insufficient resources for housing and health care [46]. Any policy solution should recognize the full range of services that seniors may need as they age. Similar social support is not available in most of the LMIC.

Long-term care Support: Aging challenges are eroding traditional eldercare in China—demographic shifts, changing family structure, industrialization, urbanization, rising population mobility, and shifting social mores. Indeed, there is a global convergence in several unsettling trends and challenges across LMICs. These include the weakening of informal family care systems for the elderly, growing needs for formal LTC of the frail and disabled who can no longer be adequately supported by family members, and mounting pressures for policy action to tackle these societal challenges [47]. Policy for LTC facilities and accessibilities varies country-by-country. While the LTCs are regulated, it is not uncommon that LTCs are privately owned and operated and the clients are expected to pay or co-pay for the services [48].

Age-friendly cities

The WHO launched the Global Age-friendly Cities (AFC) Project in 2005. According to the WHO, an age-friendly city encourages active aging by optimizing opportunities for health, participation, and security in order to enhance quality of life as people age. An age-friendly city is not just "elderly-friendly," but friendly for all ages. Eight domains of AFC are outdoor spaces and buildings, transportation, housing, social participation, respect and social inclusion, civic participation and employment, communication and information, and community support and services. For more information, please refer to the *Global Age-friendly Cities: A Guide* published by the WHO in 2007 [49].

As we have seen earlier, active or healthy aging is the process of optimizing opportunities for health, participation, and security in order to enhance quality of life as people age. In an age-friendly city, policies, services, settings, and structures support and enable people to age actively by recognizing the wide range of capacities and resources among older people; anticipating and responding flexibly to aging-related needs and preferences; respecting their decisions and lifestyle choices; protecting those who are most vulnerable; and promoting their inclusion in and contribution to all areas of community life.

COVID-19 and the elderly

COVID-related increased risk of severe illnesses

Studies have reported that the older adults were at higher risk in being infected with COVID-19, and if they get ill, they have a higher risk of death [50—52].

The higher the age, the higher the risk, with a negative effect of preexisting other diseases [51,53]. In early days of COVID-19 (on March 17, 2020), the Italian National Institute of Health reported 1625 deaths (aged 60−69: 137, aged 70−79: 578; and age 80 and over: 850). People 60 years old and over were about 96.5% of the total number of deaths in Italy, while in China they were about 80.8% of the total number of deaths [51]. The US CDC reported that over 200,000 residents and staff in long-term care facilities have died from COVID-19 [54]. Such a staggering number of elderly deaths in LTC facilities across all countries were common during the early to peak periods of COVID-19 pandemic. A study in G7 countries reported that between 20% and 70% of total deaths occurred in LTCs [55]. Data from 21 high-income countries show that while some countries have had no or very few deaths among residents in LTC facilities, other countries report that on average nearly half of all deaths linked to COVID-19 among LTC facility residents (ranging 24% in Hungary to 82% in Canada) [56].

The COVID-19 pandemic has revealed weaknesses in emergency response where LTC services has been underprioritized, resulting in the devastating impact seen across LTC services globally [57]. These events have highlighted long-standing problems in the LTC systems in most countries: underfunding, lack of accountability, fragmentation, poor coordination between health and LTC, and an undervalued workforce. The evidence also shows that once COVID-19 infection is present in LTC facilities it is difficult to control, in part due to the large number of people living close together in facilities designed for communal living and that personal care requires close proximity. The response to the pandemic must include LTC to ensure that ethnic, age, and gender groups are not marginalized.

COVID-19 increased the risk of severe illness for people with underlying conditions—all NCDs including cancer, CKD, chronic obstructive pulmonary disease (COPD), mental illnesses, heart conditions, such as myocardial infarction, heart failure, cardiomyopathy, and more, and type-2 diabetes. Risk factors such as smoking, indoor/outdoor air pollution, obesity (BMI > 32), and severe obesity (BMI > 40) exacerbated the incidence of COVID-19 related complications. Onder and colleagues [51] documented that out of 335 deaths in elderly Italians, only 3 patients (0.8%) had no pre-existing diseases, 89 (25.1%) had a single disease from the list mentioned earlier in this paragraph, 91 (25.6%) had 2 diseases, and 172 (48.5%) had 3 or more underlying diseases. The presence of these comorbidities might have increased the risk of mortality independent of COVID-19 infection [51].

Management of COVID-19 in the elderly

Complications after COVID-19 are seen mostly in those over 80 years of age with low immune function. In these patients, balanced inflammatory response is difficult because of the high responsiveness and maladjustment of the immune system. This results in a persistent, often harmful, response called a cytokine storm. With increasing age, proinflammatory cytokines increase while the immune effect decreases, making the elderly more susceptible to the virus [57]. Elderly patients are often complicated with chronic diseases, accompanied by weakness and malnutrition [58]. In addition, older patients have higher peak viral loads in the nasopharynx; therefore, the virus replicates faster [59]. At the same time, weak elderly people lack a type I interferon response, leading to a low adaptive immune response, which can cause SARS-CoV-2 to escape the human immune response [59].

When it comes to management, during the early stages of the pandemic no suitable therapy or protocol could be adopted and many people died at their own homes, hospitals, or LTC homes. In the United States alone over 370,000 people over the age of 64 died, as of February 2021, due to COVID-19 [60]. However, once the vaccines have been identified in late 2020 and implementation of vaccinations began, most of the high-income and middle-income countries prioritized vaccination for the elderly. Older adults were identified to receive vaccines under Stage 1b, immediately after the front line health workers (who belonged to Stage 1a) [61].

Pandemic and resilience

Resilience is the process of adapting well in the face of adversity, trauma, tragedy, threats, or significant sources of stress, e.g., family and relationship problems; life-altering health problems; disasters; COVID-19, etc. This empowers the humans with **crisis competence**—meaning as we get older, we get the sense that we are going to be able to handle it, because we have been able to handle challenges in the past. People get the attitude, "You know you get past it," "These things happen, but there is an end to it," "There's a life after that" and "this too shall pass" (-Mark Brennan-Ing, a senior research scientist, Hunter College's Brookdale Center for Healthy Aging).

According to Regina Koepp [62], **Five Resilience Strategies** used by older adults during the COVID-19 pandemic are as follows:

1. Focus on the quality (not quantity) of relationships.
2. Tap into wisdom and compassion.
3. Maintain regular schedules, including hobbies, chores, work, or exercise.
4. Mindfulness to focus on immediate surroundings and needs without thinking beyond the present.
5. Access mental health care and support.

Conclusion

The global population over the age of 65 is increasing, in all countries. The elderly has the right to enjoy their retirement life, after working very hard throughout their lives. Some countries have programs and support systems in place to care for the elderly, while most do not. Most elderly suffer with chronic diseases as they reach 55 years of age. With low immunity, the elderly face the double burden of both infectious diseases and NCDs. During the COVID-19, the elderly is the most affected both in terms of death and disability. The elderly in the LTC homes were victims of COVID-19. The lockdown strategies during the pandemic accelerated mental illnesses among the elderly. Since the beginning of the pandemic, older adults have the greatest risk of requiring hospitalization or dying if they are diagnosed with COVID-19. However, most, who survived first and second wave of the pandemic are quite resilient. The elderly, as a group, deserve better care and protection.

As a societal responsibility, we need to add life to years for the elderly and not merely extend their years of living. Early and timely assessment of impairment or disabilities, providing rehabilitation, and thus improving the quality of life to the aging citizens—health and healthcare including mental health—should be a priority for an equitable public health.

References

[1] Pew Research Centre. Chapter 4. Population change in the U.S. and the world from 1950 to 2050. In: Attitudes about ageing: a global perspective; January 2014. p. 40—9. https://www.pewresearch.org/global/2014/01/30/chapter-4-population-change-in-the-u-s-and-the-world-from-1950-to-2050/. [Accessed 10 April 2022].

[2] United Nations. (2022). Handbook of Statistics 2022. https://unctad.org/system/files/officialdocument/tdstat47_en.pdf

[3] An ESRI White Paper. Methodology statement: 2019 ESRI age dependency ratios. https://downloads.esri.com/esri_content_doc/dbl/us/J10301_Age_Dependency_Methodology_2019.pdf. (Accessed 10 April 2022).

[4] The UN (United Nations) DESA. World population ageing 2019. 2019. https://www.un.org/en/development/desa/population/publications/pdf/ageing/WorldPopulationAgeing2019-Highlights.pdf. [Accessed 10 April 2022].

[5] The UN (United Nations), AARP (American Association of Retired People. Ageing, older persons and the 2030 agenda for sustainable development. HelpAge International; 2017. https://www.un.org. [Accessed 12 April 2022].

[6] The UN (United Nations). Second review and appraisal of the Madrid international plan of action on ageing, 2002. Report of the secretary-general. E/CN.5/2013/6. New York: United Nations; 2012.

[7] WHO (World Health Organization). Elder abuse. Fact sheet. 2016. Available at: http://www.who.int/mediacentre/factsheets/fs357/en/. [Accessed 20 April 2022].

[8] World Health Organization. Ageing and health: key facts. Geneva; 2021. https://www.who.int/news-room/fact-sheets/detail/ageing-and-health. [Accessed 16 April 2022].

[9] World Health Organization. 10 facts on ageing and health. https://www.who.int/news-room/fact-sheets/detail/10-facts-on-ageing-and-health. (Accessed 16 April 2022).

[10] Singh AJ, Goel S, Kathiresan J, editors. Health promotion: need for public health activism. vol I & II. Germany LAP Lambert Academic Publishing; 2013.

[11] Singh AJ, Kaur S, Sharma R, editors. Advocating the cause of disability through research—some examples from North India. New Delhi: Century Publications; 2020.

[12] Health empowerment of women—a desirable strategy in 21st century hospitals. Singh AJ, Suri V, Kaur S, editorsvol. 2. New Delhi: Kalpaz Publications; 2020 (ISBN 9789353240509 & 9789353240516).

[13] IARC-WHO. Estimated number of deaths in 2020, worldwide, both sexes, ages 65+ (excl. NMSC). https://gco.iarc.fr/today/online-analysis-table?v=2020&mode=cancer&mode_pop ulation=continents&population=900&populations=900&key=asr&sex=0&cancer=39&t ype=1&statistic=5&prevalence=0&population_group=0&ages_group%5B%5D=13&ag es_group%5B%5D=17&group_cancer=1. (Accessed 25 April 2022).

[13a] Inanami H, Iwai H, Kato S, Takano Y, Yuzawa Y, Yanagisawa K, et al. Partial resection of spinous process for the elderly patients with thoraco-lumbar kyphosis: technical report. Medicina (Kaunas, Lithuania) 2021;57(2). https://doi.org/10.3390/medicina57020087.

[13b] Ko AC, Korn BS, Kikkawa DO. The aging face. Surv Ophthalmol 2017;62(2):190—202. https://doi.org/10.1016/j.survophthal.2016.09.002.

[14] Singh AJ, Kaur S, Kishore J, editors. Comprehensive textbook of elderly care. New Delhi: Century Publications; 2014. p. 427—35.

[14a] Centres for Disease Control. Public health and aging: trends in aging—United States and worldwide. JAMA 2003;289(11):1371—3. https://doi.org/10.1001/jama.289.11.1371.

[15] Pachana NA, Gallagher-Thompson D. The importance of attention to cultural factors in the approach to dementia care services for older persons. Clin Gerontol 2018;41:181—3.

[16] GBD 2019 Dementia Forecasting Collaborators. Estimation of the global prevalence of dementia in 2019 and forecasted prevalence in 2050: an analysis for the Global Burden of Disease Study 2019. Lancet Public Health January 6, 2022. https://doi.org/10.1016/S2468-2667(21)00249-8.

[17] Brodaty H, Donkin M. Family caregivers of people with dementia. Dialogues Clin Neurosci 2009;11:217—28. https://doi.org/10.31887/DCNS.2009.11.2/hbrodaty.

[18] Chockalingam A, Balaguer-Vintro I, editors. Impending global pandemic of cardiovascular diseases. Barcelona, Spain., Philadelphia, USA: World Heart Federation. Prous Science; 1999.

[19] World Health Organization. International classification of diseases. https://www.who.int/standards/classifications/classification-of-diseases. (Accessed 16 April 2022).

[20] Murray CJL, Lopez AD, editors. The global burden of disease: a comprehensive assessment of mortality and disability from diseases, injuries, and risk factors in 1990 and projected to 2020. World Health Organization; 1996. https://apps.who.int/iris/bitstream/handle/10665/41864/0965546608_eng.pdf. [Accessed 16 April 2022].

[21] GBD 2019 Demographics Collaborators. Global age-sex-specific fertility, mortality, healthy life expectancy (HALE), and population estimates in 204 countries and territories, 1950—2019: a comprehensive demographic analysis for the Global Burden of Disease Study 2019. Lancet 2020;396:1160—203. https://www.sciencedirect.com/science/article/pii/S0140673620309776. [Accessed 16 April 2022].

[22] Institute of Health Metrics and Evaluation (IHME). COVID-19 resources. https://www.healthdata.org/covid. (Accessed 16 April 2022).

[23] Marengoni A, Winblad B, Karp A, Fratiglioni L. Prevalence of chronic diseases and multimorbidity among the elderly population in Sweden. Am J Public Health 2008;98:1198—200.

[24] Haan MN, Selby JV, Rice DP, Quesenberry CP, Schofield KA, Liu J, et al. Trends in cardiovascular disease incidence and survival in the elderly. Ann Epidemiol 1996;6:348—56.

[25] Fratiglioni L, Launer LJ, Andersen K, Breteler MM, Copeland JR, Dartigues JF, et al. Incidence of dementia and major subtypes in Europe: a collaborative study of population-based cohorts. Neurologic Diseases in the Elderly Research Group. Neurology 2000;54(11 Suppl. 5):S10−5.

[26] Anderson GF, Hussey PS. Population aging: a comparison among industrialized countries. Health Aff 2000;19:191−203.

[27] UNDESA. Health inequalities in old age. https://www.un.org/development/desa/ageing/wp-content/uploads/sites/24/2018/04/Health-Inequalities-in-Old-Age.pdf. (Accessed 16 April 2022).

[28] United Nations General Assembly. Report of the secretary-general on follow-up to the second world assembly on ageing. July 26, 2012 (A/67/188).

[29] United Nations General Assembly. Report of the secretary-general on follow-up to the second world assembly on ageing, 26 July 2012 (A/67/188) and HelpAge international, "the right to health in old age: unavailable, inaccessible and unacceptable". August 2011. Available from: http://www.helpage.org/silo/files/the-right-to-health-in-old-age-unavailable-inaccessible-and-unacceptable.doc. [Accessed 16 April 2022].

[30] Marco T. Inequities in health care utilization by people aged 50+: evidence from 12 European countries. Soc Sci Med 2015;126:154−63.

[31] Equality and Human Rights Commission. Age concern and help the aged, just ageing? Fairness, equality and the life course: socio-economic inequalities in older people's access to and use of public services. 2009.

[32] Jacobzone S, Oxley H. Ageing and health care costs. Internationale Politik und Gesellschaft (International Politics and Society); 2002. Available at: http://fesportal.fes.de/pls/portal30/docs/folder/ipg/ipg1_2002/artjacobzone.htm.

[33] Mohanty SK, Chauhan RK, Mazumdar S, Srivastava A. Out-of-pocket expenditure on health care among elderly and non-elderly households in India. Soc Indic Res 2014;115:1137−57.

[34] Longev LH. Care for ageing populations globally—editorial. Lancet Healthy Longev 2021;2:e180. www.thelancet.com/healthy-longevity.

[35] Baird K. High out-of-pocket medical spending among the poor and elderly in nine developed countries HSR. Health Serv Res 2016;51:1467−88.

[36] National Development Council. Population projections for Taiwan: 2014−2060. R.O.C. Taiwan: National Development Council; 2013.

[37] Liu LF, Su PF. What factors influence healthy aging? A person-centered approach among older adults in Taiwan. Geriatr Gerontol Int 2017;17:697−707.

[38] Jowell A, Carstensen L, Barry M. A life-course model for healthier ageing: lessons learned during the COVID-19 pandemic. Lancet Healthy Longev 2020;1:e9−10.

[39] Abud T, Kounidas G, Martin KR, Werth M, Cooper K, Myint PK. Determinants of healthy ageing: a systematic review of contemporary literature. Aging Clin Exp Res 2022;8:1−9. https://doi.org/10.1007/s40520-021-02049-w. PMID: 35132578; PMCID: PMC8821855.

[40] Lu W, Pikhart H, Sacker A. Domains and measurements of healthy aging in epidemiological studies: a review. Gerontol 2019;59:e294−310. https://doi.org/10.1093/geront/gny029. [Accessed 17 April 2022].

[41] Rowe JW, Kahn RL. Successful aging 2.0: conceptual expansions for the 21st century. J Gerontol Ser B Psychol Sci Soc Sci 2015;70:593−6.

[42] Stowe JD, Cooney TM. Examining Rowe and Kahn's concept of successful aging: importance of taking a life course perspective. Gerontologist 2015;55:43−50.

[43] World Health Organization. Active ageing a policy framework. 2002. https://extranet.who.int/agefriendlyworld/wp-content/uploads/2014/06/WHO-Active-Ageing-Framework.pdf.

[44] Global Wellness Institute. What is wellness? https://globalwellnessinstitute.org/what-is-wellness/. (Accessed 18 April 2022).

[45] All About Seniors. 5 crucial healthy living tips for seniors. https://www.allaboutseniors.ca/5-crucial-healthy-living-tips-for-seniors/. (Accessed 18 April 2022).

[46] Pearson CF, Quinn CC, Loganathan S, Datta AR, Mace BB, Grabowski DC. The forgotten middle: many middle-income seniors will have insufficient resources for housing and health care. Health Aff 2019;5:851−9.

[47] Feng Z. Global convergence: aging and long-term care policy challenges in the developing world. J Aging Soc Policy 2019;31:291−7. https://doi.org/10.1080/08959420.2019.1626205.

[48] Chin CWWW, Phua K-H. Long-term care policy: Singapore's experience. J Aging Soc Policy 2016;28:113−29. https://doi.org/10.1080/08959420.2016.1145534.

[49] World Health Organization. Global age friendly cities: a guide. Switzerland: World Health Organization; 2007. p. 1−76.

[50] Peeri NC, Shrestha N, Rahman MS, Zaki R, Tan Z, Bibi S, et al. The SARS, MERS and novel Coronavirus (COVID-19) epidemics, the newest and biggest global health threats: what lessons have we learned? Int J Epidemiol 2020;49:717−26.

[51] Onder G, Rezza G, Brusaferro S. Case-fatality rate and characteristics of patients dying in relation to COVID-19 in Italy. JAMA 2020;323:1775−6.

[52] Li Q, Guan X, Wu P, Wang X, Zhou L, Tong Y, et al. Early transmission dynamics in Wuhan, China, of novel coronavirus-infected pneumonia. N Engl J Med 2020;382:1199−207. https://doi.org/10.1056/NEJMoa2001316.

[53] Liu K, Chen Y, Lin R, Han K. Review-clinical features of COVID-19 in elderly patients: a comparison with young and middle-aged patients. J Infect 2020;80:e14−8. https://doi.org/10.1016/j.jinf.2020.03.005.

[54] Chidambaram P. Over 200,000 residents and staff in long-term care facilities have died from COVID-19. Policy Watch February 2022. www.kff.org.

[55] Thompson D-C, Barbu MG, Beiu C, Popa LG, Mihai MM, Berteanu M, et al. The impact of COVID-19 pandemic on long-term care facilities worldwide: an overview on international issues. Hindawi BioMed Res Int 2020;2020:1−7. https://doi.org/10.1155/2020/8870249.

[56] De Virgiliis F, Di Giovanni S. Lung innervation in the eye of a cytokine storm: neuro-immune interactions and COVID-19. Nat Rev Neurol 2020;16(11):645−52.

[57] Office of the Auditor General of Ontario, Canada. COVID-19 preparedness and management special report on pandemic readiness and response in long-term care. 2021. https://www.auditor.on.ca/en/content/specialreports/specialreports/COVID19_ch5readinessrespon seLTC_en202104.pdf.

[58] Dandan L, Biao C, Yue L, Ximing X. Clinical characteristics of cancer patients with COVID-19 and suggestions for patients management strategy during the epidemic novel coronavirus period. Mod Oncol Med 2020;28(17):3096−8 (in Chinese).

[59] Vellas C, Delobel P, de Souto Barreto P, Izopet J. COVID-19, virology and geroscience: a perspective. J Nutr Health Aging 2020;24(7):685−91.

[60] Data visualization—COVID-19 death by age. Data as of February 17, 2021. Data visualization | COVID-19 deaths by age | The Heritage Foundation.

[61] World Health Organization. WHO sage roadmap for prioritizing uses of COVID-19 vaccines in the context of limited supply. An approach to inform planning and subsequent recommendations based upon epidemiologic setting and vaccine supply scenarios. World Health Organization; November 13, 2020. https://www.who.int/docs/default-source/immunization/sage/covid/sage-prioritization-roadmap-covid19-vaccines.pdf.

[62] Koepp R. 5 resilience strategies seniors are using during COVID-19. Psychology Today Canada. https://www.psychologytoday.com/ca/blog/the-psychology-aging/202101/5-resilience-strategies-seniors-are-using-during-covid-19.

Section III

Future of public health practice

Chapter 13

The need for a paradigm shift to ensure adequate skilled human resources for effective public health practice

Myo Minn Oo

Department of Medical Microbiology and Infectious Diseases, University of Manitoba, Winnipeg, MB, Canada

Definition of human resources for health

Human resources for health (HRH), also known as "Health Human Resources," "Health Workforce," or simply "Health Workers," is one of the most significant building blocks for health systems that serve as the primary input into the system alongside physical capital and consumables [1]. It is as such the personification of any health system [2]. Tilson and Gebbie defined the health workforce comprehensively as: "The public health workforce broadly includes all those engaged during a significant part of the time in work that creates the conditions within which people can be healthy. Specifically, the workforce comprises those who work for official public health agencies at all levels of government, community-based, and voluntary organizations with a health promotion focus, the public health—related staff of hospitals and healthcare systems, and a range of others in private industry, government, and the voluntary sector" [3]. However, the World Health Organization (WHO) provided a simplified definition for health workers, as "all people engaged in actions whose primary intent is to enhance health" [2]. In other words, HRH is involved in protecting and improving the health of communities. The formal health training and skills of health workers may vary from less than primary education to doctorate and other specialized training. In all their diversity, health workers make up the global health workforce.

Principles and Application of Evidence-Based Public Health Practice
https://doi.org/10.1016/B978-0-323-95356-6.00010-0

Various cadres of HRH

The HRH is a multidisciplinary interconnected network of teams. The team includes not only doctors but also "nurses, educators, nutritionists, social workers, engineers, and many other professionals, a large group of people working in the field as aides, sanitarians, extenders, community health workers (CHWs), and, of course, vital administrative, support, and clerical staff and a remarkable complement of volunteers" [3]. Several institutions and global organizations attempted to classify the health workforce into different cadres based on the profession, diversity, legal basis, and functions. WHO categorized HRH into three broad groups. The groups are as follows: (a) physicians or medical doctors, (b) nurses and/or midwives, and (c) others, consisting of seven subgroups of workforces. The others group includes "dentistry, pharmaceuticals, laboratory, environment and public health, community and traditional health, health management and support, and staff, including medical assistants, dieticians, nutritionists, occupational therapists, medical imaging and therapeutic equipment technicians, ophthalmic opticians, physiotherapists, personal care workers, speech pathologists, and medical trainees" [4,5]. However, lay workers like CHWs or volunteers and traditional birth attendants (TBAs) should be included in HRH since they share the major burden of work at the community level. In addition to the above groups, the definition provided by the Association of Schools of Public Health in the European Region includes professionals in other sectors like politicians, engineers, and software architects as part of HRH.

Clinicians with advanced qualifications: Physicians or medical doctors

Clinicians are professional medical practitioners who possess sophisticated skills in diagnosing and treating various medical and surgical issues [4]. They are often required to have at least 4–5 years of post-secondary education, with or without a minimum of 2–3 years of postgraduate certifications or tertiary degrees. Clinicians are licensed and highly regulated by their respective national or subnational regulatory bodies. Their designation varies within and between countries depending on the standard operating procedures of their respective health systems. Broadly, the doctors can be categorized based on (a) Degree: general medical officers (completed undergraduate degrees), postgraduates, and doctors with subspecialization (also known as superspecialists such as nephrologists, urologists, and others); (b) Level of health care: primary, secondary, or tertiary level; (c) Division: medical services, public health, or medical education; and (d) Based on employer: federal, state, or local. In medical services, the designation ranges from general medical officer to principal medical officer, assistant surgeon, to senior civil surgeon. The same applies to medical education professionals, ranging from tutors to senior professors or director-level. Prerequisites and training requirements may vary across countries. The physician or medical doctors primarily provide curative care and, to some extent, preventive and rehabilitative care.

Nurses

Nursing is the autonomous and collaborative care of individuals of all ages, families, groups, and communities, in various contexts, whether sick or healthy. It encompasses health promotion, disease prevention, and the care of the sick, disabled, and dying. Nursing also plays important roles in advocacy, promoting a safe environment, conducting research, developing health policy, managing patient and health systems, and education [6].

A *registered nurse* is a graduate who is legally authorized to practice nursing after passing an examination administered by a state board of nurse examiners or another regulatory body [4]. Nursing school usually lasts three or more years and leads to a university degree or postgraduate degree, or their equivalent. A registered nurse is knowledgeable in all aspects of nursing. Common titles include nurse practitioner, clinical nurse specialist, advance practice nurse, practice nurse, licensed nurse, diploma nurse, public health nurse, nurse clinician, and others.

Midwifes are registered health workers who are evaluated and registered by a state or comparable regulatory entity in midwifery [4]. Their primary responsibility is to provide essential healthcare to pregnant women before, during, and after delivery, as well as for newborn babies. A registered midwife is knowledgeable in all aspects of midwifery [7].

Auxiliary nurses, nurse assistants, or enrolled nurses have some secondary school training and often receive formalized on-the-job training through apprenticeships. They are trained in fundamental nursing skills but not in making nursing decisions. The length of training varies per country, ranging from a few months to 2 to 3 years. Similarly, auxiliary nurse midwives with basic nursing abilities without formal nursing decision-making training can help with mother and child health care, especially childbirth and care throughout the prenatal and postpartum periods. While they are not certified as midwives, they have some midwifery skills and are potentially useful in delivering various essential health care [7,8].

Non-clinician HRH

The other groups of HRH are an integral part of any health system, each group having a defined role. These groups include dentistry, pharmaceuticals, laboratory, environment and public health, community and traditional health, health management and support, and staff, including medical assistants, dieticians, nutritionists, occupational therapists, medical imaging and therapeutic equipment technicians, ophthalmic opticians, physiotherapists, personal care workers, speech pathologists, medical trainees, software managers, and information assistants. Together, they contribute to the diverse spectrum of HRH, from community to tertiary care, including preventive, promotive, curative, and rehabilitative services.

Lay health workforce

Lay health workers deliver community-based health services based on ad hoc requirements in the context of health intervention but do not hold any official professional or paraprofessional certificate or higher education degree [9]. They are usually kept outside the formal healthcare delivery system despite sharing a significant share of community health service delivery. The government provides incentives or honorariums, usually minimal, to these lay workers based on the delivery of specified services to the target population. Variants of lay health workers include CHWs, community or village health volunteers, TBAs, treatment supporters, promoters, and others.

Community health workers (CHWs) are frontline health workers committed to providing health care to the people within their community [10]. Their strong relationship with the community members they serve develops a sense of trust, which lends credibility to their health recommendations, diagnoses, and referrals. While their job descriptions are non-specific, the nature of the public health work they perform is determined by their education level, past training, life experience, experience working with certain groups, and other characteristics. They may advocate for health in the community, enroll and conduct outreach activities, navigate, and provide health education, counsel, and support at social and emotional levels. In addition to their training and education, they emphasize relationship-building and tailor interpersonal interactions. They are communication specialists in their own communities with a sense of cultural awareness. While having less formal education and training than nurses and physicians, this human resource group has enormous potential to extend healthcare services to vulnerable populations, such as remote communities and historically marginalized people. Accredited Social Health Activists (ASHAs) of India are a good example of CHWs. ASHAs are selected from the same community and cover approximately 1000 people. They deliver services related to maternal and child health (MCH) and others to the covered population that are linked with incentives.

Community health volunteers (CHVs) are "a diverse collection of lay persons educated to promote community health among their peers" [11]. They come from various backgrounds, based in the areas they serve, and have undergone brief training on the health issue they have volunteered to work on. It is difficult to underestimate the vital role that volunteers play in the public health profession. Many individuals who desire to help others do not require medical training to contribute to public health and safety education, contact tracing, answering phones at clinics and health hotlines, or delivering meals to the elderly. CHVs have the potential to enhance the conventional health system in the fight for universal health coverage (UHC) in low- and middle-income countries (LMICs). Their services can extend, to a certain extent, preventive, promotive, and curative health services that are on par with, if not better than, those given by professionally hired health workers.

A *traditional birth attendant (TBA)* is a lay health worker with a specific task to aid the woman during labor and who learned their expertise by delivering newborns themselves or by training with other TBAs [4]. TBAs have some biomedical training in pregnancy and delivery care. They are also called skilled birth attendants or traditional midwives in some places. Though India trained and utilized the services of TBAs in the past, the country abolished the TBA system and no longer recognizes them as skilled birth attendants.

Ad hoc workforce: Substitute health workers [12]

During emergencies, such as economic and social crises, war, or conflict scenarios or in the face of significant population growth and increased health needs, developing countries created an ad hoc group of health workers to perform the roles of qualified health professionals, including physicians, nurses, and pharmacists, despite having less pre-service training. It involves delegating of job responsibilities to a cadre of people with low qualifications or volunteers. The nature of these "substitutes" means that they are highly country-specific and the international health labor markets rarely reach them [12].

As governments establish rural health centers and attempt to shift focus away from hospitals, primary health care, and other basic care ideas may be compelled to form new cadres of the formal or informal workforce to accommodate non-traditional services. However, several concerns arise: is it viable to consider outsourcing essential skills and responsibilities to less academically trained practitioners in low-resource settings or regions that cannot afford to recruit and retain qualified health professionals? Furthermore, it is also ethically questionable to continue relying on lay or substitute health services even after the crisis has been normalized.

Global situation and management of HRH

Human resource shortages are a perennial problem worldwide, and there has been increasing demand for skilled healthcare workers. This demand often intensifies during a public health crisis or pandemic. Countries recognized the need for skilled HRH when they struggled to meet the proposed targets of the Millennium Development Goals (MDGs) [13]. MDGs emphasized health for all through quality primary health care, which necessitated the equitable distribution of HRH and the expansion of delegations to community workers. In the post-2015 MDG era, optimal HRH remains crucial for achieving the goals of Universal Health Coverage (UHC) outlined in the Sustainable Development Goals (SDGs) [14]. A significant contribution to attaining this ambitious goal is increasing and sustaining the recruitment, development, training, and retention of health workers, particularly in low- and middle-income countries (LMICs) [15]. It is estimated that achieving the proposed SDG goals would require at least 45 doctors, nurses, and midwives per 10,000 people [16]. The global strategy on HRH (2030) aims to recruit and retain at least 10 million new HRH to deliver services to the underserved population [17].

Concepts in managing HRH

Managing HRH is critical for delivering healthcare services, enhancing patient self-reported outcomes, improving health system's responsiveness to hazards or dangers, and increasing community resilience. Quantity (size, distribution, and composition) and quality (skill categories, training levels, and competency) are all essential aspects of a healthcare workforce. HRH management involves not only managing people but also improving and upgrading the knowledge, abilities, and motivations of those responsible for providing health care. The ability of a health system to employ and retain effective practitioners is affected by the exponentially rising costs of healthcare consumables. Policies addressing inflows and outflows, such as rehiring of unemployed health workers where their knowledge and skills are needed, addressing inefficiencies, and rectifying dysfunctional distribution, such as retaining health workers in underserved areas, can all be targeted for interventions. The health system needs to comprehend the HRH management strategies for such targeted interventions.

"Working lifespan" approach

The "working lifespan strategies" deals with workforce dynamics, encompassing the stage at which individuals enter the workforce, their employment period, and the point at which they leave it (Fig. 13.1) [18]. The road map for training, supporting, and keeping the workforce includes both a worker viewpoint and a systems approach. To facilitate entry into the workforce, smart investments in education and the implementation of ethical and successful recruitment strategies are necessary to build robust educational institutions, ensure academic quality, and enhance recruitment capabilities. Once individuals are part of the workforce, regular supervision, fair and reliable compensation, essential support systems, and lifelong learning opportunities will enhance the efficiency and retention of HRH. While the framework is simple, the systems view prevails. However, health systems must also consider the labor market dynamics and the personal attributes of the HRH for effective HRH management.

The Health Labor Market Framework

Comprehensive workforce planning is necessary based on the health labor market, with a reasonable understanding of the dynamics influencing

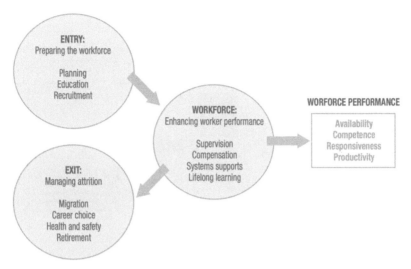

FIGURE 13.1 Working lifespan strategies [18].

workforce supply and demand at both national and global levels. Countries also need to adopt efficient strategies to optimize the supply of health personnel, addressing the highlighted challenges and achieving UHC. The Health Labor Market Framework provides a much-needed comprehensive overview of health labor market dynamics and the contributions of four types of health workforce strategies to achieve equal access to high-quality health care and UHC in the post-2015 SDGs, at national, regional, and global levels (Fig. 13.2) [14]. Firstly, the demands for health labor must align with the supply, primarily driven by education reforms. Secondly, it is crucial to create working cultures and environments within the health sector that retain competent health professionals. Factors such as wages, working conditions, safety, and career prospects play a significant role in retention. Thirdly, the outflow of competent health workers diminishes the available supply of healthcare personnel. Policies to attract health professionals back into the workforce, discourage migration, and mobilize the unemployed range from raising compensation and offering additional benefits to improving working conditions, modifying recruitment techniques, and providing training opportunities. Lastly, policy and financial decisions in education and the health labor market must align with these changing demands to achieve the required number, quality, and relevance of the health workforce. Aligned investments in HRH, addressing shortages and improving the distribution of health workers while considering labor market dynamics and education policies, should result in significant improvements in health and economic outcomes.

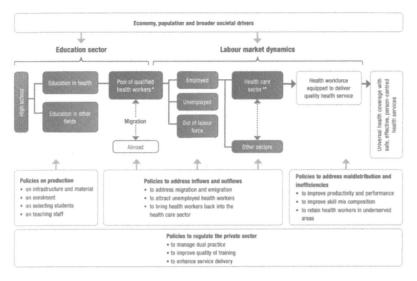

*   Supply of health workers= pool of qualified health workers willing to work in the health-care sector.
** Demand of health workers= public and private institutions that constitute the health-care sector.
Source: Sousa A, Scheffler M R, Nyoni J, Boerma T "A comprehensive health labour market framework for universal health coverage" Bull World
Health Organ 2013;91:892– 894

FIGURE 13.2   Policy levers to shape health labor markets [14].

Dimensions of HRH

For over a century, the health system has primarily focused on the availability of HRH, with a particular emphasis on doctors. However, it is important to recognize that doctors consistitute less than one-fourth of the HRH workforce, while the largest group comprises nurses and midwives. Merely quantifying the availability of HRH may not be sufficient for effective public health practice. Instead, the efficiency of HRH determines the adequate and appropriate delivery of healthcare services. Therefore, alongside availability, it is essential to consider accessibility, acceptability, functional status, and quality of services delivered to ensure the effective delivery of public health services. These dimensions are commonly, known as AAAQ: Availability, Accessibility, Acceptability, and Quality of HRH.

Availability of HRH

Availability indicates the presence of an adequate number of HRH to deliver healthcare services and improve population health. According to estimates from 2013, there was a total of 43 million HRH globally. Of them, 9.8 (22.7%) million were physicians, 20.7 (48.1%) million were nurses/midwives, and 13.0 (30.2%) million were other health workers. However, there was a significant shortage of HRH, with a total need-based shortage of 17.4 million in 2013, which accounted for nearly one-third of the total need [17]. Projections by

International agencies suggested that HRH shortages will continue to increase in the future, with variations between countries. The gap in health worker shortage between 2013 and 2030 was estimated to be highest in the WHO Western Pacific region, followed by the African region. By 2030, an additional 18 million health workers are required, especially in low- and lower-middle-income nations. Similarly, all countries will require an additional 9 million nurses and midwives by 2030 to the SDG. As an example, India currently has only 160 skilled HRH per 100,000 people, falling short of the global target of 450 per 100,000 people. In response to the adoption of the 2030 SDG, the country has integrated the HRH management with the national health policy as part of its efforts to improve the workforce [19].

Given the global shortfall in the health worforce, it is imperative to increase the education and training of health workers in areas with the greatest health needs. Furthermore, schools providing preservice education must ensure that graduates acquire the necessary skills to address local health issues, especially in impoverished regions [14,20]. Continuing job training for the health workforce, including lay health workers, the implementation of robust attrition strategies, the enhancement of health workers' healthcare delivery capacity with mobile phone-based technology, and the improvement of managerial skills and competencies have all yielded positive outcomes [21]. In many countries, the traditional and complementary medicine workforce represents a significant portion of the health workforce. For instance, Ayurveda, Yoga, Unani, Siddha, and Homeopathy (AYUSH) practitioners in India are now integrated into the routine healthcare system and provide services at all levels throughout India.

Accessibility and acceptability of HRH

Accessibility refers to equitable access in all health-related (specialist or specific care) and social (urban/rural, financial, and others) dimensions. Acceptability involves meeting the population expectations in various dimensions like age, gender, qualification or skill, and cultural awareness. In most countries, there exists an unequal and inequitable distribution of HRH within and between rural and urban areas. Despite half of the world's population residing in rural areas, only 38% of nurses and 25% of doctors work in these areas [22]. Further, specialist services are predominantly concentrated in urban areas, leading to the necessity for rural population to travel to urban areas for health care. Moreover, the distribution of HRH within rural areas (plain vs. hilly) and urban areas (proper urban vs. slum) are also imbalanced. Failure to address these shortages and maldistributions can have significant consequences for the health of billions of people worldwide [23]. While interventions have been implemented to retain the HRH in rural and other underserved areas, sustainable solutions require the development of essential facilities in these areas.

Women make up more than 70% of the global workforce in health and social services, with a significant majority working as nurses/midwives, and in other HRH roles. The acceptability of women in nursing or midwifery profession tends to be consistent across nations. For an example, in India, the country has reserved all its health worker cum auxiliary nurse midwife posts exclusively for females, both in principle and operationally. This approach has contributed to the community's acceptance of this particular cadre. It is also driven by the nations' priority on reproductive and child health, which require the delivery of health services primarily to the female population. Furthermore, this approach provides employment opportunity and empowers the women within the community. Interestingly, there are instances where cultural norms are reversed, such as in certain parts of the world, including India, where male gynecologists are renowned. In these cases, predominantly women prefer to consult male specialists over female specialists. These variations in accessibility and acceptability of HRH highlight the need for a deep understanding of local cultural contexts.

Quality of HRH

Quality in the context of HRH refers to the actual functional status of the workforce in delivering services that meet the expected standard. It goes beyond the mere presence of a qualified cadre of the workforce. The quality of HRH is determined by their competency, which encompasses their knowledge, attitude, behavior and applied skills. According to WHO, competence is defined as "the state of proficiency of a person to perform the required practice activities to the defined standards. This incorporates having the requisite competencies to do this in a given context. Competence is multidimensional and dynamic, changing with time, experience, and setting" [24]. In essence, competence is the abilities of an individual, including durability and trainability, to integrate knowledge, attitudes, and skills in performing tasks in a specific context.

The "People that Deliver" initiative outlines competence under the technical and management domains, which are applicable to healthcare service delivery as well as the health supply chain. The technical domains consists of four subdomains: assessing the need, assessing capacity, delivery of services, and resource management [24,25]. These subdomains are linked to ensure effective service delivery. The management domain, on the other hand, has two subdomains, professional and personal responsibility and accountability. This includes building competencies in areas such as communication skills, stress and time management skills, and creating a career development pathway.

Indeed, quality in healthcare delivery plays a significant role in determining the acceptability of services by the community. When HRH fails to deliver the services as expected, it can negatively impact the community's trust and acceptance of those services in the future. Hence, it is crucial for HRH to not only possess technical competence but also to be culturally aware and sensitive to the needs and values of the community they serve. By being both technically sound and culturally aware, HRH can enhance the quality of care provided and foster better acceptance and trust among the community.

Availability, acceptability, and quality of lay health workers

Lay health workers have the potential to provide interventions beyond their traditional scope when given more intense training, potentially filling the roles of certain cadres of the health workforce. These individuals have made significant contributions to population health in many countries for decades and have notably improved the implementation of various health interventions. In the context of maternal and child health (MCH), with focused training, lay health workers can play a vital role in improving MCH outcomes. They can provide antenatal care, perform safe and clean delivery procedures, act as competent birth attendants, educate mothers about exclusive breastfeeding [26], promote infant warmth and sanitary cord care, and detect life-threatening neonatal disorders. Additionally, they can contribute to enhancing vaccination programs, increasing uptake, and bringing services closer to communities [27]. Beyond MCH programs, lay health workers have effectively delivered services related to fever management, testing and treating malaria, screening common noncommunicable diseases, referral linkage and follow-up, distributing antiretroviral therapy to people living with HIV, or providing directly observed short-course therapy for tuberculosis, among other interventions [28−30].

ASHA (Accredited Social Health Activist) of India is one of the success story of the acceptability and quality of lay health workers. ASHA has been recognized for its significant contribution to the national health programs of the country and was awarded the Global Health Leaders Award during the 75th World Health Assembly. ASHA agents play a crucial role in delivering various healthcare services including those related to maternal and child health, immunization, surveillance and follow-up of communicable and noncommunicable diseases, as well as health promotion nutrition and sanitation. The acceptability of ASHA workers can be attributed to the conditions surrounding their recruitment. They are selected from the same community/village they serve, and it is a strict requirement that they are women. While a basic level of education is expected, the criterion is flexible and based on the availability of individuals in the community.

HRH in fragile, failed, or postcrisis states

The delivery of healthcare services in fragile, failed, or post-crisis states poses significant challenges yet remains. In such contexts, health systems and livelihoods of health workers are often disrupted, and the complex contextual dynamics surrounding them make it difficult to incorporate sensitive policy measures effectively [31−33]. A holistic systems thinking approach is required to swiftly and adaptively respond to health crises and address the challenges faced by the health workforce. In countries like Afghanistan, the Democratic Republic of Congo, Haiti, the Occupied Palestinian Territories, Somalia, Myanmar, and others with ongoing or recently settled civil conflicts, the government often lacks control over the engagement and quality of the health workforce. Many healthcare providers in these setting are poorly trained, lack adequate regulation, exhibit varying levels of quality. Flaws in HRH information systems, distribution, and monitoring and supervision further complicate the assessment and management of the health workforce. Additionally, a significant number of the workforce may operate "off-the-books", making it challenging to track and ensure quality service delivery. Fragile states, such as Guinea-Bissau, face specific challenges due to chronic lack of funding, commercialization of public health services, inadequate training and deployment procedures, and prolonged political instability. Guinea-Bissau has experienced socialism, civil war, and extended periods of political and military upheaval, which have further strained the country's health system over the previous 40 years, eventually becoming a hub for worldwide drug smuggling networks [34].

In fragile, failed, or conflict-affected state, the government must address six preexisting problems: recruitment, distribution, retention, performance, motivation, and monitoring and supervision. These issues must be tackled to build a resilient health workforce and rebuild the country's health system. Above all, ensure a safe working environment for all HRH is of paramount importance.

COVID-19 pandemic and HRH management

Healthcare workers have played a critical role as frontline warriors in the fight against COVID-19, despite their various vulnerabilities, including the risk illness, exhaustion, burnout, stress, harassment, and physical and psychological abuse. These vulnerabilities can be attributed to several factors such as inadequate HRH planning and policies, unhealthy working conditions, noncompliance with infection prevention and control protocols, and insufficient training. As a result, health workers have been at a higher risk of contracting COVID-19 infection. The unprecedented nature of the COVID-19 pandemic, coupled with the limited availability of healthcare workers due to isolation, infection, and even fatalities, has exacerbated the existing shortages in every country [35]. WHO estimated that healthcare workers accounted for 3% to 13% of COVID-19 infections and experienced a mortality rate of 2% due to the virus [35]. Hence, it is crucial to establish accurate, reliable, and timely data on the impact of COVID-19 on HRH. This information should be considered when planning and updating the health emergency responses and allocating resource in the future.

Inequalities in the availability of HRH during the pandemic

The persistent issue of health workforce maldistribution was rampant during the COVID-19 pandemic. Disparities in the density of physicians and nursing and midwifery personnel have significantly widened during this crisis. The gaps between the nations with the greatest and lowest densities of physicians increased by a staggering 366 times, while for nursing and midwifery personnel density, it expanded by 194 times [35]. It is importannt to note that even countries with an adequate quantity of skilled health workforce faced challenges in effectively addressing the COVID-19 pandemic. Poor management of the health workforce, often stemming from policy decisions, has been a significant contributing factor to these challenges. Insufficient coordination, inadequate resource allocation, and limited strategic planning for HRH during a crisis can impede the effective utilization of available healthcare personnel. Addressing the issue of maldistribution and improving HRH management should be a priority in order to enhance the pandemic response capacity of countries and ensure equitable access to healthcare services. This includes implementing policies that promote a more balanced distribution of healthcare workers, strengthening HRH management systems, and improving coordination and collaboration among relevant stakeholders.

Expanded roles of HRH during the pandemic

The critical element of a good health system is the ability to expand the surge capacity and flexibility of its health workforce to launch an effective response to any health emergency. During the COVID-19 pandemic, the health systems had to rapidly expand their surge capacity and flexibility to effectively respond to the crisis. This involved implementing various strategies to optimize the utilization of the health workforce. Health workers were often asked to work longer hours, staffing criteria were modified, and measures such as full-time contracting, night shift working patterns, and cancellation of vacations were implemented [36]. Technology-based interventions supporting the health workforce at all levels were helpful and invariably used in various dimensions of the prevention and management of COVID-19. One example is the "Adaptt Surge Planning Support Tool," which enabled policymakers and planners to predict the number of health workers required in hospitals, including wards and intensive care units, to cope with increasing workloads due to COVID-19. This tool utilized the COVID-19 epidemiological situation in the countries to simulate the health workforce requirements and assist in planning for the future. The flexibility of such tools allowed countries to analyze differnt staffing scenarios, taking into account the need to address skills and job shifting within the workforce in response to changing personnel availability.

Surge planning during the COVID-19 pandemic necessitated initiatives to reskill and repurpose the health workforce, as well as to increase the number of skilled healthcare professionals. This often involved providing additional education and training to doctors, nurses, and other healthcare workers from various disciplines, particularly in emergency rooms, hospital wards, and intensive care units. Health workers redeployed to ICUs, infectious disease units, or respiratory medicine wards typically underwent specialized training, such as learning how to properly use personal protective equipment or manage patients with acute respiratory failure. HRH also took on additional responsibilities related to COVID-19 such as testing, contact tracing, and monitoring [35,36]. France serves as an example of how roles were repurposed during the early stages of the COVID-19 pandemic. A 2020 law in France allowed pharmacists to take on new responsibilities, including distributing and billing masks, dispensing pulse oximeters, medications for medicated abortions, and some "off-label" products. The government also authorized them to dispense self-test kits, conduct COVID-19 testing (both rapid and PCR), and administer COVID-19 vaccines. Similarly, in India, telemedicine guidelines were released to address previous concerns medical practitioners and patients, resulting in improved community uptake and access to healthcare services.

During the COVID-19 pandemic, changes in the skill mix of health workers required both online and in-person training to develop new skills and adapt to new ways of working [35,36]. However, these changes were sometimes met with opposition from professional associations, either due to concerns about the quality and safety of care or due to vested interests. Legislation was sometimes necesssary to explain or expand medical reimbursements to accommodate these changes. In 2020, the European Union (EU), collaborated with the European Society of Intensive Care Medicine to launch a training program called COVID-19 Skills PrepAration CoursE (C19 SPACE), specifically designed for doctors and nurses who did not have previous experience in intensive care. The program covered essential topics in intensive care, such as admitting critically ill patients, providing respiratory support, managing sepsis, and understanding the operations of ICUs during the COVID-19 pandemic. More than 17,000 health professionals from across the EU enrolled in the course, and by May 2021, 12,086 individuals received certification. This training initiative not only assisted hospitals in redeploying workers during the pandemic, but also contributed to enhancing the overall preparedness of the workforce to handle future health emergencies.

Intervention and challenges in ensuring the skilled HRH

Historically, various strategies have been implementated to enhance the availability, accessibility, acceptability, and quality of HRH. Some of the strategies include creation of new courses and training programs for expected skill levels, establishment and expansion of institutes, delegation and task shifting, preferential expansion of priority cadre, and continued medical education.

Wise investments in health workforce

A healthy population is not only vital for the well-being of individuals but also plays a crucial role in driving economic growth and development. Healthcare investments have far-reaching effects that extend beyond improving population health. Research has shown that a mere one-year increase in life expectancy can lead to a substantial 4% increase in GDP per capita, underscoring the significant economic benefits of a healthier society. Recognizing the importance of the health workforce, the UN High-Level Commission on Health Employment and Economic Growth has emphasized the need for investments in global HRH to propel the global economy forward [14]. However, despite the recognized significance, global investment in the health workforce remains inadequate and sluggish. Several critical issues contribute to this challenge. Chronic underinvestment in education and training programs for healthcare professionals hinders the development of a skilled workforce. There is often a misalignment between education strategies and the actual needs of healthcare systems and populations, further exacerbating the problem. Furthermore, low prioritization and inadequate investment in preventive care, along with the challenges of deploying workforce to underserved areas and shortages and distribution issues in the global health workforce, remains unresolved. Failure to invest in skilled health professionals can have detrimental effects on healthcare systems. It leads to inefficiencies, increased healthcare inequity, higher rates of medical errors, and escalated healthcare expenditure.

Assessing the current HRH situation is important for planning investments. Future investment plans can be based on the density and distribution of the health workforce. For example, the United States regularly updates its HRH data and reported 326,602 health workers in 2012 with a worker-to-population ratio of 105/100,000 [37]. The United States has planned to invest over 7 billion US dollars in recruiting and educating public health employees to prepare for future pandemics like COVID-19 [38]. This proactive approach reflects the recognition of the value of a well-prepared and adequately supported healthcare workforce in safeguarding public health.

Political leadership and commitment to building strong resilient health systems

A strong, resilient health system strives to increase health responsiveness and efficiency while minimizing social and financial hazards [39]. Such health systems relies on the dedication and expertise of health professionals. An several LMICs, including Brazil, Ghana, Mexico, and Thailand, have demonstrated positive experiences by investing in their health workforce [40]. These countries have made consistent progress towards achieving UHC by prioritizing their health staff through political leadership, commitment, appropriate legislation and policies, actions to ensure a "fit-for-purpose and fit-to-practice" health workforce, and intersectoral collaboration [41]. These successful examples highlight the need for a shift in mindset, recognizing health staff as an investment rather than a cost. It is crucial for political leaders and policymakers in other countries to understand the signficiantce of this investment and take urgent action to increase investment, address demand and supply constraints in the health workforce, and implement structural reforms in health employment, education, and service delivery.

Reforming health service delivery and organization

In order to build resilient health systems, it is crucial for health systems worldwide to transition towards personalized long-term care and integrated, people-centered, community-based preventive health services, regardless of their progress toward universal health care. Achieving such health systems requires careful consideration of three key elements: (1) prioritizing health promotion and disease prevention, (2) improving the efficiency of HRH, and (3) strengthening linkages between health and social sectors.

Training and producing the next-generation health workforce

Aiming to train more than half a billion health workers for the health systems of the 21st century is indeed an ambitious goal. To achieve this, lifelong learning systems must move away their attention from narrow specialties to developing regionally relevant abilities that can effectively address the evolving health and social demands. Education should encompass not only technical knowledge but also emphasize multidisciplinary teamwork, ethical conduct, and other aspects of socially responsible practice. Furthermore, education and training need to be practice-oriented and tailored to the specific needs of the healthcare system. With rapid technological advancements shaping the healthcare landscape, healthcare professionals have been experiencing changes in the required skill profiles within their professions. Hence, there is a need for education and training institutions to adapt and provide opportunities for healthcare workers to continually update their skills and keep pace with these changes.

In the context of global public health emergencies, such as the COVID-19 pandemic, there is an increased demand for a larger and more skilled health workforce. This includes the need for epidemiologists who are equipped to address the unprecedented speed and widespread impact of such emergencies on the world's population. Compared to the Spanish flu outbreak in 1918, the landscape of public health readiness has drastically changed, primarily due to advancements in science and technology. Data scientists have played a crucial role in epidemic preparedness and management during the COVID-19 pandemic. Their expertise in analyzing and interpreting large datasets has enabled policymakers to make faster and more effective decisions based on evidence. Additionally, advancements in biotechnology, such as genomics,

transcriptomics, and microbiomics, have been instrumental in rapidly iden-
tifing the causal agent of the disease and obtaining the genomic sequence of
the coronavirus (SARS-CoV-2). The public availability of the whole genomic
sequence within a few days has facilitated the development of diagnostic tests,
enabling the health system to identify and isolate infected individuals, thereby
curbing the transmission of the disease. By investing in the skill development
and continuous education of the health workforce, we can enhance their capacity to respond effectively to public health crises. The integration of science
and technology advancements into healthcare practice and education will
strengthen the overall preparedness and response capabilities of the health
system, ultimately leading to better outcomes in managing and preventing
future public health emergencies.

India's introduction of Health and Wellness Centers (HWCs) is another
example of primary healthcare reform aimed at providing comprehensive
healthcare services [42]. This initiative involves converting primary healthcare
settings into HWCs, with each center being managed by a mid-level manager
trained in public health. To address the staffing needs of these HWCs, India
has implemented two strategies: offering certified training programs for
nursing or AYUSH doctors and the development of a new course called
"Bachelor's in Public Health", which was previously unavailable. These
measures aim to strengthen the primary healthcare workforce and enhance the
quality and accessibility of care provided at HWCs.

HRH retention

WHO has recommended several measures to address the availability of health
workers in remote and rural areas. These measures aim to attract, recruit, and
retain health workers in remote and rural areas. Selection of trainees from, and
delivery of training in, rural and underserved regions, financial and nonfi-
nancial incentives, and regulatory measures of service delivery restructuring
are all critical to ensuring fair deployment of health professionals.

Improved working environment

Several global and national organizations are dedicated to promoting decent
employment and a favourable working environment for all healthcare professionals. Key components of decent employment include respecting employees' rights, empowering them to influence working conditions through
meaningful interaction, and eliminating all forms of discrimination. Factors
such as low salaries, hazardous working conditions, ambiguous job responsibilities and working hours, lack of social protection and career development, and insufficient standard of living are common drivers of job
migration. Moreover, the patterns of health worker migration have grown
increasingly intricate, with significant movement occurring within regions. To

effectively understand and manage health worker mobility, maintaining an updated HRH database is essential. This database serves as a valuable tool for gaining insights into migration patterns, identifying areas of concern, and implementing targeted strategies to address workforce distribution and retention challenges. By prioritizing decent employment and establishing mechanisms to monitor and respond to health worker mobility, organizations and governments can create a supportive and sustainable environment for the health workforce. This, in turn, ensures the well-being of healthcare professionals and contributes to the provision of high-quality healthcare services to communities.

Equality and rights for women or marginalized group of HRH

The health sector has a higher proportion of women compared to any other industry. Women often play a crucial role as primary caregivers in humanitarian disasters and crises. However, gender bias persists in the payment for services, leading to women being paid less than their male counterparts, even when they possess comparable qualifications. Furthermore, healthcare workers, especially women, face significant threats such as physical, mental, and sexual harassment, as well as targeted attacks. These dangers compromise the safety and well-being of healthcare professionals. Despite their substantial presence in the health workforce, women are underrepresented in leadership and decision-making positions. This lack of representation undermines gender equality, inhibits full employment opportunities, and hinders inclusive economic development. Addressing these discriminatory practices is crucial for creating a more equitable and inclusive healthcare system that respects the rights and contributions of all healthcare professionals, regardless of gender or other social groups.

Accountability, data, and information

Assessing the labor market, economy, and population needs in relation to health is crucial for developing effective policies and making investments in the health workforce. Through labor market analysis, we can examines factors such as demand, supply, and the overall need for healthcare professionals. Having a clear understanding of the labor market dynamics helps policymakers and stakeholders comprehend the implications of educational and labor economy changes both in the present and future and enable them to make informed decisions regarding workforce planning, training programs, and resource allocation. Therefore it is essential to have reliable and up-to-date data to facilitate labor market analysis and conduct research that fills gaps in the evidence base.

Routine health facility and community reporting systems, health facility assessments, and health resource data, including health funding and workforce data, play a crucial role in developing comprehensive health service data. To enhance primary health care and UHC, these systems should be connected and interoperable to create a synergized approach to monitoring, analyzing, and administering health services. This integrated approach supports various aspects of healthcare, including patient care, facility management, and health sector planning. Knowledge of existing disparities is essential to ensure the equitable distribution of health workers within a country. Additionally, gender analysis and the collection of gender-disaggregated data are critical components of developing improved policies to address these issues.

Other HRH management strategies

The policy-level measures, investments, health system reforms, and addressing labor supply market issues are long-term interventions. However, the acute HRH shortage needs immediate attention and action. Task shifting or sharing, better incentives to lay health workers, and outsourcing through public–private partnerships (PPPs) might address such acute HRH shortages.

Task shifting and sharing

Task shifting or sharing refers to the delegation or redistribution of healthcare tasks between different HRH cadres or with the communities [43]. It is a short-term arrangement that may become a permanent practice. Task shifting involves assigning specific jobs from one HRH cadre to another, often to qualified or trained individuals who are of a lower cadre. For example, in India, auxiliary nurse midwives used to provide home-based postnatal care until two decades ago, but now this task is delegated to ASHAs and linked with incentives. Task sharing, on the other hand, increases the involvement of trained HRH within a specific category. In this case, the task is not entirely removed from one cadre but is shared between different cadres. For instance, during the COVID-19 pandemic, the C19 SPACE course trained non-ICU doctors and nurses to manage ICU care due to the high demand. Digital technologies can facilitate task shifting and sharing by enabling appropriate and timely decision-making.

Fixed and performance-based incentives to lay health workers

While lay health workers do receive incentives based on their performance, not all services are linked with incentives. For example, a lay health worker may receive incentives for providing three antenatal visits, but not for any other visits related to antenatal care. Therefore, it is important to quantify and appropriately incentivize the quality of service provided. In 2005, the Government of India introduced ASHAs [44]. Initially, their role focused on delivering selected maternal and child health services, but it has since expanded to include tuberculosis, malaria, noncommunicable diseases, and other programs. To account for the expanded scope and quality of their work, the government now provides fixed and varying (performance-based) incentives to ASHAs.

Other ad hoc HRH management strategies

PPP is one example of the ad hoc strategies for managing health workforce. It is typically established to deliver specific services that are not adequately provided by the existing HRH cadre within routine health systems. PPP can potentially be a win–win situation for the partnering entities. However, the long-term sustainability of service delivery under PPP arrangements may raise questions. Multiple countries have implemented various PPP service models [45]. In India, some functional services of PPP models in primary or secondary healthcare settings even extended to tuberculosis, kidney dialysis, and cardiac intervention.

Conclusion

Business as usual is not a viable option in the face of 21st-century health challenges, including demographic shifts, epidemiological changes, economic transitions, and the digital revolution. It is crucial to develop a health workforce that is dedicated to health promotion, disease prevention, and community-based, people-centered services. Even the wealthiest nations will face significant challenges if these issues are not effectively addressed, posing risks to social cohesion and prosperity. This is an opportunity to reevaluate traditional approaches and foster new forms of collaborative action. Countries and stakeholders must act boldly and make well-informed investments in the health workforce, aiming to improve population health and enhance preparedness for future crises.

References

[1] Kabene SM, Orchard C, Howard JM, Soriano MA, Leduc R. The importance of human resources management in health care: a global context. Hum Resour Health 2006;4:20. https://doi.org/10.1186/1478-4491-4-20.

[2] World Health Organization. World health report 2006: working together for health. Geneva; 2006.

[3] Tilson H, Gebbie KM. The public health workforce. Annu Rev Public Health 2004;25:341–56. https://doi.org/10.1146/annurev.publhealth.25.102802.124357.

[4] WHO Recommendations. Optimizing health worker roles to improve access to key maternal and newborn health interventions through task shifting. Geneva; 2012.

[5] World Health Organization. Health workforce requirements for universal health coverage and the sustainable development goals. Geneva; 2016.

[6] World Health Organization. Nursing and midwifery; 2022. https://www.who.int/health-topics/nursing#tab=tab_1. (Accessed 1 December 2022).

[7] Thompson JB, Fullerton JT, Sawyer AJ. The international confederation of midwives: global standards for midwifery education (2010) with companion guidelines. Midwifery 2011;27:409−16. https://doi.org/10.1016/J.MIDW.2011.04.001.

[8] United Nations Population Fund. The state of the world's midwifery 2021; 2021. https://www.unfpa.org/sowmy. (Accessed 1 December 2022).

[9] Lewin S, Dick J, Pond P, Zwarenstein M, Aja GN, van Wyk BE, et al. Lay health workers in primary and community health care. Cochrane Database Syst Rev 2005. https://doi.org/10.1002/14651858.CD004015.PUB2.

[10] Rural Health Information Hub. Roles of community health workers; 2022. https://www.ruralhealthinfo.org/toolkits/community-health-workers/1/roles. (Accessed 1 December 2022)

[11] Woldie M, Feyissa GT, Admasu B, Hassen K, Mitchell K, Mayhew S, et al. Community health volunteers could help improve access to and use of essential health services by communities in LMICs: an umbrella review. Health Policy Plan 2018;33:1128−43. https://doi.org/10.1093/heapol/czy094.

[12] Dovlo D. Using mid-level cadres as substitutes for internationally mobile health professionals in Africa. A desk review. Hum Resour Health 2004;2:7. https://doi.org/10.1186/1478-4491-2-7.

[13] Bhatt VR, Giri S, Koirala S. Health workforce shortage: a global crisis. 2008.

[14] World Health Organization. Global strategy on human resources for health: workforce 2030. Geneva; 2016.

[15] UN General Assembly. Transforming our world: the 2030 agenda for sustainable development. 2015.

[16] World Health Organization. World health statistics 2019: monitoring health for the SDGs, sustainable development goals. Geneva; 2019.

[17] World Health Organization. Global strategy on human resources for health: workforce 2030. WHO; 2016. p. 64.

[18] World Health Organization. The world health report 2006: working together for health. 2006. https://apps.who.int/iris/handle/10665/43432. [Accessed 1 December 2022].

[19] Dubey S, Vasa J, Zadey S. Do health policies address the availability, accessibility, acceptability, and quality of human resources for health? Analysis over three decades of National Health Policy of India. Hum Resour Health 2021;19:1−19. https://doi.org/10.1186/S12960-021-00681-1/TABLES/4.

[20] CapacityPlus. Strengthening the health workforce for improved services: results and lessons learned from CapacityPlus 2009−2015. 2016.

[21] Ayanore MA, Amuna N, Aviisah M, Awolu A, Kipo-Sunyehzi DD, Mogre V, et al. Towards resilient health systems in sub-Saharan Africa: a systematic review of the English language literature on health workforce, surveillance, and health governance issues for health systems strengthening. Ann Glob Health 2019;85. https://doi.org/10.5334/aogh.2514.

[22] United Nations. Population facts department of economic and social affairs population division (2010). New York; 2010.

[23] Dussault G, Badr E, Haroen H, Mapunda M, Mars AST, Pritasari K, et al. Follow-up on commitments at the third global forum on human resources for health: Indonesia, Sudan, Tanzania. Hum Resour Health 2016;14:16. https://doi.org/10.1186/s12960-016-0112-0.

[24] People that Deliver (PtD). Competency compendium for health supply chain management: a reference for health supply chains. Bruce; 2014.

[25] World Health Organization. Global competency framework for universal health coverage. Geneva; 2022.

[26] Gogia S, Sachdev HPS. Home-based neonatal care by community health workers for preventing mortality in neonates in low-and middle-income countries: a systematic review. J Perinatol 2016;36:S55−73. https://doi.org/10.1038/jp.2016.33.

[27] Ryman TK, Dietz V, Cairns KL. Too little but not too late: results of a literature review to improve routine immunization programs in developing countries. BMC Health Serv Res 2008;8:1−11. https://doi.org/10.1186/1472-6963-8-134.

[28] Karumbi J, Garner P. Directly observed therapy for treating tuberculosis. Cochrane Database Syst Rev 2015. https://doi.org/10.1002/14651858.CD003343.pub4.

[29] Mdege ND, Chindove S, Ali S. The effectiveness and cost implications of task-shifting in the delivery of antiretroviral therapy to HIV-infected patients: a systematic review. Health Policy Plan 2013;28:223−36. https://doi.org/10.1093/heapol/czs058.

[30] Paintain LS, Willey B, Kedenge S, Sharkey A, Kim J, Buj V, et al. Community health workers and stand-alone or integrated case management of malaria: a systematic literature review. Am J Trop Med Hyg 2014;91:461. https://doi.org/10.4269/ajtmh.14-0094.

[31] Durham J, Pavignani E, Beesley M, Hill PS. Human resources for health in six healthcare arenas under stress: a qualitative study. Hum Resour Health 2015;13:14. https://doi.org/10.1186/s12960-015-0005-7.

[32] Russo G, Pavignani E, Guerreiro CS, Neves C. Can we halt health workforce deterioration in failed states? Insights from Guinea-Bissau on the nature, persistence and evolution of its HRH crisis. Hum Resour Health 2017;15:12. https://doi.org/10.1186/s12960-017-0189-0.

[33] Witter S, Bertone MP, Chirwa Y, Namakula J, So S, Wurie HR. Evolution of policies on human resources for health: opportunities and constraints in four post-conflict and post-crisis settings. Confl Health 2017;10:31. https://doi.org/10.1186/s13031-016-0099-0.

[34] de Barros M, Gomes PG, Correia D. Les conséquences du narcotrafic sur un État fragile: le cas de la Guinée-Bissau. Altern Sud 2013;20:6−8.

[35] World Health Organization. Impact of COVID-19 on human resources for health and policy response: the case of Plurinational State of Bolivia, Chile, Colombia, Ecuador and Peru. Overview of findings from five Latin American countries. Geneva; 2021.

[36] Azizi MR, Atlasi R, Ziapour A, Abbas J, Naemi R. Innovative human resource management strategies during the COVID-19 pandemic: a systematic narrative review approach. Heliyon 2021;7. https://doi.org/10.1016/j.heliyon.2021.e07233. e07233−e07233.

[37] University of Michigan Center of Excellence in Public Health Workforce Studies. Public health workforce enumeration, 2012. Ann Arbor MI; 2013.

[38] Alfonso YN, Leider JP, Resnick B, McCullough JM, Bishai D. US public health neglected: flat or declining spending left states ill equipped to respond to COVID-19. Health Aff 2021;40:664−71. https://doi.org/10.1377/hlthaff.2020.01084.

[39] World Health Organization. Everybody's business—strengthening health systems to improve health outcomes: WHO's framework for action. Geneva; 2007.

[40] Campbell J, Buchan J, Cometto G, David B, Dussault G, Fogstad H, et al. Human resources for health and universal health coverage: fostering equity and effective coverage. Bull World Health Organ 2013;91:853−63. https://doi.org/10.2471/BLT.13.118729.

[41] Cometto G, Campbell J. Investing in human resources for health: beyond health outcomes. Hum Resour Health 2016;14:51. https://doi.org/10.1186/s12960-016-0147-2.

[42] Ministry of Health and Family Welfare (Government of India). Ayushman Bharat-Health and Wellness centers; n.d. http://ab-hwc.nhp.gov.in/. (Accessed 13 December 2022).

[43] Orkin AM, Rao S, Venugopal J, Kithulegoda N, Wegier P, Ritchie SD, et al. Conceptual framework for task shifting and task sharing: an international Delphi study. Hum Resour Health 2021;19:1−8. https://doi.org/10.1186/S12960-021-00605-Z/FIGURES/2.

[44] World Health Organization. Asha workers win global health leaders award; 2022. https://
       www.who.int/india/india-asha-workers. (Accessed 1 December 2022).
[45] The World Bank. About PPPLRC and PPPs | public private partnership; 2022. https://ppp.
       worldbank.org/public-private-partnership/about-us/about-public-private-partnerships.
       (Accessed 1 December 2022).

Chapter 14

Effective use of information technology for the quality of public health practice

Palanivel Chinnakali[1] and Swetha S. Kumar[1,2]

[1]*Department of Preventive and Social Medicine, Jawaharlal Institute of Postgraduate Medical Education & Research (JIPMER), Puducherry, India;* [2]*Department of Research, SingHealth Polyclinics, Singapore*

Introduction

Information technology has been of use in all industries and the medical field is no exception to it. It has become an important tool that forms a link between the healthcare system and consumers.

Health informatics refers to the use of principles and applications from computer and information sciences to the benefit of achieving improvement in health status and healthcare delivery. The concept of using information technology in medical practice can be traced back to the early 1950s, when early experiments led to the invention of computed tomography scans. However, the first practical use of modern-day health informatics began with exploring ways to improve the maintenance of medical records in healthcare facilities. This led to the development of electronic health records (EHRs) for maintaining patient details in hospitals and clinics. EHRs are being widely used in developed countries; however, this system has not gained much popularity in most low- and middle-income countries. In India, although EHR has been of extensive use in educational institutions and private/corporate hospitals, it is still not used to the same extent at the primary healthcare level.

Integration of health and information technology

With the advancements in technology and immense growth in science, the world has been continuously evolving and expanding in all sectors. Technological advancements have made work easy, efficient, and less time-consuming. Integration of information technology into the healthcare

Principles and Application of Evidence-Based Public Health Practice
https://doi.org/10.1016/B978-0-323-95356-6.00014-8

industry is a boon in many ways and it has a wide spectrum of use in the health sector. It has made it easy for health workers to maintain and manage patient or population information/profiles and provide quality care at reasonable prices to even people residing in inaccessible areas. This integration allows for data to be stored and maintained for an indefinite period to provide continuity of care. Compared to conventional healthcare management, this has proved to be much better in delivering high-quality performance. Thus, health information technology plays a vital role in today's health system with its uses extending from preventive education to rehabilitation. It allows healthcare providers to closely monitor and remotely control patients' health. Apart from helping in providing an accurate diagnosis, it also allows for predicting prognosis and reduces the errors caused by humans in all these procedures [1].

Importance of health information technology in public health

Public health practice area broadly includes services related to preventive, promotive, curative, and rehabilitative aspects of the health/disease across various settings. It also includes the policy-level decisions and implementation of the same. Information technology has been used in delivering all public health functions.

The major advantage of health information technology is that it allows healthcare providers to deliver quality healthcare to patients safely and securely at individual and community level. The quality improvement is from the accurate and complete data provided by this technology. The other merits are that it makes the care pathway more accessible and easier to use. It allows for service provision to all individuals including those who are inaccessible. Once the data are uploaded, they remain available for use by all cadres of healthcare providers and there is no need for duplication of data. Using the existing data, the occurrence of communicable diseases can be studied, predicted, and preventive measures can be planned using these advancements. It also plays a role in controlling and preventing chronic non-communicable conditions by promoting health among people. The vast uses mentioned below will highlight the importance that health information technology has in this era and justify its tremendous application in public health practice.

Various available information technologies

Electronic medical record/electronic health record

Electronic medical record is an electronic version of the patients' diagnosis and treatment data, which is usually confined to a healthcare provider or setting. However, EHR refers to the maintenance of the complete details of a patient in a database that can be referred to at any time to avail the medical data across healthcare setting and beyond diagnosis and treatment. The information stored corresponds to the patient's disease signs and symptoms, diagnosis, laboratory results, and medical treatments including the follow-ups. EHRs are extremely useful in the present world, especially for public health and infectious disease surveillance. It helps in maintaining and tracking a database of notifiable diseases and its reporting on a longitudinal aspect. It also allows for the analysis of important prognostic and health outcomes that can be used by the healthcare industry to plan and manage its infrastructure and investments [2].

Developed countries have already adopted a robust EHR linking with the citizen's social security number or other identifiers, and it is an indirect mandate to check insurance claims. In India, a circular was issued by the ministry in the year 2016 to create and maintain EHRs for all healthcare providers to maintain uniformity. The standard guideline for the EHRs included linking them with the "Unique Identification Number (Aadhaar)" for easy access and integration. The details required in each health record include patient identifiers, images of scans and X-rays, discharge/treatment summary, e-prescription, and digital certificates. Also, there is a provision to collect patients' mobile numbers and e-mail IDs and send their health information over mobile devices to make it easier for them to access their records. As this is still in the transition stage, many hospitals and clinics maintain patient information in a hybrid format by collecting details on a case sheet and then entering the data into the hospital information system by a data entry operator.

Advantages of EHR:

- Easy to share and handle data through multiple modes
- Can be easily spread across different registries and platforms making it accessible to all at any given time
- Patient's complete health information is available in one place and can be accessed from anywhere by the authorized person
- Electronic data saved in a standard format so communication between health providers or facilities is effortless
- Time-saving
- Cost-effective, sustainable, and eco-friendly

Challenges in EHR:

- Establishing a safe and secure technology-based environment
- Completeness and timeliness of reporting
- Accuracy in the coding of data
- Availability of dedicated staff for maintaining the electronic data
- Acceptance of the new system by the healthcare team that may be comfortable with the existing paper-based data collection

- Lack of knowledge and expertise among healthcare providers about the reporting formats which necessitates conducting periodic training for the staff.
- In places where hybrid formats are followed, inconsistency in updating or discrepancy in data with respect to the information in paper and electronic format.
- Presence of multiple platforms and interoperability of various platforms to integrate a persons' data

Telemedicine

Telemedicine is defined by the World Health Organization (WHO) as "the delivery of healthcare services, where distance is a critical factor, by all healthcare professionals using information and communication technologies to exchange the valid information for the diagnosis, treatment, and prevention of disease and injuries, research, and evaluation and for the continuing education of healthcare providers, all in the interests of advancing the health of individuals and their communities."

Telemedicine refers to the use of information technology in the field of medicine to ensure remote delivery of healthcare services even when the patient and healthcare provider are separated by distance [3]. This technology was earlier considered experimental and futuristic; however, it is the most widely used information service in the present world, especially after the COVID-19 pandemic. It has wide use in healthcare delivery and can be extensively used for providing patient education and in administrative processes.

Why telemedicine?

The goal of public health is to make health care accessible to all beneficiaries including people living in remote areas and rural villages. This population usually finds it difficult to acquire timely and quality care due to various factors like lack of resources, inadequate manpower to serve in rural areas, and nonavailability of good transportation facilities. This inadvertent problem can be overcome by using telemedicine as it helps in bridging this issue of distance and enables providing quality health care remotely [4,5].

In low- and middle-income countries, universal health coverage remains a far dream as the health system fails to provide adequate primary care in remote rural areas. Even in urban and suburban areas, there is a disparity in the availability of healthcare services. However, the development of information technology is in contrast to this and people from even the remotest of villages have internet access and have basic knowledge about computers. Thus, the healthcare system is gradually looking at telemedicine as a vital solution to providing medical care. It is considered easy to establish speciality medical services at all subcenters and primary health centers using telemedicine. The future of telemedicine is prospective and there is wide scope for satellite-based technology and fiber-optic systems to take over [6].

Uses of telemedicine

Telemedicine use is linked with the current generation of EHRs, which is useful to improve the routine public health practice. In addition to providing diagnostic, treatment, and follow-up services to patients, telemedicine can be used in other functions of public health.

(i) Epidemiological surveillance:
   Telemedicine is an important tool in the real-time monitoring of diseases, locally and globally. It can be used to deduce information on the geographic distribution and prevalence of diseases. It also helps to assess the population's health in a particular place by providing information and differentiating their levels of risk. Based on the inferences, planning of interventions can be done and preventive measures can be implemented for the anticipated epidemics.

(ii) Geographic Information System (GIS):
   GIS is the basic tool integrated with telemedicine that helps in identifying disease burden and transmission within and outside a community. It helps in understanding the spread of any communicable diseases. It differs from the conventional methods of surveillance and reporting and aids in integrating data from multiple sources, thus making it easy to conclude with GIS mapping and remote sensing. The availability of healthcare services and navigation to a public health facility are mapped for some specific services or all services. This information helps the population to identify the healthcare facility as per their need.

(iii) Health promotion and disease prevention:
   - Nowadays people are informed, educated, and motivated about health-related issues and healthy habits through telemedicine. This helps keep the community motivated to adopt a healthy lifestyle. This is less time-consuming, cost-effective, as well as reliable because the information is spread throughout the community including those living in remote areas.
   - It gives people broad choices as they are well informed about a disease condition and all available treatment options. This helps to simplify the decision-making process.
   - Usage of online mobile applications and sharing of information over the internet improves information exchange at the population level and enables the community to be self-reliant in understanding the needs of fellow beings thereby providing emotional support to them.

(iv) Monitoring and evaluation:
Telemedicine remains an important tool for the monitoring and evaluation of healthcare services. It also helps in monitoring patients receiving home care, ambulatory care, and critical care.

Challenges in using telemedicine

- User friendly?—Medical practitioners, hospital staff, elderly, and people with low literacy find it difficult to cope with the use of information technology [7]. Doctors and hospital staff are usually reluctant to spend time familiarizing themselves with telemedicine. Patients lack confidence in this approach as the traditional concept of visiting and getting consultations from doctors is bypassed which still provides greater satisfaction to the patient.
- Less comprehensive than conventional care—The complete clinical care, especially the clinical examination of the patients, is not possible with telemedicine which may be addressed in future with advanced technology.
- Cost—The initial cost to establish the traditional telemedicine system is quite high and thus it may not be easily affordable. However, the recent mobile phone-based telemedicine is quite cheap and affordable.
- Requires too much diversity—The literacy rate in India is nearly 75% (Census, 2011), but only a small percentage of people are fluent in English [9]. Therefore, with diverse languages across the country, the telemedicine facility needs to be introduced in at least all major languages making it complicated.
- Technical constraints—Telemedicine is an evolving field where the predictions and diagnosis of diseases keep changing with time and upcoming technology. To keep pace with the current trends requires a huge understanding of technologies and this may not be technically feasible for any low resource setting for introducing newer methods quite often.
- Lack of guidance—There is a lack of a proper governing body to formulate guidelines and work toward implementing the e-services in a unified manner.

Mobile-based applications, chatbots

The traditional practice of delivering healthcare services involves face-to-face interaction between the service providers and the users. However, with the advent of mobile phones and the internet, app-based service provisions have been on the rise [10]. Mobile applications are considered to be highly accessible, efficient, and cost-effective. They have wide use in the medical field in terms of patient education, self-examination and reporting, monitoring, and evaluation, and in providing feedback [11–13]. The use of mobile app programs in promoting health by reducing drinking, smoking, stress, depression, and anxiety and improving physical activity, weight loss, and dietary habits is increasingly popular [14].

In India, due to the larger population size and the lack of adequate healthcare personnel to cater to the entire population, there is a need for technological advancements like mobile-based applications (mHealth) to bridge the demand–supply gap. These mobile applications are being developed for all concerns at present. Private hospitals have personalized apps that enable users to book appointments, get prescriptions, and have interactive preventive therapies over multiple sessions. An example of this is the development of the Aarogya Setu app by the Government of India which helped provide COVID-19 care.

Dedicated mobile applications are in use for preventing mortality due to acute fatal events like myocardial infarction and stroke. For example, in stroke management, the application is helping the treatment team for triage, navigation to nearby imaging facility, early diagnosis, and management even during transport of a stroke patient. The application also alerts various members of the multidisciplinary team to be prepared in receiving a new patient as a potential case is identified via ambulance or call center [15].

Another useful and effective tool in these apps are *chatbots*. They can communicate with users in natural language mimicking a customer care executive. Users get the necessary information by chatting with these chatbots at their convenience. This facilitates knowledge sharing with patients and reduces the workload for clinicians. The chatbots are generally built on artificial intelligence (AI) and deep learning principles using or expecting the frequently observed group of problems. OneRemission1 (used in helping cancer survivors), Babylon Health (a symptom checker), and Wysa (used to interact with users to ease anxiety and depression) [16] are globally known examples. In India, various private laboratories have their applications and chatbots to help in service delivery. The Government of India introduced a WhatsApp chatbot to help in COVID-19 tracking and management.

Wearables and sensors

Wearable technologies are devices that aid in the continuous monitoring of day-to-day human activities and behavior. It also measures vital signs like body temperature, heart rate, blood oxygen saturation, physical activity, blood pressure, sleep pattern, blood sugar, and others. Examples of wearables include fitness bands, smartwatches, eyeglasses, knee caps, gloves, and other devices attachable to skin. They are embedded with sensors that transmit the information to a server via smartphone applications. This technology is an innovative solution for healthcare problems as it can be used in the prevention and management of diseases, promoting health, mental status monitoring, predicting or alarming of an acute event, and patient rehabilitation. It is considered to improve the quality of healthcare delivery while considerably reducing the frequent follow-up visits and cost of care for the patients. Integrated with other technology, it is used in practicing precision medicine for better health outcomes among patients.

A major use of wearable devices is in sports medicine where it helps athletes and coaches periodically manage their training. The advantages of these devices are that they give a regular update directly to the patients regarding their health and well-being and necessitate action. People tend to become more health conscious and are inclined toward maintaining a healthy lifestyle with the use of this technology. However, one concern with using wearable devices is the acceptance and interest of older adults as they may find it technologically challenging. Simpler devices that require minimal user manual instructions and create awareness of their usage will help in improved acceptance among users. Another area of concern is the management of data confidentiality and security which requires complex and vigilant encryption to maintain data safety [17].

Software related to health

In addition to EHR and telemedicine softwares, there are other software used routinely for efficient functioning of the health systems and delivery of healthcare services to population. The following are some of the software that are used stand-alone or integrated with other software.

Hospital management software

These help in the effective management and administration of the hospital and clinics. It includes maintaining patient registries, scheduling patient appointments, managing recalls or follow-ups, providing e-services, billing of receipts, financial management, staff monitoring, and inventory maintenance.

Medical image analysis

The best possible use of technology in the field of medicine was the development of imaging and visualization software. These work on machine learning principles and help the healthcare workers who work with enormous data by providing them with accurate findings. Apart from this, other applications include digital X-rays, 3D modeling of human anatomy, implant guiding software, and Invisalign.

Medical diagnosis software

These help doctors save time by providing a very accurate diagnosis. These are preferred as the diagnosis is based on the information provided by multiple experts from different fields. This is based on the principles of AI technology.

Remote patient monitoring software

This system enables patients to access treatment remotely at their convenience. Wearables and sensors are the preferred monitoring devices. It helps in collecting patient information remotely [18]. Patients are monitored by healthcare professionals using this software.

Health tracking apps

They are usually used in combination with wearable devices for monitoring and maintaining diet patterns, physical fitness, and the overall well-being of the users.

Artificial intelligence, machine learning, and deep learning

The technology of AI utilizes machine learning principles to simulate human reasoning and decision-making. It is based on the principle of learning by observing patterns in massive data and using the learned knowledge to make accurate predictions. Machine learning is a subset of AI which learns from the data on its own and improves the efficiency without much programming. Deep learning is a specialized machine learning which creates and uses artificial neural networks and algorithms like human brain cell functioning. The diagnosis made through AI using machine or deep learning is standardized and more accurate. It helps in reducing the workload of hospital staff, saving time, predicting prognosis, and improving efficiency.

Apart from diagnosis, AI is also used in clinical decision-making, performing surgical procedures, and wearable devices. AI-based robots like the Da Vinci surgical robot can perform complex procedures and click images through a 3D vision system [19]. Wearables generate huge data and machine learning principles are employed to read and segregate these data to make useful predictions and inferences. Further, the blockchain technology, a digital ledger, can be potentially used in EHR, inventory management, health insurance, and other areas of health care [20].

Virtual reality

Virtual reality (VR) gained popularity in the gaming industry before being considered for use in all other sectors. After exhaustive research, it has made an entry into the healthcare industry as well. DeepStream VR, Medical Realities, Google VR, Immersive Touch, and Virtually Better are some of the leading VR companies working in the healthcare industry. Some common applications of VR include the following:

- Training and public health education: VR is made use of by creating 3D interactive sessions that allow students to practice public health skills without involving living subjects. Training is given in a virtual environment that closely mimics reality.

- Diagnostics: VR facilitates the diagnosis especially in various mental health disorders (phobias). It is noninvasive and can also be combined with other techniques to establish a definitive diagnosis.
- Prevention and Treatment: VR has been used in the prevention of occupational injuries among farmers [21]. It is extensively used in pain management and in distractive therapy to engage patients with interactive games. It is also used to treat burn victims. It is considered an effective alternative to medications as they are less invasive and do not require any clinical setup. It offers relaxation and eases stress, thereby making it a good choice to treat mental illnesses, especially to deliver the cognitive behavior therapy.
- Another major use of VR for treatment purposes is its use in surgeries. Surgeries are supervised by physicians and performed by robotic devices. This is considered safer due to lesser complications as the robotic device is more accurate and the procedure is usually fast. Tele-surgeries (doctors performing surgeries on patients from a different location) are usually performed with the help of VR and are also gaining popularity.
- Physical fitness: VR is used as a substitute for drugs or surgeries and combined with cardio routines to enhance the workout experience of users.

Unmanned aerial vehicles (drones)

The use of drones in health care is getting more important especially after COVID-19. It has shown promising results in medical emergencies or disaster management, transport organs for transplantations, transport of various human samples for investigation, supply of essentials including medication, surveillance of difficult to reach areas and others [22].

Role of information technology in public health

Policymaking, implementation, and monitoring

The advantages of digital health have naturally paved the way for the government to move toward mobile governance systems. M-governance includes all strategies and methods using wireless and digital technology to perform policy making, management, and monitoring to benefit citizens, businesses, and all government units [23].

Both governance and policymaking are closely dependent on quality data which are made available using information technology in health. The data generated from the health sector through the health information system are analyzed, the gaps are identified, and solutions are compared before implementation and review. This forms the basis of public health policy. Apart from the hospital management and information system portal, valid data sources include the census, disease registers, epidemiological surveillance data, population surveys, and national health programs.

The integrated health information platform was introduced in India to maintain a uniform system of EHRs that includes the health records of patients or population at all healthcare facilities or community. It unifies data from existing electronic health systems and would be the single place to access all records of an individual. It will not only help in storing patients' health records but also analyze the data and graphically represent them for easy understanding to aid in governance [24]. Recently, National Digital Health Mission was launched which will create unique identification numbers for citizens, healthcare providers, and hospitals. The digital environment is getting tested to capture and create a comprehensive EHR in the country.

Service delivery

Preventive or promotive: Population level (community engagement)

Information and communication technology plays a vital role in health education and health promotion. Currently, the internet has been the primary source of health information in the majority of households. However, there is a disparity between the urban and rural populations as the urban areas have better internet access and digital literacy. Thus, providing health education through the internet will be beneficial in urban areas, while community engagement would be a better means of health education in rural areas [25]. Involving the community in providing health care and health information by making use of information technology has excellent reach to educate the rural communities about health. Identification of community leaders is the first step in this. Once a leader is identified, a plan needs to be formulated to effectively utilize technology and the information accessed by the people.

Curative: Including clinical decision-making algorithms and the use of technology in improving the quality of health care

The introduction of the hospital management and information system has been extremely beneficial in improving the quality of care provided by hospitals. Another important milestone in providing quality healthcare has been the introduction of the health management information system (HMIS). This is different from the hospital management and information system in that the former is concerned with the development and management of health programs, whereas the latter deals with healthcare delivery.

An important technology was developed that helped clinicians to decide on the type of management a patient required. This was known as the clinical decision support system. It is a computer-based information system that analyzes patient information and helps in decision making for managing their condition [26]. The availability of patient records in a unified platform helped all healthcare providers involved in patient care to access the records and support decision-making. The complex algorithms could weigh down the pros and cons of the different management techniques that were available based on the patient's health condition and choose the best possible solution that would provide maximum benefit to the patient. This technology is facilitating all the spectrum of clinical management of a patient, from acute to long-term care, outpatient to inpatient care, primary care, to tertiary care including reporting adverse drug reactions and management. Telemedicine, integrated or stand-alone, is facilitating the follow-up after discharge from the hospital and linking with community-based services.

Rehabilitative

Rehabilitation aims to improve the patient's quality of life by helping them achieve maximum cognitive and physical functions thereby making them lead an independent life. Importantly, the necessary assistive devices are designed as per the patients' need for better functionality. Recent technologies provide enormous musculoskeletal care and help in improving patient care at reduced costs [27]. Patient compliance is a major issue in the case of physiotherapy, and this could be handled using wearables that remind the patients about their status, give feedback on their condition, and push them to seek care if needed. Also, people with disabilities have the option of interacting over the internet to connect with similar people and provide support by sharing success stories and addressing queries [28].

IEC/behavior change communication

Information, education, and communication (IEC) involve the sharing of information through electronic or print media to impart knowledge to the audience about appropriate behavior to prevent diseases and promote health. IEC and behavior change communication (BCC) are proven tools for social change and development, as they work at the community level to identify their needs and create awareness to promote health. The advantage of IT in providing IEC/BCC is that it reaches a large number of people in less time, provides feedback, and is supported by government or NGOs as it is cost-effective and target specific [29]. It reaches out to the general public as it is conveyed as entertainment which catches the public's attention. Awareness regarding infectious diseases is spread through radio, television, and other media. During the COVID-19 pandemic, information regarding mask usage, social distancing, and vaccination was spread via phone caller tunes that reached and had an impact on the entire nation. Community health workers circulate information in the rural areas regarding hygiene and sanitation, nutrition, and other available health programs by showing videos or handing out IEC posters/pamphlets.

Prediction of emerging and reemerging diseases/outbreaks

Disease surveillance and monitoring have been advanced in recent times due to the technological advancements worldwide. It includes analyzing huge data and communicating it to the stakeholders. Studying disease dynamics and predicting epidemics have helped combat infectious disease outbreaks including the recent COVID-19 pandemic. Predictive modeling helps in identifying hotspots and changing trends in disease patterns; whereas computational modeling helps in understanding the disease progression. These techniques require large data, and this is provided with the help of the national data registries. Based on these predictions, the government issues control measures to prevent further spread and manage the affected individuals. Modeling has been effective for various diseases like Ebola, COVID-19, HIV, influenza, cholera, and West Nile virus. In the case of outbreaks, forecasting is an emerging analytical tool that helps in epidemic management. This was largely used during the Ebola outbreak [30].

Disaster management

The use of social media in disaster management has been beneficial in improving situational awareness in recent times. Examples include the Kerala (a south Indian state) floods in 2018 and 2020 that gained wide popularity through the media and sought immediate response and relief from the government. Disaster management involves the traditional four stages (risk reduction, preparedness, response, and recovery).

Risk reduction

Buildings are currently embedded with monitoring technology and sensor network systems to help local authorities be aware of and reduce disaster risk. These sensors detect even minor earth movements and water-level rises in buildings built over waterbodies. This helps in knowing the occurrence of earthquakes and floods. Satellite imaging and use of AI technologies help in analyzing stored data and creating alerts [31].

Preparedness

The community is well informed about the vulnerability and risk reduction through IEC. Digital technology helps create life-like events that can be used for field exercises to train the community in disaster preparedness. VR training systems are very helpful in this aspect. Such training helps people to understand what disasters do and they learn evacuation and management techniques.

Response

Electronic media helps in the faster spread of awareness and information. It also helps in the decision-making process. The victims have access to the internet and tend to seek help through social media. The Kerala floods are a very good example of how huge numbers of volunteers got together to help a community fight against a disaster with the help of social media.

Recovery

Recovery requires the local authorities and government to ensure returning to normal life. Information technology plays a vital role in carrying out relief work by managing the distribution of relief goods and coordinating manpower and available resources. The software can be used for loading this information and tracking the relief funds and other resources. Certain huge disasters require mobilization of the victims, and in such cases, they must be provided with certificates for authentication. Developing an information system that can process the certificate and link these data with their health records would be extremely useful in the future.

Capacity building/teaching/training

The success of any healthcare research largely depends on the availability of health information and medical knowledge. Health information technology is gifted with versatile features that provide information related to the demographic, clinical, administrative, and health status of the population. EHR and HMIS portals provide easy access to patient information that can be used for teaching and research purposes [26]. The application of information technology in medical education is abundant. Newer technology has immense scope in the field of public health; it helps learning much easier and more interesting. Digital platforms help in the transfer of educative material and enable the training of students remotely.

Health information technology in the times of COVID-19 pandemic

The COVID-19 pandemic resulted in imposing restrictions that were new to the entire world. The concept of social distancing, contactless communication, contact tracing, and lockdowns was strict in the initial days and was hard to follow. With the exorbitant rise in cases across the world and given the severity of the disease, it would have been even more difficult to handle other illnesses and ailments had it not been for teleconsultations. Identifying, reporting, and tracking COVID-19 cases were all efficiently managed with the help of health information technology. The COVID-19 pandemic indirectly expedited the use of IT in health care to a maximum extent in a short span of time. The role of IT in the times of the pandemic is discussed below.

Screening and case identification

Although the COVID-19 infection spread at a rapid pace throughout the world, the detection was insufficient and there was underreporting of cases. It was essential to detect and report all cases to curb the further spread and also to reduce the infectivity and mortality rates. Predictive machine learning models eased these problems and helped in epidemiological surveillance. The international and national COVID-19 infection tracking portals gave a real-time update on the number of cases to alert the population and to help the healthcare providers in case management. An epidemic trend monitoring was performed throughout to know the extent of infection and its fatality. Several mobile applications and software were created that captured symptom details and helped in self-monitoring among affected individuals. Workplaces and hospitals maintained daily updates of health information including body temperature readings, oxygen saturation levels, and tracing of contacts. These big data could be employed to determine whether a person was to be considered ill and required COVID-19 testing to rule out their potential of spreading the infection and prevent mortality. The pandemic even made way for the usage of infrared human body thermal imaging for the measurement of body temperature as using the conventional forehead temperature measurement gun seemed difficult and time-consuming in crowded public places.

Contact tracing

Patients infected with the disease were expected to be infectious for about 14 days and thus any contact of COVID-19 positive patients was identified and monitored for symptoms. In case of a huge number of people being contacts, it is impossible to manually keep track of everyone and monitor their symptom development. To overcome these problems, an instrument named the "close contact measuring instrument" was developed that helped in the prevention and control of infection. This tool helped in the quick and accurate identification of close contacts. Also, data mining methods like network analysis, association analysis, and knowledge graphs were utilized to determine the contacts of confirmed cases. This helped in identifying the potential cases and in curbing the spread of infection.

The subsequent problem was in categorizing the contacts into high- and low-risk contacts, for which big data analysis and digital tracking helped the healthcare workers. The digital platform stored details of passengers on public transportation and these data would be retrieved and utilized when a suspected or confirmed passenger used the public transport system. The location features and mobile phone data also played crucial roles in knowing and tracking the whereabouts of people. An application by the name of "contact tracing" was launched by Google and Apple that allowed public health agencies in tracking the contact of confirmed cases [19]. This worked on the principle of proximity via Bluetooth where the app, when installed on the user's phone, could accurately identify other mobile users who were in close contact and inform them about this.

Drug and vaccine development

Large-scale data analysis and research work had to be performed to formulate vaccines and drugs to act against the infection. Researchers predominantly used AI, cloud computing, and machine learning technologies for this purpose. AI algorithms were employed to shorten the genetic analysis and accurately detect virus mutations. The virus per se was a new one to researchers and it had to be studied and understood properly to develop drugs that could work against it. It was necessary to conduct gene sequencing and study the protein structure of the virus. Advanced digital technologies and supercomputers had to be used for this purpose.

COVID-19 and India

The Government of India launched the CoWin portal and the Arogya Setu app to help with the detection and monitoring of COVID-19 cases, booking the appointment for vaccination, reporting adverse events following vaccination, and generating and using the vaccination certificate. The Indian Council of Medical Research developed a vaccine in association with Bharat BioTech called "Covaxin" that was pronounced safe for public use after clinical trials. Digital technology played a major role in all these advancements. Hospitals used remote testing and treating facilities and the whole nation saw a digital transformation with the introduction of online classes, work-from-home options, teleconsultations, and online medicine delivery. India's fight against COVID-19 pandemic could not have been possible without the integration of information and digital technology into public health.

Health technology assessment and evidence-based adoption of information technology in public health practice

Health technology assessment (HTA) assesses the values and uses of a technology in health care through systematic assessment of the properties and consequences of the technology. In more than half of the WHO member countries, HTA is neither conducted nor the results are considered while healthcare decisions are made. However, its use will help to optimize the resource utilization and improve the efficiency of the health systems. Research is needed in this field to know evidence-based health information technology outcomes. The establishment of a standard framework for assessing the results of these would be recommended [32].

Issues and way forward in using health information technology

Although there are numerous benefits of health information technology, there are also some associated challenges:

- Lack of manpower or their capacity to effectively work on digital health platforms. Existing staff find it difficult and laborious to learn digital technology and employ it on a day-to-day basis. Although digital health is not complex, there is no adequate manpower and their skills in rural settings to carry out technological improvements.
- Big data analysis requires a huge amount of data samples to make accurate predictions. However, insufficient, inaccurate, and incomplete data hamper the accuracy and result in biased predictions which may adversely impact on the future patients' health. This insufficiency of data is referred to as data lags. This was mainly the reason why initial predictions about the COVID-19 pandemic could not effectively evaluate its course. However, it is a dynamic and quick process and learning from the previous errors are quick to rectify the future errors.
- Data fragmentation occurs mainly due to partial maintenance of data in electronic and paper formats. Derivation of manual records causes discrepancies and remains a tedious process. To avoid data entry errors, the single step data capture system needs to be adopted. A voice-to-text converter may be integrated for more effective data capture.
- Data privacy—medical record involves a lot of personal information and there is a problem of data hacking if there is no proper safety and privacy. Users get worried about data leakage and hence it is necessary to use adequate safety measures to prevent these issues.
- There are no clear guidelines or frameworks and a nodal governing body to coordinate the integration of digital health. A responsible and comprehensive regulatory body needs to be established to particularly deal with digital health and its related systems.
- The digital transformation in healthcare is happening in ad-hoc manner and solitarily for each program. A comprehensive digital structural and operational environment need to be created in the healthcare systems at all levels to fulfill the maximum demands.
- Regular feedback from all types of end-users at various levels needs to be institutionalized and the same needs to be considered for amendments for better usage.

References

[1] Shahmoradi L, Habibi-Koolaee M. Integration of health information systems to promote health. Iran J Public Health 2016;45(8):1096−7.

[2] Willis SJ, Cocoros NM, Randall LM, Ochoa AM, Haney G, Hsu KK, et al. Electronic health record use in public health infectious disease surveillance, USA, 2018−2019. Curr Infect Dis Rep 2019;21(10):32.

[3] Brown N. A brief history of telemedicine. Telemed Inf Exch 1995;105:833−5.

[4] Bashshur RL, Armstrong PA, Youssef ZI. Telemedicine: explorations in the use of tele-communications in health care. Springfield, IL: Charles C Thomas; 1975.

[5] Bashshur R, Lovett J. Assessment of telemedicine: results of the initial experience. Aviat Space Environ Med 1977;48:65−70.

[6] Bedi BS. Telemedicine in India: initiatives and perspective, eHealth 2003. Addressing the Digital Divide; 2003.

[7] Dasgupta A, Deb S. Telemedicine: a new horizon in public health in India. Indian J Community Med 2008;33(1):3−8. https://doi.org/10.4103/0970-0218.39234.

[8] Deleted in review

[9] Saxena G, Singh JP. E-medicine in India: hurdles and future prospects. In: Paper presentation at an international seminar organized at the international institute of professional studies. Devi Ahilya University; Conference Presentation; 2003.

[10] Riley WT, Rivera DE, Atienza AA, Nilsen W, Allison SM, Mermelstein R. Health behavior models in the age of mobile interventions: are our theories up to the task? Transl Behav Med 2011;1:53−71. https://doi.org/10.1007/s13142-011-0021-7.

[11] Fukuoka Y, Gay CL, Joiner KL, Vittinghoff E. A novel diabetes prevention intervention using a mobile app: a randomized controlled trial with overweight adults at risk. Am J Prev Med 2015;49:223−37. https://doi.org/10.1016/j.amepre.2015.01.003.

[12] Turner-McGrievy GM, Beets MW, Moore JB, Kaczynski AT, Barr-Anderson DJ, Tate DF. Comparison of traditional versus mobile app self-monitoring of physical activity and dietary intake among overweight adults participating in an mHealth weight loss program. J Am Med Inf Assoc 2013;20:513−8. https://doi.org/10.1136/amiajnl-2012-001510.

[13] Buller DB, Berwick M, Lantz K, Buller MK, Shane J, Kane I, et al. Evaluation of immediate and 12-week effects of a smartphone sun-safety mobile application: a randomized clinical trial. JAMA Dermatol 2015;151:505−12. https://doi.org/10.1001/jamadermatol.2014.3894.

[14] Rabbi M, Pfammatter A, Zhang M, Spring B, Choudhury T. Automated personalized feedback for physical activity and dietary behavior change with mobile phones: a randomized controlled trial on adults. JMIR mHealth uHealth 2015;3. https://doi.org/10.2196/mhealth.4160.

[15] Andrew BY, Stack CM, Yang JP, Dodds JA. mStroke: "mobile stroke"-improving acute stroke care with smartphone technology. J Stroke Cerebrovasc Dis 2017;26(7):1449−56. https://doi.org/10.1016/j.jstrokecerebrovasdis.2017.03.016.

[16] Inkster B, Sarda S, Subramanian V. An empathy-driven, conversational artificial intelligence agent (Wysa) for digital mental well-being: real-world data evaluation mixed-methods study. JMIR mHealth uHealth 2018; 23;6(11):e12106. https://doi.org/10.2196/12106.

[17] Wu M, Luo J. Wearable technology applications in healthcare: a literature review. Online J Nurs Inf 2019;23(3).

[18] Caban M. The top 10 types of healthcare software. Objectivity blog. 202. Available from: https://www.objectivity.co.uk/blog/top-10-types-of-healthcare-software/. (Accessed 7 November 2022).

[19] Wang Q, Su M, Zhang M, Li R. Integrating digital technologies and public health to fight Covid-19 pandemic: key technologies, applications, challenges and outlook of digital healthcare. Int J Environ Res Public Health 2021;18(11):6053. https://doi.org/10.3390/ijerph18116053.

[20] Velmovitsky PE, Bublitz FM, Fadrique LX, Morita PP. Blockchain applications in health care and public health: increased transparency. JMIR Med Inf 2021;9(6):e20713. https://doi.org/10.2196/20713.

[21] Namkoong K, Chen J, Leach J, Song Y, Vincent S, Byrd AP, et al. Virtual reality for public health: a study on a VR intervention to enhance occupational injury prevention. J Public Health 2022:fdab407. https://doi.org/10.1093/pubmed/fdab407.

[22] Qureshi A, Fakhar-I-Adil M, Arshad H, Aslam A, Deeba F. Applications of medical drones in public health: an overview. J Human Anat 2021;5(1):000152. https://doi.org/10.23880/jhua-16000152.

[23] Vij N. Role of information technology in policy implementation of maternal health benefits in India [Dissertation]. New York: Syracuse University; 2016. Available from: https://surface.syr.edu/etd/648.

[24] Chauhan H, US S, Singh SK. Health information and its crucial role in policy formulation and implementation. J Health Manag 2021;23(1):54−62. https://doi.org/10.1177/0972063421994.

[25] Crilly JF, Keefe RH, Volpe F. Use of electronic technologies to promote community and personal health for individuals unconnected to health care systems. Am J Public Health 2011;101(7):1163−7. https://doi.org/10.2105/AJPH.2010.300003.

[26] Sinha RK. Impact of health information technology in public health. Sri Lanka J Bio-Med Inf 2010;1(4):223−36. https://doi.org/10.4038/sljbmi.v1i4.2239.

[27] Stevens IL. How technology is reshaping rehabilitation in the Middle East. Omnia Health; 2020. Available from: https://insights.omnia-health.com/technology/how-technology-reshaping-rehabilitation-middle-east. [Accessed 7 November 2022].

[28] Winstein C, Requejo P. Innovative technologies for rehabilitation and health promotion: what is the evidence? Phys Ther 2015;95(3):294−8. https://doi.org/10.2522/ptj.2015.95.2.294.

[29] Kar P. Role of information education & communication (IEC) in rural development. TVC 2019. Available from: https://www.thevisualhouse.in/blog/role-of-information-education-communication. [Accessed 7 November 2022].

[30] Christaki E. New technologies in predicting, preventing and controlling emerging infectious diseases. Virulence 2015;6(6):558−65. https://doi.org/10.1080/21505594.2015.1040975.

[31] Sakurai M, Murayama Y. Information technologies and disaster management—benefits and issues. Prog Disaster Sci 2019;2:100012. ISSN 2590-0617.

[32] Christopoulou SC, Kotsilieris T, Anagnostopoulos I. Assessment of health information technology interventions in evidence-based medicine: a systematic review by adopting a methodological evaluation framework. Healthcare 2018;6(3):109. https://doi.org/10.3390/healthcare6030109.

Chapter 15

Sustaining population benefit using evidence-based public health

Rama Shankar Rath[1] and Ayush Lohiya[2]

[1]*Department of Community Medicine and Family Medicine, All India Institute of Medical Sciences (AIIMS), Gorakhpur, Uttar Pradesh, India;* [2]*Department of Public Health, Kalyan Singh Super Speciality Cancer Institute, Lucknow, Uttar Pradesh, India*

Introduction

Public health professionals test various interventions based on population priority and assess the efficacy of the interventions in a limited setting. Following successful outcomes, they test the same interventions in real-world settings with or without some modifications and evaluate the effectiveness. The aim is to integrate successful interventions within the existing setting for continuous and extended population benefit. Such translation of research findings into policy or practice in health care takes more than a decade. Since testing of interventions consumes heavy resources in public health research (operational/implementation/health systems research), the effective interventions need to be continued in the routine setting to address the priority problems. The integration and continuation of activities or benefits are closely linked with the terminology "sustainability."

Definition of sustainability

Sustainability is defined in many ways, broadly based on the views of public health practice or funders. Sustainability was defined as the "capacity of the project to continue to deliver its intended benefits over a long period of time" [1]. Claquin et al. defined sustainability as the capacity to maintain service coverage at a level that will provide continuing control of health problems [2]. United States Agency for International Development (USAID) stated that a program is sustainable when it is able to deliver an appropriate level of benefit for an extended period of time after major financial, managerial, and technical assistance from an external donor is terminated [3].

Principles and Application of Evidence-Based Public Health Practice
https://doi.org/10.1016/B978-0-323-95356-6.00016-1

Moore et al. have identified three important constructs for defining sustainability. The identified components are the time component, intervention component, and outcome or benefit component [4]. The time component is one of the key factors in defining sustainability. Though no cut-off period was defined for assessing or reporting sustainability, it varied from immediate termination of external funding to more than 25 years. Different stakeholders have different perception of sustainability.

From the individual/beneficiaries perspective, it is the continued service provision and or health benefits from the program. In contrast, the health system looks it in two ways: firstly, to self-sustain the program by including it in the routine organizational structure and secondly, to empower the community to help the organizational structure for continued benefits after the funding dies off [5]. These two are the other important aspects of sustainability identified by Moore et al. [4]. The population health benefit is generally assessed in terms of traditional implementation outcomes like availability, accessibility, acceptability, uptake or coverage, appropriateness, and cost-effectiveness of the intervention.

Lennox et al. identified two more important constructs of sustainability, i.e., capacity building and recovering costs [6]. The capacity building should be done at both levels, within the institution/organization and outside institutions by establishing collaborations for problem-solving. One can also view this point as the broadened organizational structure, whereas a few may think of it as a separate entity. The recovering cost implies that the health benefits should outweigh the cost of the input resources [7].

Despite different views, all these definitions include the converge of few common parameters, i.e., "the ability of the program or intervention to continuously provide the health benefit over a period of time with organizational capacity building and interorganizational collaborations by using minimum inputs." All constructs mentioned here must be kept in mind before defining sustainability. Thus, sustainability can be viewed as a bicycle with one wheel as the organizational and community level changes over time, and the other wheel (with collaborations) helping the rider (population) to reach a destination (sustained benefit) with minimum efforts (input).

Sustainability is sometimes interchangeably used with institutionalization, routinization, incorporation, and other similar terms. However, one cannot simply equate sustainability with the above-mentioned terms. Institutionalization or routinization or incorporation talks about the continuation or survival of the intervention within the organization framework (a part of the sustainability) but remains silent about the benefits related to the same. Rather institutionalization, routinization, or incorporation are part of and are the processes of achieving the much broader concept of "sustainability" [8,9].

Levels of sustainability

Although continued program activities and health benefits are the core components of sustainability, the activities and health benefits may not remain the same as tested during the efficacy or effectiveness research. Depending on the prevailing context, the health benefits tend to reduce immediately or over time. Hence, researchers proposed a classification of levels of sustainability. LaPelle et al. demonstrated four levels (Levels 1—4 or No, Low, Moderate, and High) of sustainability based on the scope of the services and creative use of resources [10].

Factors influencing sustainability

The sustainability of an intervention depends on many factors. These factors may be categorized into three broad domains. (Fig. 15.1) Firstly, the factors related to the intervention design and its implementation. Secondly, the factors related to the organization or the agency implementing the intervention. Lastly, the factors related to the community environment which is likely to affect and affected by the intervention.

Factors related to the intervention design and its implementation

The intervention factors are the most important determinants of sustainability:

a. Intervention/project negotiation process:
The intervention/project objectives must be aligned with the needs / objectives of the executing agency. The funding agency and the executing agency must find a suitable middle point in framing the objectives of the intervention and the proposed outcome.

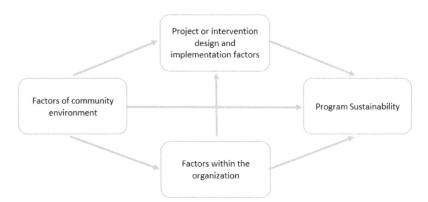

FIGURE 15.1   Determinants of sustainability of public health interventions.

b. Intervention/project effectiveness:

The effectiveness of the intervention/project is one of the key determinants of the project implementation. In the absence of effective intervention, it is worthless to expect the sustainability of the intervention.

c. Intervention/project duration:

The duration of the intervention/project is inversely related to the project sustainability. Intervention/project duration of 3 years or less is negatively associated with sustainability, whereas interventions with a period of 5 years or more are more favorable for long-term sustenance [8].

d. Intervention/project financing:

The sustainability of the financing beyond the funding agency support is a key determinant. Majority of the studies showed that the internal funding is necessary for sustainability. The mechanism of internal funding could be increasing the service charges. Community-based funding, diversification of services, and a sliding scale of service costs are important components of sustainable financing [11,12]. Financial sustainability is of two types: supply-side sustainability and demand-side sustainability. Supply-side financial sustainability is linked to the external funding source, whereas demand-side financial sustainability depends on internal funding. The demand-side financial sustainability depends on the willingness-to-pay (WTP). On the other hand, excessive supply-side stability may lead to poor intervention/program sustainability [13].

e. Intervention/project type:

Studies suggest that health service research is more likely to be sustained as compared to the preventive health services. It could be due to the immediate health benefit offered by the health service research, which is very rare in preventive health research [14].

f. Training:

Integration of the training component is more likely to provide the sustenance benefit due to the continued effect of the trainers [15].

Factors within organizational settings

a. Organizational strength

The organization or the implementing agency which is well integrated, whose institutional goals are oriented with the program goals, and has strong leadership, high skill level, and maturity is more likely to provide sustainability [12,14].

b. Integration with the existing program

Standalone vertical programs are generally the best programs to achieve goals. But the vertical programs has its own set of problems. These programs are costlier due to the requirement of specific resources to achieve the goals. Vertical programs also create a parallel set of human resources and machinery which may lead to an internal tussle between the existing health machinery and vertical program machinery [12,14].

Integrating the program with the existing program increases the chances of sustainability by providing funding directly to the organization and using the sociocultural platform of the existing program [12,16].

c. Program leadership

Program leadership is one of the important aspects of program sustainability. The leaders act as advocates, champions, or linking agents of the intervention with the existing program [17,18].

Factors of the community environment

a. Socioeconomic and political considerations

There are two schools of thought regarding the socio-economic status (SES) and political considerations. One believes that the problem environment is less likely to sustain the program, whereas other advocate that the problem environment enables better community associations [19,20].

b. Community participation

Community participation at various levels facilitate the sustainability of the intervention [21]. Community participation acts in three major ways, firstly by positively influencing community behavior, secondly, by promoting a sense of ownership, and lastly, by a widespread change in the community and program sustainability [22–26].

Planning or evaluation of sustainability

The sustainability of an intervention is essential to ensure a long-term impact on the population. Hence, public health professionals must focus on the sustainability from the planning stage of any program. They should consider the political support, stability of the current funding, availability of the future funding, organizational capacity and partnership, community needs, and community participation while planning to implement any program. The sustainability planning model (SPM) provides a simple and systematic insight to ensure sustainability while planning and implementing such new public health interventions.

Sustainability planning model (SPM)

The SPM is based on the "Intervention (Program) Theory" which depends on the "sustainable factors" [27]. The sustainable factors are of two types, capacity-building factors and sustainable innovation (Fig. 15.2).

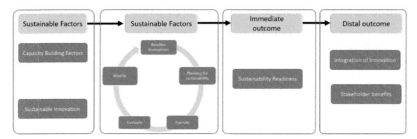

FIGURE 15.2 Sustainability planning model process.

Capacity building is one of the key determinants of the sustainability of any program or innovation. Capacity building includes various components:

1. Development of ***structures and linkages*** to sustain innovation.
2. ***Champion roles and leadership action*** that can facilitate the innovation and help in sustaining the program at each level.
3. ***Resources*** specifically the alternate sources of funding, trained and committed human resources, physical, infrastructural, and technological resources, which can facilitate sustainability.
4. ***Administrative policies*** that facilitate and assure that innovation is part of the system will likely promote sustainability. Combining power, incentives, rewards, and certification makes the system more likely to adopt the innovation faster and for a longer term than other methods [28].
5. ***Expertise*** in sustaining innovation which includes the development of champions, and experts who can help in sustaining the program, is key to long-term sustainability of the program.

The sustainable innovation includes the following:

1. ***Alignment*** between the innovation and the needs of innovation stakeholders: Innovation cannot be the standalone component. It must be aligned with the organizational requirement and the needs of the stakeholders. Innovations that are aligned with the stakeholders' needs are more likely to be sustained. For example, the development of the accredited social health activist (ASHA) worker cadre at the primary care level filled the gap between the community and the health system which was required for various health programs.
2. ***Positive healthy relationship*** between various stakeholders, including the community, must be there in order to increase the commitment and ease of implementation of the changes.
3. ***Quality and integrity of innovation*** is important in developing the initial success of providing benefits to the community, thereby increasing the confidence in both the implementers and the community.
4. ***Ownership among stakeholders*** is one of the positive motivator for sustainability.

The sustainabile factors depend on "sustainability action," i.e., a series of actions to achieve the outcomes. The action starts at the planning phase, when the implementation of the innovation begins. An evaluation of the innovation performance follows the implementation. In certain conditions, the innovation does not perform according to the needs of the stakeholder/program manager. In such cases, the required modifications must be done according to the needs of the stakeholder/program manager. The final outcome of the SPM is integrating innovation into the intended program, which will ultimately benefit the stakeholders. The innovations and modifications should be continued until it becomes suitable for integration into the routine program.

The sustainability of public health innovation is the ultimate goal of public health practitioners. Planning for sustainability should be done a priori, not retrospectively after a lot of resources and time has gone into the innovation itself.

Sustainability evaluation models or frameworks

Sustainability is sometimes conceptualized as a ***process*** of achieving the public health outcome. In contrast, sometimes it is conceptualized as an ***outcome*** of the investment of funds, human resources, and time for the interventions. Many sustainability evaluation models, theories, or frameworks are available, similar to the number of definitions of sustainability. Normalization Process Theory (NPT) and Dynamic Sustainability Framework (DSF) are the most accepted and commonly used approaches guiding sustainability.

Normalization Process Theory (NPT)

NPT looks at the sustainability of complex interventions beyond the process of normalization. It facilitates understanding the social processes to adopt and operationalize the new or modified practices. The core propositions of the theory (Fig. 15.3) are (a) complex interventions which become a part of the organization as people are working with it and (b) the implementation process and its operational mechanisms. The four operational mechanisms are:

1. Coherence (Competencies holding the practices together)
2. Cognitive participation (of individuals and groups)
3. Collective action (Engagement of individuals based on their interaction and action)
4. Reflexive monitoring (Informal and formal monitoring of actions by the participants)

The third proposition is that continuous investment by a group of people is required to carry the action in time and space [30]. Though NPT considers the different socio-environment contexts, it still assumes a fixed intervention design. These concepts can be used to evaluate an intervention before starting a trial for its long-term sustainability [31].

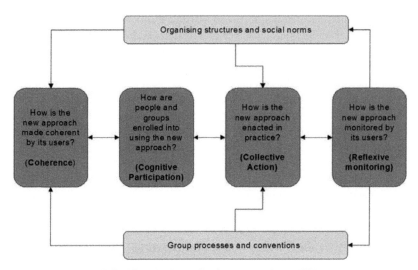

FIGURE 15.3 Normalization process theory [29].

Dynamic Sustainability Framework (DSF)

In traditional biomedical research, pharmaceutical products are tested in controlled clinical settings, wherein a linear process is followed for intervention development and assessment. However, the linear approach might not be applicable in a pragmatic setting, especially for public health interventions. The public health interventions tend to be more complex and are implemented across communities having varying levels of heterogeneity. Hence, one should expect a varied or even smaller effect of public health interventions from efficacy to effectiveness to sustainability. This smaller effect could be due to voltage drop or program drift [32].

Voltage drop is the drop in the effectiveness of the interventions due to complexity of the real world. Program drift is the decrease in program effectiveness due to deviation from the intervention protocol by the stakeholders. Both these phenomenon suggest that the optimal and expected benefit of the intervention can occur only if the modification is done according to the local needs or time. It initiates the concept of a dynamic model of sustainability, i.e., the DSF [33].

The DSF depends on three important pillars (Fig. 15.4). The first is Intervention, the second is Practice Settings, and the third is the Ecological System. The intervention has few components delivered through the human resource in the background of certain delivery platforms to achieve specific outcomes, practically oriented toward achieving the more significant public health outcomes. The success of the intervention is in assimilation into the system for which it was planned. Notably, the broader ecological system of the practice setting determines the fitting of the intervention to the setting.

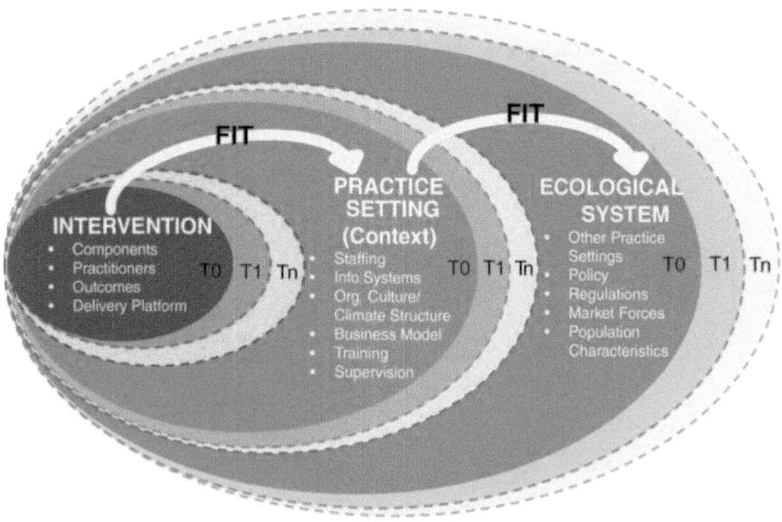

FIGURE 15.4   Dynamic sustainability framework [32].

This framework suggest the need for changes that make the intervention more suitable for the setting. It also suggest the changes to be made in the intervention which makes it easy for the setting to adopt the intervention for optimal outcomes. These changes may occur knowingly or unknowingly by the end user. It is the main concept behind the continuous quality improvement of the intervention at the local level.

The framework also suggests that the intervention is bound to change depending on local needs. Hence, the intervention should not be tagged in the optimization frame according to the policymakers before the implementation. There should be scope for making certain changes in the intervention according to local needs [34,35]. The local needs and context are bound to change continuously, hence the intervention (Fig. 15.5).

In contrast to NPT and DSF models, there are other models and frameworks that emphasize sustainability after implementing and evaluating a public health intervention, namely Re-AIM (Effectiveness, Adoption, Implementation, and Maintenance) and PRECEDE-PROCEED models, where the intervention design and its adaptations are given lesser importance [36].

Case studies on the sustainability of population benefit

23 years sustainability of home-based newborn care in a resource-limited community setting

Society for Education, Action and Research in Community Health (SEARCH), a nongovernment organization, conducted a field trial on home-based newborn

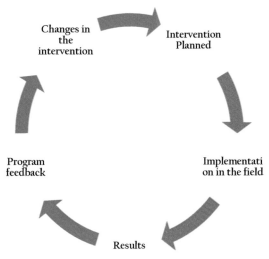

FIGURE 15.5   Intervention adaptation.

care in 39 villages of rural Maharashtra, India, from 1995 to 1998. The community health workers visited the newborns at scheduled time points to screen, diagnose, and refer or treat sepsis at home. Following the trial, the authors assessed the coverage, diagnostic fidelity, and adoption (home-based treatment) of the intervention between 1996 and 2019, i.e., 23 years [37]. During the sustainability assessment period, the authors observed 93.8% coverage, 89% diagnostic fidelity, and 88.4% adoption of the intervention. In addition, a decreasing trend of sepsis and neonatal mortality was observed during the sustainability period. Further, SEARCH replicated and scaled up the home-based new born care (HBNC) in other parts of India through partnership and external funding, following which the intervention was scaled up at the national level in 2011. Since then, the ASHA, a community health worker, has been providing HBNC as part of routine maternal and child healthcare services of the National Health Mission.

The long-term sustainability of HBNC and positive health outcome could be due to (a) acceptance of SEARCH by the community and community participation, (b) the expertise and organizational objective of SEARCH, (c) the involvement of community health worker—a female from the local community, (d) regular supervision, (e) periodic training of community health workers, (f) performance-based incentives, (g) partnership with various organizations and communities, and (h) long-term funding. Further, the national-level scale-up could be due to (a) the priority of the national health policy to reduce neonatal mortality, (b) the presence of a similar organizational structure and resources, (c) the cost-effectiveness of the intervention, and (d) adaptation of the intervention (other services also delivered in addition to HBNC).

Sustainability of interventions and benefits in a hospital setting

The cardiovascular center (CVC) of Maastricht University Hospital tested health counseling intervention for all patients with cardiovascular diseases between 1998 and 2001. The hospital continued the health counseling as part of the post-implementation trial till 2003 under the project Heart-beat Limburg. Based on the post-implementation trial, the hospital wanted to integrate health counseling while managing cardiovascular patients in CVC under the project Heartbeat 2 (2003−06) [38]. For continuing health counseling as a routine practice, they established a linkage system between CVC and the public health system, where the latter acted as an external change agent. The change agent discussed regularly with CVC regarding the need for sustaining health counseling, the mode of sustaining it, the persons responsible, and the level of sustainability. Due to the linkage system, the patients reported a short- and long-term increase in physical activity and stress reduction, though smoking cessation did not sustain in the long term. The authors attributed the unwillingness of CVC due to the lack of clear vision and related budgetary issues for low sustainability.

Sustainability of worksite intervention and benefits

In 2001, Danish researchers tested a worksite canteen intervention, i.e., training of canteen staff and goal setting by them to increase the fruit and vegetable intake among the customers. After 4 months of "No" contact period, they assessed the fruit and vegetable intake in the canteens and observed an increase of 95 g per customer compared to baseline [39]. All except one canteen maintained an increased uptake of fruits and vegetables (95 g per customer) after 5 years of intervention. The authors mentioned that the involvement of management from the beginning of the intervention, involvement and training of all canteen staff, empowerment of canteen staff, networking between the canteen managers, and shared goal setting as the reasons for the sustainability of the interventions in their study setting. Further, the authors identified five broad factors linked with sustainability in their study: modification of the program over time, presence of a champion, fitting the program with the organizations' mission, and benefits to stakeholders.

Sustainability of COVID-19 vaccination during the pandemic

Following the development and testing of safety and efficacy, international agencies and national governments rolled out the vaccination program against COVID-19. COVID-19 vaccination protected millions of lives worldwide. The sustainability of COVID-19 vaccination could be linked with (a) identification of the risk population step by step, (b) equitable allocation and timely deployment of the vaccine, (c) networking and collaboration with international agencies and pharmaceutical industries, (d) multisectoral involvements, (e) risk communication, and (f) availability of a cost-effective delivery system [40].

Conclusion

Sustainability is one of the essential aspects of every innovation in public health practice. Non-sustainable public health interventions are a waste of resources. Public health professionals must consider the sustainability of the intervention and population benefit at the time of planning of intervention. They should link the intervention with an available sustainability model or framework to systematically proceed from efficacy to effectiveness to sustainability. Foreseeing the voltage drop and program drift, the intervention should be adaptable to the prevailing practice setting of the broad ecosystem to produce optimal health outcomes.

References

[1] Michael B, Cheema S. Case studies of project sustainability: implications for policy and operations from Asian experience. World Bank; 1990. https://documents.worldbank.org/en/publication/documents-reports/documentdetail/601151468760543920/Case-studies-of-projec t-sustainability-implications-for-policy-and-operations-from-Asian-experience. [Accessed 9 December 2022].

[2] Claquin P. Sustainability of EPI: utopia or sine qua non condition of child survival. Arlington, VA; 1989.

[3] Agency for International Development. Sustainability of development programs: a compendium of donor experience. Washington, DC: Center for Development Information and Evaluation; 1988.

[4] Moore JE, Mascarenhas A, Bain J, Straus SE. Developing a comprehensive definition of sustainability. Implement Sci 2017;12:110. https://doi.org/10.1186/s13012-017-0637-1.

[5] Bodkin A, Hakimi S. Sustainable by design: a systematic review of factors for health promotion program sustainability. BMC Publ Health 2020;20:964. https://doi.org/10.1186/s12889-020-09091-9.

[6] Lennox L, Maher L, Reed J. Navigating the sustainability landscape: a systematic review of sustainability approaches in healthcare. Implement Sci 2018;13:27. https://doi.org/10.1186/s13012-017-0707-4.

[7] Brinkerhoff Derick, Goldsmith Arther. Promoting the sustainability of development institutions: A framework for strategy. World Development 1992;20(3):369−83. https://doi.org/10.1016/0305-750x(92)90030-y.

[8] Steckler A, Goodman RM. How to institutionalize health promotion programs. Am J Health Promot 1989;3:34−43. https://doi.org/10.4278/0890-1171-3.4.34.

[9] Goodman RM, Steckler A. A model for the institutionalization of health promotion programs. Fam Community Health 1989;11:63−78.

[10] LaPelle NR, Zapka J, Ockene JK. Sustainability of public health programs: the example of tobacco treatment services in Massachusetts. Am J Public Health 2006;96:1363−9. https://doi.org/10.2105/AJPH.2005.067124.

[11] Abel-Smith B, Dua A. Community-financing in developing countries: the potential for the health sector. Health Policy Plann 1988;3:95−108. https://doi.org/10.1093/heapol/3.2.95.

[12] Gertler PJ, van der Gaag J. The willingness to pay for medical care. Johns Hopkins University Press for the World Bank; 1990.

[13] Horn M. Operations research: helping family planning programs work better. 1st ed. Wiley-Liss; 1991.

[14] Bossert TJ. Can they get along without us? Sustainability of donor-supported health projects in Central America and Africa. Soc Sci Med 1990;30:1015–23. https://doi.org/10.1016/0277-9536(90)90148-l.

[15] Jackson C, Fortmann SP, Flora JA, Melton RJ, Snider JP, Littlefield D. The capacity-building approach to intervention maintenance implemented by the Stanford Five-City Project. Health Educ Res 1994;9:385–96. https://doi.org/10.1093/her/9.3.385.

[16] Streefland PH. The continuity of vaccination programmes: reflections and a case from Gujarat, India. Soc Sci Med 1989;29:1091–8. https://doi.org/10.1016/0277-9536(89)90021-x.

[17] Monahan JL, Scheirer MA. The role of linking agents in the diffusion of health promotion programs. Health Educ Q 1988;15:417–33. https://doi.org/10.1177/109019818801500404.

[18] Orlandi MA. The diffusion and adoption of worksite health promotion innovations: an analysis of barriers. Prev Med 1986;15:522–36. https://doi.org/10.1016/0091-7435(86)90028-9.

[19] Shea S, Basch CE, Wechsler H, Lantigua R. The Washington Heights-Inwood Healthy Heart Program: a 6-year report from a disadvantaged urban setting. Am J Public Health 1996;86:166–71. https://doi.org/10.2105/ajph.86.2.166.

[20] McKnight J. Regenerating community. Social Policy; 1987. p. 54–8.

[21] Rifkin SB. Lessons from community participation in health programmes: a review of the post Alma-Ata experience. Int Health 2009;1:31–6. https://doi.org/10.1016/j.inhe.2009.02.001.

[22] Kinne S, Thompson B, Chrisman NJ, Hanley JR. Community organization to enhance the delivery of preventive health services. Am J Prev Med 1989;5:225–9.

[23] Bracht N, Kingsbury L, Rissel C. Health promotion at the community level: new advance. 2nd ed. Thousand Oaks: SAGE Publications, Inc.; 1999. p. 83–104. https://doi.org/10.4135/9781452204789.

[24] Flynn BS. Measuring community leaders' perceived ownership of health education programs: initial tests of reliability and validity. Health Educ Res 1995;10:27–36. https://doi.org/10.1093/her/10.1.27.

[25] Wallerstein N. Powerlessness, empowerment, and health: implications for health promotion programs. Am J Health Promot 1992;6:197–205. https://doi.org/10.4278/0890-1171-6.3.197.

[26] Robertson A, Minkler M. New health promotion movement: a critical examination. Health Educ Q 1994;21:295–312. https://doi.org/10.1177/109019819402100303.

[27] Johnson K, Hays C, Center H, Daley C. Building capacity and sustainable prevention innovations: a sustainability planning model. Eval Progr Plann 2004;27:135–49. https://doi.org/10.1016/j.evalprogplan.2004.01.002.

[28] Lawrence TB, Winn MI, Jennings PD. The temporal dynamics of institutionalization. AMR (Adv Magn Reson) 2001;26:624–44. https://doi.org/10.5465/amr.2001.5393901.

[29] Ong BN, Morden A, Brooks L, Porcheret M, Edwards JJ, Sanders T, et al. Changing policy and practice: making sense of national guidelines for osteoarthritis. Soc Sci Med 2014;106:101–9. https://doi.org/10.1016/j.socscimed.2014.01.036.

[30] May C, Finch T. Implementing, embedding, and integrating practices: an outline of normalization process theory. Sociology 2009;43:535–54. https://doi.org/10.1177/0038038509103208.

[31] Murray Elizabeth, Treweek Shaun, Pope Catherine, MacFarlane Anne, Ballini Luciana, Dowrick Christopher, et al. Normalisation process theory: a framework for developing, evaluating and implementing complex interventions. BMC Med. 2010;63(8). https://doi.org/10.1186/1741-7015-8-63.

[32] Chambers DA, Glasgow RE, Stange KC. The dynamic sustainability framework: addressing the paradox of sustainment amid ongoing change. Implement Sci 2013;8:117. https://doi.org/10.1186/1748-5908-8-117.

[33] Cohen DJ, Crabtree BF, Etz RS, Balasubramanian BA, Donahue KE, Leviton LC, et al. Fidelity versus flexibility. Translating evidence-based research into practice. Am J Prev Med 2008;35:S381−9. https://doi.org/10.1016/j.amepre.2008.08.005.

[34] Ebi K. Climate change and health risks: assessing and responding to them through "adaptive management.". Health Aff 2011;30:924−30. https://doi.org/10.1377/hlthaff.2011.0071.

[35] Stange KC, Breslau ES, Dietrich AJ, Glasgow RE. State-of-the-art and future directions in multilevel interventions across the cancer control continuum. J Natl Cancer Inst Monogr 2012;2012:20−31. https://doi.org/10.1093/jncimonographs/lgs006.

[36] Walugembe DR, Sibbald S, Le Ber MJ, Kothari A. Sustainability of public health interventions: where are the gaps? Health Res Policy Syst 2019;17:8. https://doi.org/10.1186/s12961-018-0405-y.

[37] Bang A, Baitule S, Deshmukh M, Bang A, Duby J. Home-based management of neonatal sepsis: 23 years of sustained implementation and effectiveness in rural Gadchiroli, India, 1996−2019. BMJ Glob Health 2022;7:e008469. https://doi.org/10.1136/bmjgh-2022-008469.

[38] Jansen M, Harting J, Ebben N, Kroon B, Stappers J, Van Engelshoven E, et al. The concept of sustainability and the use of outcome indicators. A case study to continue a successful health counselling intervention. Fam Pract 2008;25(Suppl. 1):i32−7. https://doi.org/10.1093/fampra/cmn066.

[39] Thorsen AV, Lassen AD, Tetens I, Hels O, Mikkelsen BE. Long-term sustainability of a worksite canteen intervention of serving more fruit and vegetables. Public Health Nutr 2010;13:1647−52. https://doi.org/10.1017/S1368980010001242.

[40] Vu TS, Le M-A, Huynh NTV, Truong L, Vu GT, Nguyen LH, et al. Towards efficacy and sustainability of global, regional and national COVID-19 vaccination programs. J Glob Health 2021;11:03099. https://doi.org/10.7189/jogh.11.03099.

Chapter 16

Precision medicine and public health practice

Gomathi Ramaswamy

All India Institute of Medical Sciences (AIIMS), Bibinagar, Telangana, India

Introduction

Precision literally refers to being accurate, impeccable, ensuring the highest quality, or being exact with what is expected. Precision in the diagnostic test means how precisely a test will give similar results when repeated multiple times. Precision in patient management refers to making the correct diagnosis and delivering the right intervention strategy to the right person. Clinical medicine aims to provide individualized patient care based on the standard universal protocol assuming one size will fit all. However, it incorporates individual conditions based on clinical examination, considers the presence of other conditions, and/or laboratory results and tries to provide personalized medicine. The current era of medicine is about the target-to-treat, which cures diseases in most infectious disease cases or by managing the symptoms and signs as in noncommunicable diseases (NCDs). Though the standard universal treatment protocol is followed, the success rate is not the same for everyone or every condition. This fact can be observed even if the individuals are from the same geographical region or the same family. It indicates that each disease might behave in a certain way in individuals based on their genetic, phenotypic, environmental, and lifestyle factors.

Traditionally, the disease management of an individual or a group follows the detailed assessment of the agent, host, and environment. In the case of clinical management, the patients are treated based on the assessment, and in the case of public health, the population is stratified into risk categories for implementing various preventive interventions. Over a period, advanced assessments are introduced in clinical management, predominantly focusing on agents and, to a certain extent, the environment. For example, tuberculosis (TB) was managed based on the presence of TB bacilli in the sputum smear (smear positive or negative) and place of disease (pulmonary or extrapulmonary) and previous intake of TB medicines (new or previously treated). Later,

Principles and Application of Evidence-Based Public Health Practice
https://doi.org/10.1016/B978-0-323-95356-6.00006-9

microbial drug resistance testing—based treatment was introduced among the selected group of patients (previously treated or patients with poor sputum conversion) with TB. Currently, all TB patients are universally undergoing drug resistance testing, and treatment is initiated based on the resistance pattern. Importantly, drug resistance testing is rolled out to peripheral health institutes where TB diagnosis is routinely happening. Similarly, the socio-behavioral and environmental exposures related to TB are assessed in detail and necessary interventions are provided like smoking cessation, alcohol cessation, provision of antidiabetic medication and incentives to improve nutritional intake. However, the host factor is limited to age, sex, and the presence of comorbidities and the link of host genes with the disease occurrence and treatment outcome is yet to be studied and incorporated into routine clinical and public health management if found effective.

Similar focus on the agent or environment neglecting the host genetic factors is observed in most of disease prevention and management. They might significantly improve the individual's health if they are studied and integrated with the treatment protocol. The study of genetic factors will help improve the understanding of diseases and their management for better clinical and population health outcomes. In 2015, the former president of the United States of America launched the precision medicine initiative in a research mode to revolutionize treatment practices and improve people's health [1]. The precision medicine initiative is expected to do genome sequencing of one million US people. Similarly, the United Kingdom has started a genome sequencing project and is expected to recruit 100,000 population. The initiative will give access to healthcare providers to rich and resourceful data to make decision-making tailored to the individual within a community.

Public health has evolved from prevention and health promotion strategies to health system strengthening, healthcare financing, globalization, international health, and cost-effective interventions with comprehensive intersectoral strategies targeting lifestyle, socio-economic, and environmental determinants. Though it is traditional, public health promotes an individual's health through population health. For the same, precise data are needed at individual and population levels, which are generated now with newer and more advanced technologies like wearable technologies, and track data from various sources like social media, electronic health records (EHRs), environment, and others. In the last few years, precision medicine and precision public health have been largely discussed. In this chapter, we discussed the role of precision medicine in public health, the challenges, and the way forward to improve population health while administering precision medicine to improve population health.

Precision medicine

What is precision medicine and related terminologies

Dr. William Powderly, the Larry J. Shapiro Director of the Institute for Public Health, says, "Precision medicine is a medical model that accounts for variability in genes, environment, and lifestyle to tailor disease treatment and prevention to the individual" [2]. In other words, we can say the provision of the right medicine to the right person at the right time is also a right for every human in the world. Precision medicine considers the genomics and metabolomics of individuals along with societal determinants to construct decisions for every individual's treatment. The terms personalized medicine, genomic medicine, targeted therapy, and client-specific approach sound similar to precision medicine. Though personalized medicine and precision medicine are used interchangeably in many contexts, the latter is an appropriate and correct terminology. The President's Council of Advisors on Science and Technology refers to personalized medicine as the "tailoring of medical treatment to the individual characteristics of each patient. It does not mean the creation of drugs or medical devices that are unique to a patient, but rather the ability to classify individuals into subpopulations that differ in their susceptibility to a particular disease or their response to a specific treatment" [3]. Personalized medicine is more about narrowing down to treatment specific to the patient and precision medicine to plan the management of the disease considering the genetic, environmental, and personal lifestyle determinants.

Factors influencing precision medicine

Precision medicine will require robust information and technology systems with strong data management capacity. For example, genome sequencing of a human generates more than 200 gigabytes of data [4]. Precision medicine is a tailored approach for an individual's treatment, prevention, or diagnosis by integrating phenotypes, genetic, and other biological information, from imaging to laboratory tests and health records. Such incorporation of technology aids in gaining greater knowledge. Availability of such technology in the country, accessibility to needy populations, cost of such investigation, time, trained personnel, maintenance system, and positive political environment are a few key factors that determine the possibility of utilization of precision medicine. The studies have reported that the approximate cost of running a genome sequencing was 95 million dollars in 2001. But now, it is just less than 1000 dollars, thanks to rapid technological advances and data availability. The awareness of the people, novelty in diagnosis and treatment, the success of such innovations, incremental cost, health insurance, availability and accessibility of the services, and spending capacity of the people will also influence the utilization of precision medicine [5,6]. Further, greater reduction and elimination of certain diseases and evidence-based clinical practices warrant a detailed study of the host factors, especially genetic ones that still need to be explored for many diseases. Further, advancing technology facilitates transforming traditional clinical care with precision medicine.

Role of precision medicine in individual and population health

Precision medicine incorporates omics-, genomics, transcriptomics, proteomics, and metabolomics, for individual-specific decision-making. Genomics is the study of the structure, function, and expression of all the genes of an organism. It contains approximately 3.2 billion bases and 40,000 protein-coding genes. Transcriptome is the mRNA (ribonucleic acid) present in a cell that facilitates protein synthesis. A proteome is a protein that is expressed in cells or tissues of an organism and there are >100,000 proteins available. Metabolomes are low molecular weight compounds (metabolites, n > 5000) in the cells that participate in metabolic reactions [8]. All these omics are linked with each other and with the environment and socio-behavioral exposure in maintaining health or disease occurrence (Fig. 16.1) [7]. A disease-specific

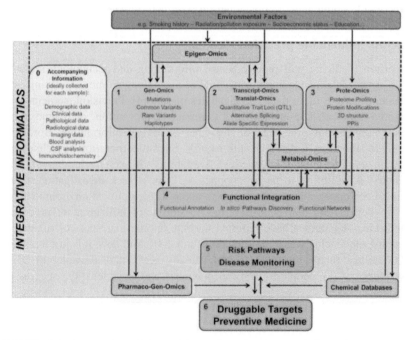

FIGURE 16.1 An overview of the precision medicine-based approach for health and disease management [7].

precision medicine-based approach will bring more light to managing diseases of hereditary origin, such as cancers, diabetes, hypertension, congenital disorders, certain psychiatric disorders, and other rare diseases. Human genomics will tailor the health strategies for the prevention of disease, screening, investigations, and treatment plan specific to the individual. Such strategies will also help devise population-based interventions for similar groups. For example, screening for newborn metabolic disorders can be an individual-specific intervention targeted toward newborns at high risk for metabolic disorders. The screening for newborns can be implemented as a population-level strategy targeting all newborns in the country or as a targeted intervention for certain newborns based on the country's policy and financial status. Among the panel of tests for metabolic diseases, a newborn can get tested for selected tests based on the applications of precision medicine [4,9].

Evidence available and applications

The term precision medicine may appear new, but some of the principles of precision medicine are already in practice. Genetic sequencing or other omics are currently used in risk categorization, diagnosis/tracking the infection, and treatment of diseases. Its application in communicable diseases, especially whole genome sequencing (WGS) of the organism, helped in tracking the disease transmissibility, severity, and resistance pattern. WGS and the use of a geographic information system helped conduct a social network analysis (person-to-person transmission or hospital-based transmission) of XDR-TB in South Africa. Similar applications have been reported for influenza, human immunodeficiency virus, and others. In NCDs, several precision medicine applications have been reported, especially on cancers. For example, (a) BRCA1 and BRCA2 mutations in assessing the risk of breast and ovarian cancers and prophylactic mastectomy and salpingo-oophorectomy and (b) epidermal growth factor receptor mutation among nonsmoking Asian women with adenocarcinoma lung and targeted therapy using erlotinib and gefitinib for improved progression-free survival [10].

Another example (general to all patients) is the EHR, which provides detailed patient information to the treatment provider. A hospital management information system integrated with the clinical decision support system (CDSS) offers critical patient information such as existing diseases, contraindications, allergies, and current disease status. It can also pull similar other patients' detail from any part of the world and helps the treatment provider make wise decision specific to the patient. However, such CDSS facilities are available in developed countries with technological advances [11]. In developing countries, though EHR is available at tertiary care centers, primary and secondary health centers still rely on papers. After the launch of the Global Strategy on Digital Health 2020−25, countries have committed to digitalizing their health system primarily to create a digital health record and unique identification details for every individual. During the initial stages, this digital health record can be used for insurance, census, and recording of the health of the individuals. Countries have a long way to go to integrate CDSS and precision medicine into routine practice. Integration of precision medicine, such as genome sequencing for diagnosis, identification of sensitive drugs, antimicrobial resistance, treatment responders, and nonresponders, will bring major transformations in medicine. Precision medicine could also help educate individuals and understand their genomic profile. Such a population-based approach will translate genomic application from an individual to a larger population and aid in informed decision-making through a precision-based approach [12].

Limitations of precision medicine

Precision medicine aims to identify the mutation in genomic sequence, which might be due to environmental or other causes that would result in phenotypic alterations in the near future. However, all genomic mutations might not result in phenotypic changes or might result in minor phenotypic alterations. So how to decide whether the mutation is minor or major and results in small or large effects? The genetic research is majorly based on discrete/mendelian mutations or continuous/polygenic mutations identified by Karl Pearson. The mutations resulting in large effects are identified easily, and the small effects, which could be due to environmental effects, lifestyle, etc., are largely non-identified. However, it requires a larger sample from large population groups to unravel the gene–gene and gene–environment interaction. Identifying and interpreting the genetic variations and linking it with the diagnosis and treatment of a patient needs specific skills that are lacking across healthcare providers (in both clinical and public health professionals). Further, genetic variations-based therapies are available only for minimal and selected conditions among thousands of diseases. Of these therapies, only some are found to be more effective and have a great impact on the clinical outcome of the patients compared to the traditional clinical approach. Another important limitation is the affordability of these therapies, which is expected to increase health inequity to a greater extent in the current scenario. The complexity is a larger challenge in precision medicine, but it could be a blessing in disguise when the technology evolves [13,14].

Convergence of precision medicine and public health

Intersecting areas of precision medicine and public health

For any identified disease, the principles of precision medicine and public health in patient management are entirely different. The public health focus is on primordial, primary, secondary, and tertiary care for the larger group of the population. Precision medicine, especially at this point, is about treatment with specific drugs or interventions based on the individual's genetic, phenotypic, and other factors specific to the individual. Robust planning and stratification are required to make public health interventions successful. It requires political commitment, the supportive financial status of the country, efficient program officers, planning, implementation, monitoring, and evaluation to be successful [15]. Precision medicine is a medical model and more of individual-level decision-making. But from a bird's eye view, precision medicine is intertwined with public health in multiple socio-economic, cultural, and life-style factors. These factors have to be considered during decision-making in both public health and precision medicine. Public health and precision medicine sound like separate entities, but they are interlinked [16,17]. The derived terminology is precision public health. Here, the n of 1 (individual) is linked with n of many (population) and vice versa that help in better risk stratification and population management (Fig. 16.2) [9]. Further, precision in public health practice refers to the application of emerging methods and technologies to assess the risk and disease and develop and implement evidence-based public health interventions that will prevent disease, promote health, and reduce health inequity [18].

The use of genomic data in public health practice is in the early stage since the application in individual's clinical care is also not well developed to date. Hence, existing and cost-effective public health interventions need to be implemented with improved efficiency before the widespread use of genomic

PRECISION PUBLIC HEALTH

FIGURE 16.2 The precision public health cycle [9].

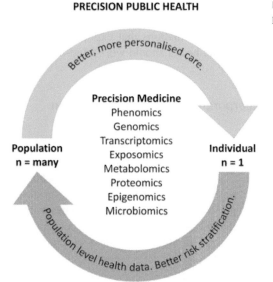

Better, more personalised care.

Precision Medicine
Phenomics
Genomics
Transcriptomics
Exposomics
Metabolomics
Proteomics
Epigenomics
Microbiomics

Population
n = many

Individual
n = 1

Population level health data. Better risk stratification.

data in public health practice since two-thirds of the diseases are attributed to socio-economic and cultural factors. Despite the limitations, precision public health helped tackle the recent COVID-19 pandemic worldwide.

Pandemic and precision public health

The World Health Organization declares an outbreak as a pandemic when it is spread worldwide or affects or crosses international boundaries affecting a larger population. The causative agent for most pandemics was either influenza or coronavirus. The recent COVID-19 pandemic is similar to the Asian flu that occurred in 1957–58, which killed around 10 million people. The mortality due to COVID-19 is estimated at 6.8 million at the end of 2022 [19]. However, compared to the previous pandemics, many advanced technologies are now available for diagnosis, monitoring, and management. However, there is also a downfall: advanced technology is associated with frequent travel and globalization, leading to the quick spread of disease before it is diagnosed. Handling a pandemic is largely a public health activity. It requires precision public health to identify the strain and its manifestation in different age groups, gender, with any disease conditions. In the COVID-19 pandemic, the identification of different COVID-19 strains such as alpha (B.1.1.7), beta (B.1.351), gamma (P.1), delta (B.1.617.2), and omicron (B.1.1.529) and its origin helped to identify the virulent or highly transmissible strain and currently circulating strain and hence the necessary precautionary steps [20]. The Geographic Information System based analysis provided information about the current and future hotspots and predicted the effects of various public health interventions. Precision medicine also helped to identify the susceptible people and plan for isolation, quarantine, and as a larger step for public health safety, a nationwide lockdown. The pandemics are one of the situations that explicitly show the world that precision medicine and public health should work hand-in-hand as precision public health [21].

Economic implications of incorporating precision medicine in public health

The European Union's Horizon 2020 Advisory Group defines PM as the "characterization of individuals' phenotypes and genotypes for tailoring the right therapeutic strategy for the right person at the right time, and/or to determine the predisposition to disease and/or to deliver timely and targeted prevention." Such a tailored strategy with genetic sequencing comes with a cost. As precision medicine is a relatively newer and emerging concept, the data on the economics of routine care are limited. Based on a systematic review by Kasztura et al. the precision approaches are exorbitantly high and not cost-effective compared to routine care. The studies exploring the willingness-to-pay thresholds reported varied levels of rate—approximately USD 20,000/

QALY in studies from the United Kingdom or Europe and USD 200,000/ QALY in studies from the United States of America. Hence, the cost-effectiveness based on QALY might be acceptable and feasible in certain countries. In others, it may not be acceptable, even in a few high-income economies. WHO identified best buy NCD intervention only if it costed ≤I\$ 100 per DALY averted. The prevalence of the disease condition, cost of the tests including genomic sequencing, gross domestic product of the country, treatment options available, and companion tests, the success rate of the intervention such as mortality, disability etc., and exploitation of available common and effective treatments have implications on utilizing the precision medicine in public health. With more and more data available and advances in biomedical engineering and medical technology, precision medicine has the potential to be a cost-effective and successful strategy in the future [6,22].

Decision-making in public health with precision

Precision medicine revolutionizes the medical field by incorporating individual lifestyle, genetic, and environmental factors to provide appropriate treatment. Though population-based strategies are widely adopted and implemented, the success of the program or procedure depends on how precisely it can address the individual's health condition. For example, iron-folic acid supplementation is the widely implemented strategy for the widely prevalent health condition—anemia. However, suppose individual-level factors, such as vitamin B12 deficiency, folic acid deficiency, hemoglobinopathies, malarial infections, fluorosis, worm infections, etc., which are specific to the person apart from iron deficiency, are not addressed. In that case, it is unlikely that anemia can be controlled. Maybe that is why, among all the countries with a high prevalence of anemia, only one country (Tunisia) is in line to achieve the WHO target of a 50% reduction in the burden of anemia among women of the reproductive age group [23]. Hence, even though public health strategies are planned at the population level, precision-based approaches are crucial to achieving the desired effect. This can be applied more specifically to diseases with multifactorial causation and multiple risk factors.

Challenges and way forward

At this point in time, the list of challenges outweighs the opportunities of precision medicine, especially in public health practice. First, in terms of prevention, if precision medicine identifies high-risk individuals based on genetic, environmental, and behavioral factors, what could be the response of the individuals to prevent the occurrence of disease? A study by Marteau et al. reported that there were no changes in lifestyle in the group of individuals with familial hypercholesterolemia who participated in a randomized control trial. The authors provided information on the risk factors for preventing and controlling hypercholesterolemia. However, they could not find any statistically significant difference [24]. A perception of the common law of fate might set in among those identified as high risk leading to less motivation or mental health disorders. There is a lot of inequity in both developed and developing economies for most health conditions, and COVID-19 has worsened this situation. Precision medicine, a superior technological health intervention, will have a major challenge in overcoming health inequity worldwide [10,25].

Role of big data in achieving health for all through precision medicine

Applicability of one size for all through precision-based public health approaches

The epidemiologist does research intending to understand the problem in the large population, such as the distribution and determinants of the disease. Such population-based studies focus on a larger group of people and aim to determine how many people are affected and what determinants are associated with the disease. So it is a common perception that population-based studies approach a condition at the population level and provide a solution, one size for all. The population-based medicine explores the socio-economic, cultural, and individual factors which can precisely influence the health of a specific group of people. This knowledge will aid the treating physician when a patient with such characteristics and disease of interest comes across [26]. The physician will be able to identify the factors from the population-based studies and apply this in deciding the patient's management plan. However, one should consider the issue of ecological fallacy in interpreting population-level data at the individual level.

Digital health—The source of big data

The granular data from different programs, planned demographic health surveys, and patient-level data from the digitalization of the healthcare system provide a large volume of data—in other words, "big data." The analysis of such big data will help in assessing the effectiveness of any program or intervention, implementation gaps, and areas of improvement. However, adding genomic sequencing data with the existing big data will require a different landscape of introspection, analysis, interpretation, and application. The data collected from routine health programs are usually incomplete, and appropriate planning is necessary to handle such data to plan precision-based interventions. If utilized properly, the predictive analysis could effectively pinpoint key challenges, barriers, and opportunities and tap potential resources and effective implementation strategies. The predictive analysis of large data from the program, genome, and socio-cultural data would serve as a great source to devise CDSS for communicable, noncommunicable, genetic, and metabolic diseases [27].

Personally identifiable information in health-related data—Whether a cause of concern?

A large quantity of data is being collected actively (by the government/ agency—DHS, census) and passively (provided by the individual) at various healthcare system levels, DHS, census, vital registries, etc. This formal or informal complex data structure is called big data and is the resource source for future planning [28]. However, the individual-specific and targeted strategies of precision medicine would require more detailed personal data to develop CDSS or artificial intelligence (AI). It leads to the question-'Whether precision medicine jeopardizes one's privacy and compromises security?'. Even big champions like Google and Facebook became vulnerable to hackers and data breaches at one point. Hence, it is crucial to think about to what extent personal information and digital identity will be used in the public domain of precision medicine. One's privacy can become unprotected at any stage of big data analysis, such as during data collection, cloud storage, data preservation, and processing. Recent advances such as differential privacy (obtaining personal data by excluding personal identifiers), identity-based anonymization, blockchain technology, and others can be explored to identify the best solution for the country. Such exercise should be in-built with any precision medicine approach. As precision medicine is still evolving, data security and privacy should be given equal importance in precision medicine structures [29].

Challenges and opportunities in future

Though big data and AI will help build data repository for many disease conditions, more data could still be needed. Genetic sequencing and mapping should be made available for various race groups and subgroups. Even among family members, there could be discordance in genes in a few individuals and could result in diseases that are not part of the family and are not picked by precision medicine. Belsky et al. explored and generated a genetic risk score using single-nucleotide polymorphisms from genome-wide association studies. The study was conducted as a population-based cohort study exploring the developmental and biological characteristics of asthma. Individuals with higher genetic risk developed asthma earlier in life and life-course-persistent asthma symptoms than those with a lower genetic risk. There was, however, no overlap in the genetic risk score between individuals with asthma and those without; half of the healthy individuals had between 14 and 25 asthma risk alleles without being affected by asthma by 38 years of age, which was the follow-up period at the time of the study. It was evident from this study that the genetic risk of asthma at an individual level cannot be predicted by genetic information alone [29]. Similarly, two strategies, namely family history-based and population-based breast and ovarian cancer risk assessment using BRCA 1 and BRCA 2 mutations, have been tested [30,31]. The family history–based strategy is more cost-effective and focused on high-risk populations. However, we should not forget the "Prevention Paradox" concept, where the majority of the cases may come from low- to moderate-risk populations. Though the latter strategy seems quite precise, it is not cost-effective from the public health point of view most of the time.

Summary

In summary, precision medicine will transform modern medicine by integrating genetic medicine to make the individual-specific diagnosis, treatment, and prevention decisions. As a dynamic and ever-growing field, public health practice also embraces modern technologies as part of precision public health. Multicountry transdisciplinary research is required to understand and reap the maximum benefit of precision medicine and for further scale-up in public health. The ongoing precision medicine initiatives in the United States, United Kingdom and other countries may facilitate the rational and cost-effective use of omics data in public health practice. Though there are hurdles, it is a rough road that leads to greatness. Precision public health may bring a solution for chronic diseases, which are a huge challenge with the current management strategy. Hence, this is the right time for public health experts to upgrade their skills in dealing with newer and advanced technologies and data, incorporating the epidemiological and program aspects to develop a new era of precision public health.

References

[1] The White House. The precision medicine initiative; 2015. https://obamawhitehouse. archives.gov/precision-medicine. (Accessed 15 February 2023).

[2] dos Santos Alves R. Alex Soares de Souza et all. In: Igarss 2014. Institute for Public Health (IPH); 2014. p. 1–5.

[3] President's Council of Advisors on Science and Technology (U.S.). HSDL—priorities for personalized medicine. Report of the President's Council of Advisors on Science and Technology; September 2008.

[4] Velmovitsky PE, Bevilacqua T, Alencar P, Cowan D, Morita PP. Convergence of precision medicine and public health into precision public health: toward a big data perspective. Front Public Health 2021;9:561873. https://doi.org/10.3389/FPUBH.2021.561873.

[5] Hall MA. King v. Burwell—ACA Armageddon averted. N Engl J Med 2015;373:497–9. https://doi.org/10.1056/nejmp1504077.

[6] Kasztura M, Richard A, Bempong NE, Loncar D, Flahault A. Cost-effectiveness of precision medicine: a scoping review. Int J Public Health 2019;64:1261–71. https://doi.org/10.1007/s00038-019-01298-x.

[7] Manzoni C, Kia DA, Vandrovcova J, Hardy J, Wood NW, Lewis PA, et al. Genome, transcriptome and proteome: the rise of omics data and their integration in biomedical sciences. Brief Bioinform 2018;19:286−302. https://doi.org/10.1093/BIB/BBW114.

[8] Horgan RP, Kenny LC. 'Omic' technologies: genomics, transcriptomics, proteomics and metabolomics. Obstet Gynaecol 2011;13:189−95. https://doi.org/10.1576/TOAG.13.3.189.27672.

[9] Bilkey GA, Burns BL, Coles EP, Mahede T, Baynam G, Nowak KJ. Optimizing precision medicine for public health. Front Public Health 2019;7:42. https://doi.org/10.3389/FPUBH.2019.00042.

[10] Ramaswami R, Bayer R, Galea S. Precision medicine from a public health perspective. Annu Rev Public Health 2018;39:153−68. https://doi.org/10.1146/annurev-publhealth-040617-014158.

[11] Temesgen Z, Cirillo DM, Raviglione MC. Precision medicine and public health interventions: tuberculosis as a model? Lancet Public Health 2019;4:e374. https://doi.org/10.1016/S2468-2667(19)30130-6.

[12] Roberts MC, Fohner AE, Landry L, Olstad DL, Smit AK, Turbitt E, et al. Advancing precision public health using human genomics: examples from the field and future research opportunities. Genome Med 2021;13:1−10. https://doi.org/10.1186/s13073-021-00911-0.

[13] Goetz LH, Schork NJ. Personalized medicine: motivation, challenges and progress. Fertil Steril 2018;109:952. https://doi.org/10.1016/J.FERTNSTERT.2018.05.006.

[14] Vasconcellos VF, Colli LM, Awada A, de C Junior G. Precision oncology: as much expectations as limitations. Ecancermedicalscience 2018;12. https://doi.org/10.3332/ECANCER.2018.ED86.

[15] Yadav AK, Sagar R. Precision medicine in public health in India: nascent but poised in the right direction. Indian J Public Health 2021;65:414−7. https://doi.org/10.4103/ijph.ijph_1728_21.

[16] Khoury MJ. From precision medicine to precision public health. Bethesda, Maryland: National Institutes for Health; 2022. p. 1−20.

[17] Meurer JR, Whittle JC, Lamb KM, Kosasih MA, Dwinell MR, Urrutia RA. Precision medicine and precision public health: academic education and community engagement. Am J Prev Med 2019;57:286−9. https://doi.org/10.1016/j.amepre.2019.03.010.

[18] Naumova EN. Precision public health: is it all about the data? J Public Health Policy 2022;43:481. https://doi.org/10.1057/S41271-022-00367-5.

[19] World Health Organization. Coronavirus disease (COVID-19); 2022. https://www.who.int/emergencies/diseases/novel-coronavirus-2019?adgroupsurvey=%7Badgroupsurvey%7D&gclid=Cj0KCQjwnvOaBhDTARIsAJf8eVM3jE533XPQCbz0icjv1vn6MlQ4WgYk_-4X9t3vm2zdtke3-ab902saApypEALw_wcB. (Accessed 30 October 2022).

[20] World Health Organization. Tracking SARS-CoV-2 variants; 2022. https://www.who.int/activities/tracking-SARS-CoV-2-variants. (Accessed 15 February 2023).

[21] Oran DP, Topol EJ. Prevalence of asymptomatic SARS-CoV-2 infection. A narrative review. Ann Intern Med 2020;173:362−8. https://doi.org/10.7326/M20-3012.

[22] Veenstra DL, Mandelblatt J, Neumann P, Basu A, Peterson JF, Ramsey SD. Health economics tools and precision medicine: opportunities and challenges. Forum Health Econ Policy 2020;23. https://doi.org/10.1515/FHEP-2019-0013.

[23] Kinyoki D, Osgood-Zimmerman AE, Bhattacharjee NV, Schaeffer LE, Lazzar-Atwood A, Lu D, et al. Anemia prevalence in women of reproductive age in low- and middle-income countries between 2000 and 2018. Nat Med 2021;27:1761−82. https://doi.org/10.1038/s41591-021-01498-0.

[24] Marteau T, Senior V, Humphries SE, Bobrow M, Cranston T, Crook MA, et al. Psychological impact of genetic testing for familial hypercholesterolemia within a previously aware population: a randomized controlled trial. Am J Med Genet A 2004;128A:285−93. https://doi.org/10.1002/AJMG.A.30102.

[25] Taylor-Robinson D, Kee F. Precision public health—the Emperor's new clothes. Int J Epidemiol 2019;48:1−6. https://doi.org/10.1093/ije/dyy184.

[26] Khoury MJ, Engelgau M, Chambers DA, Mensah GA. Beyond public health genomics: can big data and predictive analytics deliver precision public health? Public Health Genom 2019;21:244−9. https://doi.org/10.1159/000501465.

[27] Khoury MJ, Armstrong GL, Bunnell RE, Cyril J, Iademarco MF. The intersection of genomics and big data with public health: opportunities for precision public health. PLoS Med 2020;17:1−14. https://doi.org/10.1371/journal.pmed.1003373.

[28] Sun S, Ching AH. Social systems matter: precision medicine, public health, and the medical model. East Asian Sci Technol Soc 2021;15:439−66. https://doi.org/10.1080/18752160.2021.1938440.

[29] Jain P, Gyanchandani M, Khare N. Big data privacy: a technological perspective and review. J Big Data 2016;3. https://doi.org/10.1186/s40537-016-0059-y.

[30] Hanley GE, McAlpine JN, Miller D, Huntsman D, Schrader KA, Blake Gilks C, et al. A population-based analysis of germline BRCA1 and BRCA2 testing among ovarian cancer patients in an era of histotype-specific approaches to ovarian cancer prevention. BMC Cancer 2018;18:1−8. https://doi.org/10.1186/S12885-018-4153-8/TABLES/3.

[31] Raza SA, Salemi JL, Zoorob RJ. Historical perspectives on prevention paradox: when the population moves as a whole. J Family Med Prim Care 2018;7:1163. https://doi.org/10.4103/JFMPC.JFMPC_275_18.

Chapter 17

Public health practice—A futuristic perspective

Patricio V. Marquez[1,2] and Arun Chockalingam[3,4,5]

[1]World Bank Group, Washington, DC, United States; [2]Johns Hopkins University Bloomberg School of Public Health, Baltimore, MD, United States; [3]Medicine and Global Health, Faculty of Medicine, University of Toronto, Toronto, ON, Canada; [4]Health Sciences, York University, Toronto, ON, Canada; [5]Global Health, National Heart, Lung and Blood Institute at the National Institute of Health, Bethesda, MD, United States

Introduction

The focus of this chapter is to present a perspective on the public health challenges and opportunities that lay ahead. COVID-19 has exposed not only global public health vulnerabilities, but it thrived on inequality, exploiting deep disparities across societies, and normalizing inequity—for testing, vaccines, and antivirals [1]. The accumulated lessons during this period clearly point out the urgency of building resilient health systems under the universal health coverage (UHC) paradigm that includes the full spectrum of essential, quality health services, from health promotion to prevention, treatment, rehabilitation, and palliative care across the life course, and that offers protection to people from the impoverishment consequences of paying for health services out of pocket. In addition to making emergency policy and clinical decisions with the available evidence during the pandemic, it also made the countries to realize the importance of public health and social determinants of health to create a healthier population [2].

The COVID-19 pandemic exposed global public health vulnerabilities

The COVID-19 pandemic is a stark remainder of the ongoing and future challenge of newly emerging diseases, those that have never been recognized before, and reemerging or resurging infectious diseases, those that have been around for decades or centuries, but have come back in a different form or a different location [3].

Principles and Application of Evidence-Based Public Health Practice
https://doi.org/10.1016/B978-0-323-95356-6.00008-2

TABLE 17.1 Serious viral outbreaks over the past 100 years.

Sn	Name	Virus type	Year began	Global deaths
1	Spanish Flu	Orthomyxovirus	1918	50,000,000
2	Asian Flu (H2N2)	Orthomyxovirus	1957	1,100,000
3	Hong Kong Flu (H3N2)	Orthomyxovirus	1968	1,000,000
4	HIV	Retrovirus	1981	32,700,000
5	SARS-CoV-1	Coronavirus	2002	774
6	Influenza (H1N1)	Orthomyxovirus	2009	284,000
7	MERS	Coronavirus	2012	875
8	Ebola	Filovirus	2014	11,310
9	Zika	Filovirus	2015	N/A
10	Ebola	Filovirus	2018	2300
11	SARS-CoV-2	Coronavirus	2019	6,312,377 (as of June 14, 2022)

Adopted from US Government, 2021; data for SARS-CoV-2 are from Our World in Data, 2022 https://ourworldindata.org/covid-deaths

As shown in Table 17.1, serious viral outbreaks have occurred periodically since the early 1900s, caused by pandemic pathogens which span five virus families [4]. Two of these outbreaks have caused more deaths than COVID-19: the Spanish Flu, which appeared in two deadly waves in 1918 and 1919, after the devastating consequences of World War I, and the HIV pandemic, that has been around for more than 40 years. In addition, many other new viruses have been emerging in recent decades, including the severe acute respiratory syndrome (SARS), avian influenza in humans, Ebola, Marburg hemorrhagic fever, Nipah virus, and the current COVID-19 pandemic, which have triggered major international concerns, raised new scientific challenges, caused major human suffering and deaths, and imposed enormous economic damage. In the case of the COVID-19 pandemic, its full impact has been much greater than what is indicated by reported deaths. While the cumulative number of confirmed deaths since the beginning of the pandemic stands at 6.2 million, a recent paper estimates that 18.2 million people died worldwide because of the pandemic [5], as measured by excess mortality [6], over that period.

Adding to these public health challenges, yellow fever and epidemic meningococcal diseases have made a comeback in the last quarter of the 20th century [7], while antimicrobial resistance (AMR), which is resistant to drugs to treat infections caused by microbes (e.g., TB), parasites (e.g., malaria), viruses (e.g., HIV) and fungi (e.g., Candida), is looming as another public health risk that has the potential to wreak similar havoc in a globalized world [8,9].

Since these public health crises are not something new, the question that we need to ask is what have countries learned to be ready to deal with these threats?

Earlier pandemics had provided the global community with clear warnings about the danger of fast-moving zoonotic diseases—those transferred from animals to humans, which have increased in frequency and severity, causing significant health, social, and economic damage. Each time such a disease struck, resources were hastily poured into fighting the outbreak, and calls were made to build back better, so that communities and countries should be better equipped to detect and deter subsequent crises.

Unfortunately, in many instances, it seems that once the sense of fear has diminished and the outbreak controlled, it is back to the common practice of undervaluing the importance of having in place resilient health systems. Dealing effectively with these crises requires the capacity to conduct constant and expanded (including animal health) disease surveillance, prompt diagnosis and case confirmation, and robust research to understand the basic biology of new organisms and our susceptibilities to them, as well as to develop effective and safe countermeasures to control them [3].

It was reasonable to expect, therefore, that national governments and international agencies would have learned lessons from the past—perhaps most clearly from the 2014–16 Ebola epidemic in West Africa [10]. By early 2020, health systems might have been ready to manage a new infectious disease outbreak, with robust public health infrastructure for disease prevention, early detection, surveillance, and response capacity across the medical care continuum. Appropriate year-on-year budget allocations might have become commonplace to ensure the medium- and longer-term sustainability of crisis-response systems. Yet, while many countries had developed pandemic preparedness plans in the wake of Ebola, most of these plans were never resourced. Only in a few countries had governments, often with international support, established solid capacity for infectious disease control. The long-standing cycle of "panic and neglect" meant that, after the surge of global fear inspired by each new infectious disease outbreak, funds and attention quickly shifted elsewhere, once the immediate threat was contained. Most nations had fallen back into patterns of insufficient expenditures for public health and disease prevention, as other priorities absorbed attention and resources.

In short, while progress had been made in some areas, the vision of robust outbreak preparedness did not match the reality in the vast majority of countries. In fact, as the early waves of COVID-19 hit, even countries highly rated for preparedness struggled to bring the novel coronavirus under control [11]. Indeed, the rapid and uncontrolled spread of the novel coronavirus was facilitated by system-wide failures across countries that left governments blind to the threat and unable to mount effective responses [12].

Moving forward, the health and economic threat posed by these crises should become a driver for supporting the continuous development of public health institutions, systems, capacities and practices, building upon accumulated experiences and new scientific knowledge and technological know-how. And this threat stands to be more menacing in the face of inaction and misplaced social and economic priorities when one considers that the unprecedented global shock caused by COVID-19 could still leave long-lasting scars that reduce economic prospects compared with their precrisis trends, particularly in emerging market economies which are likely to endure greater losses because they had relatively less access to vaccines and their pandemic-support packages were smaller, and for many economies, the outbreak of the war in Ukraine, and economic crises in Sri Lanka, Pakistan and other countries is adding to the challenges [13,14]. The risks posed by rapid urbanization and the conglomeration of people living in large megacities with limited access to basic services [15,16], climate change that is driving the emergence of new pathogens such as the coronavirus and the reemergence of known pathogens such as Ebola [17], social conflict and wars that not only breed social dislocation and the displacement of large numbers of people, but also that hinder public health action, and massive disruption of global supply chains in interconnected economies, stand to further compound these challenges in the future [18].

Adopting a new understanding of global health security

While different proposals have been advanced and efforts initiated to support better pandemic governance and financing at the global level, there is an equal need to focus on and support national efforts at the country level. To this end, a report by the Global Challenges Foundation argues that *"global leaders must resist two great temptations: the desire to build new institutions (instead of strengthening existing ones like the WHO), and the tendency to securitize health instead of implementing strong public health measures, surge capacity to accommodate heightened pandemic requirements while ensuring access to routine health care, and enabling healthy populations"* [19].

To avoid the risk and unintended consequences of overglobalizing, overengineering, and oversecuritizing health that underplay the underlying broader determinants of health, the report proposes a new understanding of global health security (GHS) based on three interlocking functions at the national level: (i) resilient healthcare systems with built-in surge capacity (including for primary health care [PHC]); (ii) resilient public health core capacities that meet International Health Regulations standards; and (iii) proactive investments toward supportive environments, well-being, and healthy populations. The best functional model can be evolved based on the evidences created in the past and future evidence-based practices.

In the following sections of this chapter, different aspects of these inter-locking functions are discussed offering a futuristic perspective of public health practice.

Repositioning public heath practice

The definition of public health given by Charles-Edward Amory Winslow acknowledges the interdependent link between collective and individual based actions, given their impact on the health of the population [20]. While the function of public health action is to promote and protect the health of people and the communities where they live, learn, work, and play, the function of medical care services is to treat people who are sick [21].

As such, public health includes the organization of comprehensive, coordinated, and integrated services based on a defined population [22], including population or individually based actions to promote health, prevent disease, treat disease (diagnosis, treatment, palliative care, and rehabilitation), and provide the short-, medium-, and long-term care required [23]. These actions also include interventions to influence the environmental, social, economic, cultural, and political conditions that impact the health of the population [24,25].

During the past 150 years, as it has been widely documented, two factors have shaped the modern public health system: first, the growth of scientific knowledge about sources and means of controlling disease and second, the growth of public acceptance of disease control as both a possibility and a public responsibility [26]. Indeed, the perception of illness in society evolved from acceptance and limited public action to confront them, to a situation where scientific discoveries helped understand the sources of contagion and means of controlling disease, and more effective interventions against health threats were developed. Public organizations and agencies were also formed to employ newly discovered interventions against health threats, including sanitation, hygiene, immunization, regulation, health education, and personal health care [27,28].

The history of public health in the Americas illustrates the evolution of public health practice over the past centuries, particularly about the sustained efforts over time in sanitation, hygiene, and disease control directed at old infectious disease scourges that brought death, social disruption, and economic ruin to the countries [29,30]. In the countries of this region, the adoption of uniform quarantine regulations during the latter years of the 19th and the early 20th centuries, at different international conferences, under the aegis of the newly established Pan American Sanitary Bureau in 1902, that preceded the establishment of the World Health Organization, helped remove barriers to steam navigation by codifying new preventive measures into specific health legislation and programs based on the great microbiological discoveries of Pasteur, Koch, and Klebs that had revolutionized public health practice in Europe. Later in the early decades of the 20th century, the establishment of national health departments, the forerunners of the present-day ministries of health, was geared to combat the infectious diseases that hampered maritime trade in major port cities and diminished labor productivity in the countries' export-oriented economies, resting on the acceptance and application of the germ theory in disease causation that was supported by leading public health specialists such as the Brazilians Oswaldo Cruz and Carlos Chagas, the Cuban Carlos J. Finlay, as well as Walter Reed, and others in the United States. As a result of these efforts, great inroads were made in the conquest of many of the diseases that had warranted quarantine, particularly yellow fever, that represented the principal scourge of international trade throughout the colonial period, up to the beginning of the 20th century. In subsequent periods, the activities of the national health departments expanded with the support of the International Health Commission of the Rockefeller Foundation to undertake land sanitation programs centered on the control of hookworm infection and malaria, with the aim of improving the productivity of workers in exporting regions, such as those growing bananas and coffee. The support of the Rockefeller Foundation was also important for the establishment in 1918 of the School of Public Health, University of São Paulo, following the example of the Johns Hopkins School of Hygiene and Public Health in Baltimore that was founded in 1916 also with a grant from the Foundation. Additionally, the Foundation provided fellowships for the training of Latin American public health cadres (for example, epidemiologists, biostatisticians) to staff the departments, manage public health programs, and conduct public health research. During the 1930s and 1940s, changing economic and socio-political conditions led to the elevation of the health departments to the ministry level and the concurrent expansion of their activities to include the provision of personal health services. Similarly, the development of medical care programs under social insurance schemes was directly linked to the process of industrialization.

The observed and ongoing transitions in demography, epidemiology, nutrition, climate change, and economy across the world warrant a significant reform in public health practice which is yet to happen [31].

Improving public health practice in the future

Globally, over the past century, people have become healthier and wealthier, and live longer lives [32]. Indeed, over the second half of the 20th century and the early decades of the 21st century, life expectancy increased, child mortality has fallen, but more importantly, life expectancy has increased at all ages [33]. More than 40 years have passed since the last naturally occurring case of smallpox, marking the "death of a disease" that for more than 3000 years killed hundreds of millions of people or left them permanently scarred or blind [34]. Vaccines have drastically reduced the occurrence of polio and measles [35]. Besides the public health gains, the economic burden of disease was also reduced. As observed in the landmark 1993 World Bank Group World Development Report "Investing in Health," these successes have come about in part because of growing incomes and increasing education around the globe and in part because of governments' efforts to expand health services, which, moreover, have been enriched by technological progress [36].

Despite these remarkable improvements, enormous health problems and inequalities remain, requiring a sustained effort to addressing the well-being of all countries. A more widespread adoption of evidence-based public health strategies can contribute to the improvement of population health [37]. The type of evidence that is critical to guide action concerns the causes of diseases and the magnitude of risk factors, the relative cost and effectiveness of specific interventions, and the adoption and adaptation of interventions to specific contextual conditions.

The improvement of decision making and public health practice in the future requires that it is anchored on an evidence-based approach in public health to meet health goals while using effectively available resources [38]. Similar to systematic reviews of clinical trials on therapeutics, the evidence need to be created on the effect or impact of various public health practices across the countries [39]. Public health guidance based on evidence is also critical to help countries ensure value for money in limited-resource environments through careful selection and funding of procedures and drugs [40]. Further, public health needs to expand its practice area beyond the core health sector based on evidence. Moving forward, it should be clear that scientific evidence is not only a critical tool to improve treatment and spending decisions in the health sector—it also helps build capacity to adapt new knowledge and technologies that ultimately benefit people [40].

Building resilient health systems

The COVID-19 pandemic, as previous public health crises, has demonstrated that inequitable and fragmented health systems that cannot meet the needs of the population under normal circumstances cannot cope effectively with epidemics and other health emergencies [41]. As a result, the need to strengthen public health functions within health systems to make them more resilient to changing needs and threats; it is increasingly recognized as a future priority [42].

In a health system, resilience is defined as the ability to prepare, manage (absorb, adapt, and transform), and learn from shocks. These shocks are often sudden and extreme natural, financial, and 'other' acute disturbances, such as the COVID-19 pandemic [43]. In building resilience in health systems, efforts should focus not only on absorbing unforeseen shocks precipitated by emerging health needs, but also on ensuring continuity in health improvement, sustaining gains in systems functioning and fostering people centeredness, while delivering high-quality care [44,45]. The shock experience and its management during the pandemic provide valuable information not only for improving the current system but also in relation to better handling of another shock in the future. The inclusion and integration of other systems of medical practice which is backed by evidence potentially impact the population health positively [46].

The critical role of essential public health functions

In addressing the systemic deficiency of health systems, an area that requires priority attention is the strengthening of essential public health functions (EPHFs). As illustrated in Fig. 17.1, these functions are the capacities of the health authorities, at all institutional levels, to act with civil society to strengthen health systems and ensure the full exercise of public health, by acting on the social factors and determinants that impact the health of the population. This implies that country governments need to adopt policies and target investments to develop and maintain institutional capacity at the national and local levels to perform EPHFs [47].

Detecting outbreaks and health risks early

Disease surveillance is a critical public health function, but as COVID-19 has made clear, the world woefully underinvests in this function [48]. It is the continuous, systematic collection, analysis, and interpretation of health-related data that serve as an early warning system for impending outbreaks that could

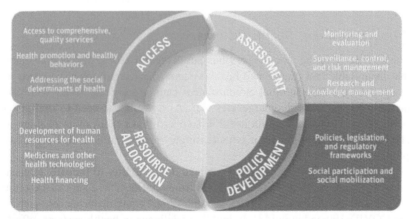

FIGURE 17.1 Essential public health functions within an integrated approach to public health. *Adopted from PAHO 2020.*

become public health emergencies; supports prompt diagnosis, and robust research to understand the basic biology of new organisms and our susceptibilities to them, as well as to develop effective countermeasures to control them [49]. These systems also enable monitoring and evaluation of the impact of an intervention, helps track progress toward specified goals; and monitors and clarifies the epidemiology of health problems, guiding priority-setting and planning and evaluation of public health policy and strategies. An "always-on" early warning system is essential to detecting disease outbreaks quickly before they spread, cost lives, and become difficult to control.

As human have extensive contact with the environment and wild life known to harbor vast numbers of viruses, many of which have not yet spilled into humans, better surveillance of pathogen spillover and development of global databases of virus genomics and serology, along with better management of wildlife trade, and substantial reduction of deforestation, is vital to prevent and minimize the impact of future epidemics and pandemics. Country experiences show that a coordinated effort is needed to identify the strengths and weaknesses in veterinary and public health systems, and develop interventions to sustain good practices and bridge gaps to tackle the animal and human health dimensions of infectious diseases of animal origin [50].

As humans have extensive contact with the environment and wildlife known to harbor vast numbers of viruses, many of which have not yet spilled into humans, better surveillance of pathogen spillover and development of global databases of viral genomics and serology, along with better management of wildlife trade, animal health management, and substantial reduction of deforestation, are vital to prevent and minimize the impact of future epidemics and pandemics [51].

Veterinary services have a key role to play in addressing pandemic risks at the source. By improving animal health and welfare, they fulfill the fundamental mission of protecting human populations from animal diseases, and vice versa. This responsibility is increasingly shared by the broader community because emerging diseases are, to a large extent, driven by numerous factors, such as climate change, land use, agricultural expansion, food systems, urbanization, conflicts and war, and population mobility in an ever more interconnected world [52].

Besides passive and active public health surveillance methods, as demonstrated during the COVID-19 pandemic, strengthening public health surveillance through wastewater testing is another essential investment for anticipating and dealing with future health threats [53]. Use of validated point of care testing tools may improve the real-time surveillance and quick prediction of future outbreaks for control [54]. Building capacity in countries to conduct thorough surveillance of the use of all newly authorized drugs, both brand-name and generic, is another critical "public good" investment to ensure that drugs work correctly and that their health benefits outweigh their known risks [51].

In all cases, reporting systems with standard procedures and definitions are required to ensure that no cases are missed. Moreover, every case during an epidemic needs to be regarded as a public health emergency and investigated immediately; its chain of transmission used to determine the extent of virus circulation in the community. Increased laboratory helps with case confirmation. Gaps in the logistics required for collecting specimens and transporting them from the field to the laboratory need addressing. Laboratory support networks once established can help analyze all types of samples, and reference laboratories can provide more sophisticated tests, including genetic characterization of virus isolates. Information and communication strategies in any type of disease outbreak help governments and regional and global institutions with the delivery to timely notification to countries of any outbreaks so that traveler's advisories can be issued and response plans are implemented. All these activities required to be carried out by a strong, well-trained, and motivated cadre of health workers to be able to timely detect, identify, confirm, and inform about new pathogens and people's susceptibilities to them, as well as to develop effective and safe countermeasures to control them [3].

It should also be clear, as advised by Brian Bremner in a brilliant new book, that preventing more pandemics in the future will hinge on developing a better understanding the billions of tiny pathogens circulating in nature [55]. That is, as he noted, if we are going to successfully handle future viral and bacterial outbreaks, we will need to head off pathogens quietly circulating in nature or within wildlife and livestock animal carriers well before they escape into human populations. That requires sustained, long-term investment in a biosurveillance system that tracks microbial risks at something approaching the molecular level. Bremner also observes that recent technological developments, such as artificial intelligence (AI) (e.g., AI modeled on the neural networks in human brains led to the discovery of a structurally unique antibiotic effective against dangerous strains of the bacterium *Escherichia coli* and drug-resistant tuberculosis) and bioengineering, are also disrupting animal disease vectors (e.g., genetically modified mosquitoes have been released into the air to suppress the population of an invasive species of mosquito called *Aedes aegypti*, which transmits Zika and dengue fever and is rapidly spreading worldwide as the climate warms) [56]. As shown by the 1-year turnaround of COVID messenger RNA and viral vector vaccines by companies such as Pfizer, BioNTech, Moderna, and AstraZeneca, a stunning achievement in public health history, better intelligence on microbes, is also key for vaccine development. Indeed, accumulated knowledge about coronaviruses from earlier epidemics of SARS and Middle East respiratory syndrome contribute to this development.

'Do no harm': The role of pharmacovigilance in implementing countermeasures

Building capacity in countries and at the regional level to conduct thorough pharmacovigilance, the surveillance of the use of all medicines and vaccines, with particular attention to newly authorized drugs, is another critical 'public good' investment to ensure that the drugs work correctly and that their health benefits outweigh their known risks [57]. The reporting of adverse effects and reactions after the use of a drug is a critical tool to: (i) inform decision-making in a health system; (ii) educate and provide guidance to health providers; (iii) help address public safety concerns about new medicines and vaccines; and (iv) stimulate prompt policy and regulatory actions. The overall goal in undertaking this work is to improve patient care and safety with the use of any kind of medication, improve public health and safety in relation to medication use, contribute to the assessment of benefits, harm, effectiveness, and risks of medicines, and encourage safe, rational, more effective and cost-effective use of drugs in countries [58].

The integration of pharmacovigilance with clinical or public health practice and its further development and strengthening are vital tasks going forward to contribute to building resilience in health systems. This is of particular importance as a decades-long barrage of antibiotics, antivirals, antifungals, and antiparasitics into the biosphere has triggered biochemical adaptations that make infections harder to treat. Pharmacovigilance data can serve, for example, as a trigger to identify AMR [59]. Their overuse, underuse, or use of antibiotics in inappropriate infections contribute to the increase of AMR, which is associated with a number of clinical complications, unnecessary prolongation of hospital stays, and, in some cases, death of the patient. This situation represents a high expenditure in the health system, and it will be increasing in the following years, if not effectively addressed [9].

Addressing the interrelation between communicable diseases and NCDs and injuries

Given resource constraints, and some shared determinants, characteristics, and interventions, there is also scope for integrated approaches focusing on functions (prevention, treatment, and care) rather than on disease categories [29]. Action against noncommunicable diseases (NCDs) and injuries is needed now, and must take place alongside continued efforts to address communicable diseases and maternal and child health, and to reach the Sustainable Development Goals in health by 2030 [60]. One set of actions cannot wait for the other. Nevertheless, there are opportunities to take advantage of the commonalities between these disease groups and build on existing work. For example, the medical care for all the above set of diseases or conditions is potentially provided at one place. However, the public health programs are running separately for each condition vertically without any linking or integration.

While great attention has been placed on the dichotomy between communicable diseases and maternal, perinatal, and nutritional causes of morbidity and mortality, and more recently, on the growing relative importance in the world of NCDs and injuries, as leading causes of disease and death, less attention has been paid to the extent to which these conditions are interrelated [61]. The same underlying social conditions, such as poverty and insanitary environments and behavioral risk factors, are associated with the onset of both communicable diseases and NCDs, and there are close relationships between these disease groups in terms of causation, comorbidity, and care. Frequently, both communicable diseases and NCDs coexist in the same individual, and one can increase the risk or impact of the other. Some infections cause or are related to NCDs; for example, cervical cancer, a leading killer of women in low- and middle-income countries, is caused by the human papilloma virus. Treatment of communicable diseases can also increase NCD risk: Antiretroviral drug therapy for HIV is saving lives, but as the HIV-infected population ages, cardiovascular disease prevalence and mortality increase significantly as shown in recent research. NCDs or their risk factors can also increase the risk of infection; for example, smoking and diabetes each increase the risk of tuberculosis, and comorbidity of tuberculosis and diabetes can worsen outcomes for both diseases. And, as observed during the COVID-19 pandemic, individuals with preexisting chronic diseases presented an increased higher risk of developing severe COVID-19 disease and death once infected with the coronavirus.

Also, many maternal illnesses and behaviors affect children, including tobacco and alcohol use, anemia, and over- and undernutrition. Gestational diabetes is a strong predictor of future illness, both of the mother, who may develop diabetes and cardiovascular diseases later in life, and the child, who also becomes at risk. Poor maternal nutrition before and during pregnancy together with smoking tobacco during pregnancy contribute to poor intrauterine growth, resulting in low birth weight which in turn predisposes to NCD risk later in life. Thus, poverty of many countries in the world, aggravated by the economic downturn and fiscal constraints due to the COVID-19 pandemic, may result in an epidemic of cardiovascular diseases in middle age for those who survive. The problem is compounded by HIV/AIDS: for example, low birth weight and malnutrition are more frequent in HIV-infected children.

Given the reality of resource constrains, more so now after COVID-19, integration and resource-sharing in the health system becomes an inescapable imperative [29]. It is important, therefore, to acknowledge and address the potential risks of setting up yet another vertical program in resource-constrained countries—and to promote integration and resource-sharing where feasible. For example, at the primary care level, maternal and child health programs could include combined interventions to alleviate malnutrition and reduce smoking in pregnant women, promote healthy nutrition and smoke-free homes in families, identify and manage hypertension and diabetes in pregnancy, and increase the uptake of breastfeeding, monitor birth weight, and screening of congenital anomalies in newborn. Collaboration with reproductive and sexual health programs could promote the use of condoms and safe sex practices and raise awareness of early signs and symptoms of breast and cervical cancer. The scope of immunization programs could be expanded to include not only vaccine-preventable diseases among children but also improved access to human papillomavirus vaccines to prevent cervical cancer and hepatitis B vaccination to prevent liver cancer and now COVID-19 vaccination, with particular attention to people with preexisting conditions and the elderly. Models already exist for collaboration with tuberculosis control programs to benefit patients with noninfectious respiratory symptoms in primary care facilities, such as asthma and chronic obstructive pulmonary disease. Screening for hypertension and elevated blood sugar levels can be administered among people diagnosed with HIV infection or among eligible population during delivery of reproductive and child health services at community level. Leveraging the resources, experience and models of existing HIV/AIDS programs could benefit other chronic conditions; and a greater emphasis on strengthening health systems through universal coverage, stronger PHC, integrated chronic care delivery, and community-based interventions is likely to be valuable for any condition.

Also, NCDs, child health, and HIV are inextricably related to mental disorders [62]. A fundamental rethinking is needed to overcome the false dichotomy between physical health and mental health in health services delivery. The social cost of mental illness and substance use disorders—which compound the impact of poor physical health—are terribly high for individuals, families, communities, and the economy. Since the impact of mental health is pervasive and relevant to not only health but to other sectors, like education and labor, investing in mental health would significantly contribute to more general efforts to reduce poverty and share prosperity. Indeed, many non-health—related global concerns have clear linkages to mental illness, such as enduring poverty, natural disasters, wars, and refugee crises. They provide entry points to link priorities and collaboration with relevant actors in order to increase investment in mental health, align funding, service provision, and population needs. Examples include the following: (i) integration of mental health interventions as part of school-based and youth programs is critically important because 75% of mental and substance use disorders begin before age 25, and suicide is a leading cause of death among youth, (ii) integrating mental health into wellness programs in the workplace can help mobilize private companies to invest in health promotion and disease prevention activities; (iv) programs in fragile and conflict-affected settings can be leveraged to mainstream integrated physical and mental health interventions alongside other social services to address the needs of displaced populations and refugees; and (v) microcredit schemes can be used to provide low-interest, small business loans, training, and mentorship to entrepreneurs with a history of mental health or addiction challenges (including former prisoners), facilitating their reintegration into the community [63]. As one size will not fit all, the strong regulatory approaches to prevent and curb the substance use disorders used in some places of the world may be better studied and adapted if needed [64].

In sum, much illness and inefficient resource use can be avoided in countries—diseases and disabilities are frequently preventable—but comprehensive and systematic approaches need to be applied which build on existing resources and experience and capitalize on the interlinkages between communicable diseases, NCDs, maternal and child health, mental health, and socio-economic development.

Redefining primary healthcare

To address preexisting distortions and underlying weaknesses in the delivery of health services, a three-layered approach for post-COVID-19 investments that integrates public health system functions and PHC has been proposed [65]. It is based on the 'Prevent, Detect, and Respond' model of health security, enabling prioritization and integration of health security investments within the service delivery systems. The first layer focusses on pandemic risk reduction and emphasizes PHC systems that can monitor and detect emerging disease outbreaks at the community level. The second layer, which operates primarily during the early phase of an outbreak, focuses on the identification and protection of at-risk populations, scaling up of testing and contact tracing, epidemic intelligence, risk communication, and public health measures. Layer three includes surge responses and secondary and tertiary hospital interventions that require advanced case management. Expenditures in the first two precrisis layers are more effective in strengthening resilience to epidemics, and their secondary impacts than investments to fight a full-fledged epidemic in the final layer. The above layers also applicable to prevention and management of NCDs where primary care is provided at first layer and complications are managed at layer two or three.

In practice, the adoption of this integrated approach would help overcome some of the dichotomies that have persisted for decades in the global health debate: prevention versus treatment, primary versus specialized care, vertical versus horizontal strategies, and social determinants of health versus health services [66]. By helping settle the disputes between vertical and horizontal approaches to healthcare, and adopting a "diagonal" perspective, an integrated approach works by prioritizing disease control and enhancing the capacity of health systems to effectively deal with the population health needs and be able to be ready to respond effectively to public health crises [67,68]. This consideration is particularly important given the real threat of emerging and remerging infectious diseases alongside the raising importance of NCDs and injuries as leading causes of death and disability in a rapidly aging world.

Three priority action areas have been suggested to remake PHC systems [69]. As outlined in Box 17.1, they are posited to help reform how PHC is designed, financed, and delivered.

Critical enabling tools for the reaming of PHC are deconcentrated or decentralized decision-making and management structures and processes, as well as the use of evidence-based clinical protocols to guide care coordination [70]; health management information systems to coordinate the on-time flow of patient and administrative and financial information across facilities [71]; and new incentive frameworks that link resource allocation or payments to the production of quality services and good health outcomes, helping reduce avoidable costs of untimely, uncoordinated, expensive, and substandard care [72].

BOX 17.1 Priorities for reimagining primary health care

1. **Adopting a multidisciplinary team-based approach.** Team-based care models have emerged as the state-of-the-art primary health service delivery platform, offering integrated, responsive, continuous, and community-oriented care. Depending on local needs, the team could consist of community health workers, registered nurses, and general practice or family medicine specialists working collaboratively. This approach improves patient care, prevents disease, and protects community health and has important benefits for emergency preparedness, response, and resilience.

2. **Reforming healthcare workforces.** In many countries, the primary healthcare workforce lacks sufficient numbers, competencies, and deployment to deliver quality team-based care. Changes are needed to training, deployment, management, evaluation, and compensation of healthcare workers.

3. **Financing for primary healthcare systems.** Funding primary health care yields high returns and promotes sustainability. Countries should develop tailored plans to invest in their primary healthcare systems and donors should increase funding for primary care. Modeling suggests that an additional US$200 billion per year will be required through 2030 for low- and middle-income countries to improve their primary care systems. Providers should be incentivized through "value-based" payments to innovate and deliver high-quality care while reducing costs, encouraging team-based approaches, rewarding quality, and promoting accountability. Donors should shift from investing in specific programs to investing in systems.

Source: *Barıs E, Silverman R, Wang H, Zhao F, Pate MA. Walking the talk: reimagining primary health care after COVID-19. Washington, DC: World Bank; 2021. https://openknowledge. worldbank.org/handle/10986/35842*

Healthcare innovations that promote care coordination among hospitals, physicians, nurses, therapists, and home care providers in accordance with evidence-based care protocols and that reimburse services using annual or capitated fees for members of an assigned population can be scaled up to promote collaborative structures centered on ambulatory, community-based, primary care services [73]. These arrangements have the potential to reduce costly emergency room visits and inpatient services through early detection and treatment of chronic diseases and by keeping people healthy and out of the hospital.

It should be clear, however, as suggested in a recent book, *Improving Health Care in Low- and Middle-Income Countries* [74], that the development of robust health systems over the medium- and long-term also requires sustainable quality improvement efforts that are institutionalized through policies, organizational and institutional arrangements, performance management processes, and adequate and predictable funding allocations. This will support the establishment of a "quality improvement culture" that is ingrained at all levels of the health system. The engagement of frontline teams, following clearly defined standards of work for the delivery of safe, effective, and efficient health care, that is monitored and adjusted as a team responsibility, would be key to sustain performance and contribute to improving the health conditions of the population.

The current status of public health practice in occupational settings especially in low- and middle-income countries needs focus and linkage with routine population-based or facility-based healthcare services. At the workplace, the employers could adopt and scale up measures to protect workers from occupational risks. Also, they could offer on-site clinics as part of workplace health and wellness programs, covering both physical and mental health conditions, to help workers and their families access health promotion counseling to encourage physical activities, diet regimens, and vaccination and to provide secondary prevention services such as screening for high blood pressure and blood sugar levels, and psychosocial support for anxiety and depressive disorders and alcohol and substance use disorders [75].

Reforming public health and medical education

The training and skills of public health professionals vary within and between countries. Similarly, their availability at different levels of health care also varies like community health workers at PHC level to person with postgraduate degree or doctoral degree in tertiary care level. The community health workers who provide primarily the preventive and promotive healthcare services at community level are usually trained inadequately, on ad hoc basis and in isolated manner, due to which their comprehension and delivery of services varies across the world. They are most of the time overburdened with multiple tasks and poorly paid. Hence, there is a need to create or amend the existing set of basic public health skills and adequately train the community health workers for effective delivery of public health services. As present, in some countries at national and subnational level, a separate public health cadre covering the core principles or functions of public health needs to be implemented to avoid the human resource scarcity especially during the management of outbreaks or emergencies [76,77]. Further, improving the public health and medical education programs—which are hampered in many countries by outdated conceptual and methodological structures and practices—is a critical future task to support the development of resilient health systems. As digital advances can support and accelerate public health achievement (e.g., AI-enabled frontier technologies are helping to save lives, diagnose diseases, and extend life expectancy), while posing privacy, security, and inequality risks [78], inclusion of the concepts of modern technological exposures such as AI and big data can help prepare the health workforce to harness and manage new technologies.

The challenge is shared globally, as different countries are struggling to sufficiently staff their health systems with well-trained, deployed, managed and motivated physicians and nurses to provide quality medical care, and competent staff to manage service delivery and carry out essential public health work such as disease surveillance.

With few exceptions, such as the 2010 Lancet commission report [79], medical, nursing, and public health education reform has failed to appear in the international health agenda. This needs to change because the extraordinary progress in medical knowledge during the last 50 years, coupled with the introduction of new technologies, drugs and procedures, and the promise of more profound and rapid changes in the future catapulted by the "genome revolution" and evidence from different disciplines, clearly points out that medical and public health education programs cannot remain static. They need to continuously change with these developments and serve as the "conduit" for channeling new knowledge to reform medical and public health institutions and practices. Importantly, information technology skill is inevitable, and a special focus is needed to impart the same among the healthcare professionals.

Education reform requires well-planned and systematic efforts, including the revision of often fragmented medical education curriculum by defining aims, outcomes, and structure of the whole program, for each year, and for core modules [80]. Also, in many countries, there is the need to adapt new learning and training materials in the local language; introduce laboratory training (e.g., bedside teaching, using equipment) to develop the clinical skills of students; replace oral examinations with test-based assessments to objectively measure student performance; support training to improve the knowledge and teaching skills of professors; and introduce national licensing examination for recent graduates to determine who is fit to practice medicine. Similar efforts need to be undertaken for postgraduate medical training through the introduction of residency programs for specialists. The general and specialist medical curriculum needs to ensure that the relevant and necessary public health skills are taught and acquired by the medical professionals during their academic training. Similarly, the effort to reform medical education will need to be accompanied in the future by similar reforms in nursing and public health education.

International organizations and donors need to support this effort not only to help ensure that future physicians and nurses, as well as public health specialists, are well prepared to tend the health needs of the population, but to sustain ongoing healthcare organization and financing reforms.

Leveraging new technologies

The introduction and use of new technologies are also key to the effective and efficient delivery of PHC health services, helping transition from low-quality PHC services to high-quality care for all. For example, the critical importance of new and repurposed pharmaceutical products, both vaccines and therapies, during a public health crisis was made evident during the COVID-19 pandemic. Indeed, the development and scaled-up administration of the new COVID-19 vaccines, which have been at the center of the global response, were life-saving and occurred under accelerated delivery timelines—no vaccines in history have been developed and received emergency use authorization as fast as the COVID-19 vaccines [81,82].

The vaccination effort has also required building public confidence through public education campaigns and community mobilization, using social media and other communication tools [83].

Digital care options through teletherapy and new apps have seen explosive growth during the COVID-19 pandemic, serving a vital role during lockdowns [84]. The use of smartphones and specialized apps can help keep people healthy, via text message reminders about medication schedules; keeping track of lab results and vital signs; and monitoring progress in achieving personal health goals; e.g., monitoring every day the number of steps taken and comparing them with the steps taken the previous day or week, calorie or nutritional intake, and others. They offer alternative service delivery models that help overcome obstacles that hinder access to care, such as transportation barriers, stigma associated with visiting mental health clinics, personnel shortages, hospital-acquired infections/conditions, and high costs. These platforms, especially in mobile formats, can offer remote screening, diagnosis, monitoring, and treatment, facilitate remote training for nonspecialist healthcare workers, and enhance online peer-to-peer support and self-care [85]. Importantly, the advanced point of care testing technology will help to rapidly diagnose and control the disease especially the communicable diseases.

Also, thanks to new technologies, procedures that used to require lengthy hospitalization now can be performed in an outpatient facility, with more convenience and safety for the patient [86]. And, the development of precision medicine approaches for disease treatment and prevention that takes into account individual variability in genes, environment, and lifestyle for each person [87] stands to further revolutionize medical and public health practices. While the role of precision medicine in day-to-day healthcare is still relatively limited, examples can be found in several areas of medicine. For example, a person who needs a blood transfusion is not given blood from a randomly selected donor; instead, the donor's blood type is matched to the recipient to reduce the risk of complications. It is hoped that as the adoption of this approach expands to many areas of health and health care in the coming years, and it will allow care providers and researchers to predict more accurately which treatment and prevention strategies for a particular disease will work in which groups of people, in contrast to a one-size-fits-all approach, in which disease treatment and prevention strategies are developed for the average person, with less consideration for the differences between individuals.

Addressing bottlenecks to build vaccine production capacity in developing countries

The COVID-19 pandemic exposed the large disparity in access to vaccines between developed and developing countries. To deal with this problem, a joint proposal by India and South Africa was advanced in 2020 requesting the World Trade Organization (WTO) to waive certain provisions of the Trade-Related Aspects of Intellectual Property Rights (TRIPS) Agreement to support the pandemic response [88]. The proposal recognizes that other intellectual property rights may also hinder rapid access to affordable medical products beyond patents. In addition, it states that many countries, especially developing countries, may face institutional and legal difficulties when using flexibilities available in the TRIPS Agreement. Specifically, it raises particular concern for countries with insufficient or no manufacturing capacity and the cumbersome and lengthy process for the import and export of pharmaceutical products. It, therefore, calls for the WTO Members to agree to waive some of the obligations under the TRIPS Agreement on the protection and enforcement of patents, copyright, and related rights, industrial designs, and protection of undisclosed information during the COVID-19 pandemic. This proposal, however, has not been approved at the WTO given the opposition from several European countries and the pharmaceutical industry to a waiver of rules protecting the intellectual property behind the vaccines.

Going forward, besides addressing the legal aspects related to the TRIPS Agreement, it is imperative that attention be placed on identifying options to deal with other potential constraints in developing countries, such as availability of therapeutics for locally prevalent diseases, available production capacity, trained personnel with required knowledge and skills, and supply chains of raw materials and other critical inputs [89]. In the current globalized world, the diseases of developing countries cannot be ignored. Potential options include the provision of expert support from vaccine or therapeutic makers to new plants in developing countries to manage the technology transfer process, financial support, both from bilateral and multilateral sources, and easing of export controls so that the pharmaceutical industry is not restricted in where they send their vaccines/therapeutics or ingredients to make them.

The role of big data in public health

Big data powers modern innovation. In the health sector, big data is transforming public health through applicable, scalable technologies and treatments [90]. The stores of digital information have the potential to offer exceptional access to research and care as well as to expedite clinical trials to advance the development of new drugs and therapies and policy decisions. However, there are privacy and security risks that need to be considered. Understanding big data's role in public health as it continues to evolve should be a priority going forward.

Big data and transforming health care

The multiples of billion terabytes (10 to the power of 21 or 1021) of data collected from around the world, deidentified and handled responsibly, may lead to new insights into medical conditions, treatments, and prevention methods. Already, these tools are being applied in public health solutions. From Internet of Things (IoT) medical devices [91] to machine learning pandemic-tracking features, big data has empowered new technologies to allow for greater transparency and responsiveness of care [92]. IoT in medicine has the power to bridge patient data with their physicians in a seamless, insightful connection, bringing preventative treatments into the future. With the power of wearable devices and sensors spanning vast networks, the potential for data collection is enormous. Public health professionals with access to these data can then predict and track the patterns such as the spread of a pandemic or the correlations between symptoms and risk factors. Big data has already assisted public health in fighting COVID-19, protecting medical data, and so much more.

Data availability facilitates a better understanding of communicable and NCDs and their outcomes. It also allows for greater accessibility of health care through better telemedicine [93]. Through data analytics, care professionals can leverage diagnosis and treatment of certain conditions remotely, take better care decisions that cut costs, and improve efficiency [94]. This allows patients who are vulnerable or who live far from care to more safely, timely, and cost-effectively access the resources they need. Predictive analytics can help streamline every aspect of care, including the supply chain for medical supplies and pharmaceuticals. However, the developing countries are yet to build or strengthen their information technology infrastructure to capture and utilize such big data. The role of data in health care will only grow in the future, making the application and understanding of these tools now a must for public health and healthcare providers. A specialist data scientist group may dominate the public health practice sooner worldwide.

Big data strengthens machine learning by feeding it new variables and corresponding patterns. With all the potential of such a system in healthcare, it should be no surprise that big data has already made waves across the healthcare sector. Fueled by AI and machine learning capabilities, big data has produced a variety of use cases in the medical field. Expanding the accessibility of care to strengthening medical cybersecurity, AI can maximize the potential of care providers to reach and understand their patients [95].

For example, data scientists at Crisis Text Line (a service supported by the Substance Abuse and Mental Health Services Administration (SAMHSA), an agency within the U.S. Department of Health and Human Services, serves people across the United States experiencing any type of crisis and provides free, 24/7 emotional support and information through texting with a live, trained specialist) have used machine learning to assess how high risk for suicide a person might be based on the words and emojis they use. They use these data to assist those most at risk as quickly as possible to maximize the good they can do with the resources they have. Tracking patterns, symptoms, and rates of infection are all additional features offered by big data. During the COVID-19 pandemic, the value of these applications was evidence in resources like the Mayo Clinic's tracking tool for pandemic data [96]. Enhanced by real-time data and predictive analytics, this tool offered key insights into the spread of the virus across US counties. Caution must be applied, as the risk of cybercrime has exploded in the wake of the pandemic, with the healthcare field seeing record-level data breaches in 2020 [97]. At the same time, with the help of machine learning, data breaches can be modeled and analyzed by an AI that can then learn from the datasets and offer protection.

Improving financing to achieve universal health coverage

As a "new normal" is constructed after the COVID-19 experience, it is important to focus on ways to redouble efforts to achieve UHC globally. The pandemic has showed with great clarity that underfunded, institutionally weak, and hospital-centric health systems pose major public health, economic, and national security risks for all countries across the income spectrum.

Moving forward, improved efficiency and equity in the use of available health resources can help ensure more rapid progress toward UHC, and generate higher priority for health in government spending decisions [98]. Areas for countries to focus on include the prioritization of investments in PHC, and bolstering public health and health security functions while defining a set of guaranteed health services, including health promotion and disease prevention, as well as EPHFs, including disease surveillance, outbreak response, monitoring and evaluation, and governance.

The commitment to public health objectives and resilient health systems also needs to be frmly entrenched in the development programs of governments and supported with the necessary budgetary allocations in a sustainable way. While donor contributions help, particularly in low-income countries, governments can mobilize additional tax revenue as a share of GDP to fund and sustain public health and essential medical care services via better tax administration (including value-added taxes), tackling tax avoidance and evasion (e.g., taxing financial capital flows), broadening the tax base by removing cost-ineffective tax expenditures (e.g., fuel subsidies), and increasing excise taxes on unhealthy products (including on tobacco, alcohol, and sugar-sweetened drinks) [99]. This is needed, as clearly articulated in a recent article in Foreign Affairs by Nobel Laureate in Economics, Joseph E. Stiglitz and two of his colleagues [100], because:

> *The state requires something simple to perform its multiple roles: revenue. It takes money to build roads and ports, to provide education for the young and health care for the sick, to finance the basic research that is the wellspring of all progress, and to staff the bureaucracies that keep societies and economies in motion. No successful market can survive without the underpinnings of a strong, functioning state.*

Prohealth taxes, in particular, offer a potential win−win−win outcome, as they can contribute to reduce health risks, expand fiscal space for UHC, and enhance equity [101].

Indeed, taxes on tobacco, alcohol, and sugar-sweetened beverages are effective but underused policies of disease prevention and health promotion, that could also help mobilize additional government revenue to fund investments and programs that benefit the entire population and enhance equity [101]. The evidence across a wide range of countries shows that a 50% increase in cigarette price typically leads to a 20% decline in cigarette consumption. Lowering consumption reduces tobacco-attributable sickness and death: about half of this effect comes from current smokers quitting and the other by reducing smoking initiation among young people. Taxing alcohol and sugar-sweetened beverages helps to reduce consumption and prevent the onset of related chronic diseases such as cardiovascular diseases, cirrhosis of the liver, obesity, and diabetes. Moreover, taxation to increase the price of alcohol products, along with strict enforcement of drunk-driving laws, can help reduce the high human and economic cost of road traffic injuries, fatalities, and domestic violence. Emerging evidence shows positive health impacts from sugar-sweetened beverages taxes by reducing consumption and hence helping to control the growing obesity epidemic and its impact on NCDs.

Besides generating public health benefits, these health taxes, can substantially boost government revenues. This is of critical importance during

COVID-19, as policymakers must maintain their public health responses while also mobilizing domestic revenue for investment in future pandemic preparedness and other essential health services. Health tax increases would have the additional advantage of reducing future health care costs by curbing the growth of the NCDs that tobacco, alcohol, and sugar-sweetened beverages can cause. But the possibilities for taxation are broader: sugar production and imports, fossil fuels (or carbon), and industrial or vehicle emissions. Also, of importance is reducing expensive subsidies that now exist on fossil fuels and often on unhealthy food production or unhealthy child dietary supplements. The phasing out of these of fossil fuel subsidies, for example, which impose large fiscal costs while adding to negative environmental and health impacts, could also help expand fiscal space for health.

Projections presented in a World Bank Group report [98] for the G-20 meeting in Osaka, Japan, in 2019 showed that the substantial UHC financing gap in low- and lower-middle-income countries (now exacerbated by COVID-19) can be attenuated by excise tax increases on tobacco, alcohol, and sugar-sweetened beverages. These calculations showed that a 50% increase in prices for these products could generate additional revenues of approximately US$24.7 billion in 54 low- and middle-income countries by 2030. Importantly, the revenue raised can additionally benefit poorer households when it is used progressively.

Improved financial protection can also be helped by reducing reliance on out-of-pocket payments for health through prepaid and pooled sources, with subsidies for people who cannot afford to contribute.

International finance institutions, such as the World Bank Group and the International Monetary Fund, as well as donor governments, bilateral agencies, and private sector entities, can play a vital role in support of government efforts in low- and middle-income countries to develop resilient health systems, and to modernize and strengthen tax administration systems to mobilize additional domestic resources for health and other priority investments that benefit the entire population.

For example, at second Global COVID-19 Summit in May 2022, cohosted by the United States, Belize, Germany, Indonesia, and Senegal, convened over partners and organizations from around the world, significant financial commitments were made to accelerate collective efforts to support COVID-19 vaccination efforts, enhance access to tests and treatments, protect the health workforce, and finance, and build health security for future pandemics and other health crises [102]. The new financial commitments totaling US$3.2 billion includes nearly US$2.5 billion for COVID-19 and related response activities and US$712 million in new commitments toward a new pandemic preparedness and GHS fund at the World Bank Group. The establishment of The Pandemic Fund by the World Bank Group in mid-2022, with support from the G20, recognized the urgent need for coordinated action to build stronger health systems, and mobilize additional resources for pandemic prevention, preparedness, and response.

Multisectoral action for health

Addressing the interlinked social, economic, and environmental factors that influence the health of a population at the local, national, and global levels also requires investment and actions undertaken by non-health sectors [103].

As noted before, the risk of spillover is greater when there are more opportunities for animals and humans to make contact (e.g., in the trade of wildlife, in animal farming, or when forests are cleared for mining, farming, or roads) [104]. Stopping spillovers and preventing future pandemics, therefore, require that decision-makers focus on the adoption of multisectoral actions that reduce the risk of animals and people exchanging viruses, including (i) protecting tropical and subtropical forests, since changes in the way land is used, particularly tropical and subtropical forests, might be the largest driver of emerging infectious diseases of zoonotic origin globally [105]; (ii) banning or strictly regulating commercial markets and trade of live wild animals that pose a public health risk; (iii) improving biosecurity when dealing with farmed animals through better veterinary care, enhanced surveillance for animal disease, improvements to feeding and housing animals, and quarantines to limit pathogen spread; and (iv) adopting effective and sustained interventions to improve people's health and economic security, particularly in hotspots for the emergence of infectious diseases, and taking into account that people in poor health—such as those who have malnutrition or uncontrolled HIV infection—can be more susceptible to zoonotic pathogens, and in immunosuppressed individuals such as these, pathogens can mutate before being passed on to others [106].

The adoption of intersectoral policies is also needed to build less vulnerable and more resilient populations. These policies fall into four broad categories: taxes and subsidies, regulations and related enforcement mechanisms, built environment, and information [107]. These policies are designed to reduce the population level of behavioral and environmental risk factors, and the implementation of many of these policies, such as tobacco, alcohol, and sugary drinks taxes and regulations, and road safety measures, requires action beyond the health sector, and can deliver broader social benefits in addition to their benefits for health. Selective, evidence-based actions to reduce health risks and disease and injury in countries will help address the disease burden and achieve a more sustainable improvement in health outcomes, more efficient use of resources, and enhanced equity among the population.

As more countries pursue the goal of UHC, which is high on the global health agenda, there is a need to focus not only on how to expand access to needed medical care and financial protection for all, but also to rethink how public policy is structured and geared to "nudge" people to improved health-related

decision-making and actions, and to adopt fiscal and regulatory measures to tackle social, economic, and environmental factors that contribute to ill health, premature mortality, and disability [108].

Indeed, governments can play a catalytic role in tackling the main health risk factors at the population level as part of an integrated health agenda, since (i) there are substantial societal costs resulting from, for example, second-hand smoke and alcohol-induced injuries and fatalities [109,110]; (ii) people are not always fully aware of the health (and other) consequences of unhealthy life-style choices such as smoking, alcohol abuse, physical inactivity, and poor diet; they may also be misled by information provided by the food, alcohol, and tobacco industries; and (iii) children and adolescents (and even adults) tend not to take into account the future consequences of their current choices, irrespective of whether they are informed about them [111].

There is ample accumulated evidence from across the globe that shows that fiscal measures (e.g., higher excise taxes for tobacco, alcohol, and sugary drinks), combined with regulatory measures (e.g., smoking bans in public places to prevent the negative effects of secondhand smoke) [112,113], or measures by the police (e.g., enforcement to deter drunk driving) [114], or the adoption of workplace mental health and well-being strategies [115,116], are critical tools in the public health arsenal that lead to lower rates of lung cancer, heart attacks, road traffic deaths, reduced use of related high-cost treatment and trauma care services, as well as prevent or minimize other indirect economic costs such as absenteeism, presentism, and low productivity due to mental ill-health and substance use disorders [117,118]. However, none of the above measures are really under the direct control of public health sector.

Similarly, evidence from different countries can be used for designing effective policies to empower people to make informed decisions and do what is in their best interests. In *Poor Economics* [119], a book by Nobel Laureates in Economics, Abhijit Banerjee and Esther Duflo of MIT, evidence is presented on how firmly held beliefs by the poor, who often lack critical information (e.g., how HIV is transmitted or prevented), contribute to decisions and behaviors that put them at risk of or contribute to the spread of communicable diseases. But the authors also argue that information alone will not do the trick. What is needed are those "policy nudges," such as free services or rewards as done under conditional cash transfer programs (e.g., nation-wide programs in Mexico and Brazil), which encourage people to demand and utilize preventive and treatment services (e.g., prenatal care and institutional deliveries, taking pills over the course of treatment to prevent the onset of multidrug-resistant TB).

Insurance arrangements and healthcare organizations and payment innovations that are already used in different countries can be leveraged to further advance this public health agenda as well [120]. For example, insurance companies, by charging lower premiums for those who quit smoking, lose weight, overcome sedentarism by increasing physical activity, and pass screening tests for artery-clogging cholesterol, high blood pressure, and high sugar levels, provide an incentive for individuals and families to assume responsibility for their health.

At a more macro level, for example, the redesign of the urban space to facilitate pedestrian and cyclist mobility by rolling out new cycling networks and improving pedestrian walkways can generate significant health benefits by reducing health risks associated with sedentary lifestyles while making cyclists and pedestrians safer [121]. Supporting pedestrian- and bike-friendly policies, programs, and investments in cities, therefore, offers the potential to contribute to the reduction of traffic congestion and pollution, but also to improve the health of the population at different age groups.

As highlighted in the 2022 G7 Health Communicate [122], the importance of combating climate change to protect health is a global public health imperative. Climate change is already affecting the health of people, animals, and ecosystems globally, and projections indicate increasingly negative health impacts. The aim going forward must be to better protect people around the world from the health impact of climate change. This includes using early warning systems more effectively for health protection, integrating climate adaptation measures into the training of health professionals, and promoting cooperation between G7 public health institutes on climate and health.

Taking forward the 'Health in All Policies,' the health implications of every public policy need to be discussed at subnational, national, and international level. To foster the discussion and drawing the strategic actions, the public health professionals need to be involved at all stages of policy development and implementation [123].

Acceleration and application of public health research

In the era of evidence-based medicine, it is still argued that the policy decisions are made in public health empirically without considering the available evidence. Though it is partially true, the best solution may not be readily available or applicable to the complex public health environment as it is used in clinical practice [124]. For example, the medical practice primarily focused on clinical outcome. However, the public health practice looks for its cost-effectiveness, availability, acceptability, accessibility, scalability, and others. The public health interventions used to control the COVID-19 pandemic are one of the recent examples where the policy and practice decisions are taken empirically since it is a completely a new public health problem. However, it cannot be ignored the magnitude and the fast-tracked creation of evidence during the same period for better evidence-based clinical and public health decisions. Though the quantification of the effect of an intervention in public health is a chronic challenge, the public health researchers need to provide a solution using newer technology considering the complex environment of its application. Further considering the complexity in public health practice, the generation of evidence needs to be changed to practice-based rather than in a controlled environment [125]. In this regard, the conduction of operational/implementation/health systems/policy research must be part of routine public health practice. For example, operational research conducted routinely in TB or HIV programs of various countries has improved the effectiveness of the interventions implemented in those programs [126]. Building the research capacity of public health professionals, data accessibility to researchers and creation of research teams, conducting research beyond the health sector, and adequate funding to public health research are the other chronic issues which can be addressed for acceleration and application of research findings.

Charting a course of action

As the saying goes, a crisis poses challenges but also offers opportunities to learn and evolve. All of us in the global health community have an obligation not only to learn from the COVID-19 pandemic and what has worked before but to avoid, paraphrasing the Harvard philosopher George Santayana [127], being condemned to face unprepared similar crises in the future.

The observations and perspectives offered in this chapter are not new. Since 2015, we have witnessed a significant movement toward the goal of UHC, highlighting the need to accelerate progress toward affordability of care and access to basic services. But to achieve these objectives and to ensure the financial sustainability of health and social support systems, which can be severely undermined by the uncontrolled outbreaks of disease and the rise of healthcare costs, it is imperative that a stronger push toward UHC include efforts to build both health systems resilience and population-level resilience, giving equal importance to emergency preparedness, health promotion, disease prevention, and treatment and care. Our collective goal should be the promotion of healthy, resilient, less vulnerable populations and not simply more, better financed, health services for the sick. Viewed in these terms, the attainment of UHC across the world should be seen as a key investment to contribute to address the deep-seated economic and social inequalities that were revealed and magnified during the COVID-19 pandemic [128].

Moving forward, we should be mindful that as public policy is an inherently political process, a clear understanding and careful but active modulation of the interrelationships among individuals, organized interest groups, and governments in a given context and period of time are of critical importance to influence the design and enactment of policies, laws, and regulations that shape a health system [129]. This implies that, in advancing future proposals to strengthen and build resilience in health systems, first and foremost there is a need to clearly articulate how they will contribute to the achievement of broader social goals in a country, or more broadly stated, to construct societies that are more inclusive, where health care is a social good available to all on equal terms, rather than a private consumption good like other services such as food and housing, that is only fully afforded to the rich. The unification of vision and social goals as the frame of reference for health reform proposals is key to align the interests of different actors and mobilize their support for their adoption and sustainable implementation over time.

The codification into laws of health as a right of citizenry is a vital step that is required to transform health systems as it helps to translate specific policy proposals into legal mandates, institutionalizing a framework that ensures the directionality and continuity of the reform process across political cycles, and helps withstand recurring bouts of political opposition and changed social and economic interests in a country.

Hence, to conclude, we will do well to remember the wise words of Henry Sigerist, the early 20th century director of the Johns Hopkins University Institute of History of Medicine, who observed that: "*It is quite obvious that the means and methods used in the prevention of disease are those provided by medicine and science. And yet, whether these methods are applied or not, does not depend on medicine alone, but to a much higher extent on the philo-sophical and social tendencies of the time*" [130]. The realization of the public health goals discussed above, therefore, has to be a shared social responsibility as the determinants of good physical and mental health cut across all areas of human activity.

References

[1]   Horton R. Offline: Bill Gates and the fate of WHO. Lancet 2022;1853:399. https://doi.org/10.1016/S0140-6736(22)00874-1.

[2]   OECD. Building resilience to the Covid-19 pandemic: the role of centres of government. Tackling Coronavirus Contributing to a Global Effort 2020:25. https://www.oecd.org/coronavirus/policy-responses/building-resilience-to-the-covid-19-pandemic-the-role-of-cen tres-of-government-883d2961/. [Accessed 1 July 2022].

[3]   Fauci AS. Emerging and reemerging infectious diseases: the perpetual challenge. Acad Med 2005;80:1079—85. https://doi.org/10.1097/00001888-200512000-00002.

[4]   The White House. American pandemic preparedness: transforming our capabilities. Report. 2021. p. 27. https://www.whitehouse.gov/wp-content/uploads/2021/09/American-Pandem ic-Preparedness-Transforming-Our-Capabilities-Final-For-Web.pdf?page=29. [Accessed 1 July 2022].

[5]   COVID-19 Excess Mortality Collaborators. Estimating excess mortality due to the COVID-19 pandemic: a systematic analysis of COVID-19-related mortality, 2020—21. Lancet 2022;399:1513—36. https://doi.org/10.1016/S0140-6736(21)02796-3.

[6]   Checchi F, Roberts L. Interpreting and using mortality data in humanitarian emergencies—a primer for non-epidemiologists. Network Paper Number 52 Humanitarian Practice Network 2005:41. https://odihpn.org/publication/interpreting-and-using-mortality-data-in-humanitarian-emergencies/. [Accessed 1 July 2022].

[7]   World Health Organization. The world health report 2007: a safer future: global public health security in the 21st century. Geneva: World Health Organization; 2007.

[8]   Marquez PV. Antimicrobial Resistance: a new global public health "ticking bomb"?. https://blogs.worldbank.org/health/antimicrobial-resistance-new-global-public-health-ticking-bomb. [Accessed 1 July 2022].

[9]   Jonas OB, Irwin A, Berthe FCJ, Le Gall FG, Marquez PV. Drug-resistant infections: a threat to our economic future (vol. 2): final report. HNP/Agriculture Global Antimicrobial Resistance Initiative; 2017. p. 172. https://documents.worldbank.org/en/publication/documents-reports/documentdetail/323311493396993758/final-report. [Accessed 1 July 2022].

[10]   Kelly E, Marquez PV. Health, nutrition and wash. In: United Nations, World Bank Group, European Union, African Development Bank, editor. Recovering from the Ebola crisis. World Bank Group; 2021. p. 128.

[11]   Horton RC. The COVID-19 catastrophe: what's gone wrong and how to stop it happening again. 2nd ed. Cambridge: Polity; 2021.

[12]   Gottlieb S. Uncontrolled spread: why COVID-19 crushed us and how we can defeat the next pandemic. 1st ed. Harper Collins; 2021.

[13]   Andaloussi MB, Christiansen LE, Habib A, Malacrino D. Healing the pandemic's economic scars demands prompt action—IMF Blog. IMFBlog; 2022. https://blogs.imf.org/2022/05/17/healing-the-pandemics-economic-scars-demands-prompt-action/?utm_medium=email&utm_source=govdelivery. [Accessed 1 July 2022].

[14]   Editorial Board. Sri Lanka and Pakistan, brothers in crisis | East Asia Forum. East Asia Forum; 2022. https://www.eastasiaforum.org/2022/04/25/sri-lanka-and-pakistan-brothers-in-crisis/. [Accessed 1 July 2022].

[15]   Potsiou C, Doytsher Y, Kelly P, Khouri R, McLaren R, Mueller H. Rapid urbanization and mega cities: the need for spatial information management. Int Fed Surv 2010:25. https://www.fig.net/resources/monthly_articles/2010/march_2010/march_2010_potsiou_etal.pdf. [Accessed 1 July 2022].

[16]   Edward G, Cutler D. Survival of the city: living and thriving in an age of isolation. 1st ed. New York: Penguin Press;; 2021.

[17]   Beyer RM, Manica A, Mora C. Shifts in global bat diversity suggest a possible role of climate change in the emergence of SARS-CoV-1 and SARS-CoV-2. Sci Total Environ 2021;767:145413. https://doi.org/10.1016/j.scitotenv.2021.145413.

[18]   International Monetary Fund. World economic outlook: war sets back the global recovery. Washington, DC; 2022.

[19]   Kickbusch I, Kheng Khor S, Heymann D. A new understanding of global health security: three interlocking functions. Global Challenges Foundation; 2021. p. 10. https://globalchallenges.org/wp-content/uploads/2021/11/A-new-understanding-of-global-heath-security.pdf. [Accessed 1 July 2022].

[20]   Winslow CEA. The untilled fields of public health. Science 1920;51:23—33. https://doi.org/10.1126/SCIENCE.51.1306.23.

[21]   American Public Health Association. 10 essential public health services; 2020. https://www.apha.org/what-is-public-health/10-essential-public-health-services. [Accessed 1 July 2022].

[22]   Frenk J. Public health: a field of study and an arena for action. Salud Publica Mex 1988;30:246—54.

[23]   Pan American Health Organization. 53rd directing council. 66th session of the regional committee of WHO for the Americas. Strategy for universal access to health and universal health coverage. Agenda itemvol 43; 2014. p. 25. https://www.paho.org/hq/dmdocuments/2014/CD53-5-e.pdf. [Accessed 1 July 2022].

[24] Emergency Preparedness ADGO, Integrated Health Services. Essential public health functions, health systems and health security-developing conceptual clarity and a WHO roadmap for action. Geneva; 2018.

[25] Martin-Moreno JM, Harris M, Jakubowski E, Kluge H. Defining and assessing public health functions: a global analysis. Annu Rev Public Health 2016;37:335−55. https://doi.org/10.1146/annurev-publhealth-032315-021429.

[26] Institute of Medicine (US) Committee for the Study of the Future of Public Health. A history of the public health system. The future of public health. Washington, DC: National Academic Press; 1988.

[27] Chave S. The origins and development of public health. In: Holland WW, Detels R, Knox G, editors. Oxford textbook of public health, vol. 1: history, determinants, scope, and strategies. New York: Oxford University Press; 1984.

[28] Fee E. Disease and discovery. A history of the Johns Hopkins school of hygiene and public health. 1st ed. Baltimore: John Hopkins University Press; 2016. p. 1916−39.

[29] Marquez PV, Farrington JL. No more disease silos for sub-Saharan Africa. BMJ 2012;345:e5812. https://doi.org/10.1136/bmj.e5812.

[30] Nunes ED. Social thinking in health in Latin America: revisiting Juan César García. Cad Saúde Pública 2013;29:1752−62. https://doi.org/10.1590/0102-311x00020613.

[31] Labonté R, Mohindra K, Schrecker T. The growing impact of globalization for health and public health practice. Annu Rev Public Health 2011;32:263−83. https://doi.org/10.1146/annurev-publhealth-031210-101225.

[32] Deaton A. The great escape: health, wealth, and the origins of inequality. New Jersey: Princeton University Press; 2013.

[33] Roser M. It's not just about child mortality, life expectancy improved at all ages. Our World Data; 2020. https://ourworldindata.org/its-not-just-about-child-mortality-life-expectancy-improved-at-all-ages. [Accessed 6 July 2022].

[34] Henderson DA. Smallpox: the death of a disease. The inside story of eradicating a worldwide killer. 1st ed. New York: Prometheus Books; 2009.

[35] Marquez PV. How to defeat COVID-19? Some lessons from the global smallpox eradication and the polio-free effort in the Americas. Blog; 2021. http://pvmarquez.com/smallpoxpoliovaccination. [Accessed 6 July 2022].

[36] World Bank. World development report 1993: investing in health. New York: Oxford University Press; 1993. https://doi.org/10.1596/0-1952-0890-0.

[37] Brownson RC, Fielding JE, Maylahn CM. Evidence-based public health: a fundamental concept for public health practice. Annu Rev Public Health 2009;30:175−201. https://doi.org/10.1146/annurev.publhealth.031308.100134.

[38] Jacobs JA, Jones E, Gabella BA, Spring B, Brownson RC. Tools for implementing an evidence-based approach in public health practice. Prev Chronic Dis 2012;9:E116. https://doi.org/10.5888/pcd9.110324.

[39] Marquez PV, Chalkidou K, Cutler D, Doyle N. Redesigning health care in ECA: some lessons from the UK. ECA Knowl Brief; 2010. https://web.worldbank.org/archive/website01419/WEB/0__C-139.HTM.

[40] Chalkidou K, Marquez P, Dhillon PK, Teerawattananon Y, Anothaisintawee T, Gadelha CAG, et al. Evidence-informed frameworks for cost-effective cancer care and prevention in low, middle, and high-income countries. Lancet Oncol 2014;15:e119−31. https://doi.org/10.1016/S1470-2045(13)70547-3.

[41] Bascolo E, Houghton N, Del Riego A, Fitzgerald J. A renewed framework for the essential public health functions in the Americas. Rev Panam Salud Publica 2020;44:e119. https://doi.org/10.26633/RPSP.2020.119.

[42] Pan American Health Organization. 55th directing council. 68th session of the regional committee of WHO for the Americas. Resilient health systems. Agenda itemvol 45; 2016. p. 22. https://www.paho.org/hq/dmdocuments/2016/CD55-9-e.pdf. [Accessed 6 July 2022].

[43] European Observatory on Health Systems and Policies, Sagan A, Greer SL, Webb E, McKee M, Muscat NA, et al. Strengthening health system resilience in the COVID-19 era. Eurohealth 2022;28:4−8.

[44] Haldane V, Ong S-E, Chuah FL-H, Legido-Quigley H. Health systems resilience: meaningful construct or catchphrase? Lancet 2017;1513:389. https://doi.org/10.1016/S0140-6736(17)30946-7.

[45] Legido-Quigley H, Asgari-Jirhandeh N. Resilient and people-centred health systems: progress, challenges and future directions in Asia. New Delhi: World Health Organization, Regional Office for South-East Asia; 2018.

[46] Farmanova E, Baker GR, Cohen D. Combining integration of care and a population health approach: a scoping review of redesign strategies and interventions, and their impact. Int J Integr Care 2019;5:19. https://doi.org/10.5334/ijic.4197.

[47] Pan American Health Organization. The essential public health functions in the Americas: a renewal for the 21st century. Conceptual framework and description. 1st ed. Washington, DC: OrganizacPan American Health Organization; 2020. https://doi.org/10.37774/9789275122648.

[48] Gates B. How to prevent the next pandemic. 1st ed. New York: Penguin Random House LLC; 2022.

[49] Centers for Disease Control and Prevention. Public health emergency preparedness and response capabilities: national standards for state, local, tribal, and territorial public health. 2018. p. 176. https://www.cdc.gov/cpr/readiness/00_docs/CDC_PreparednesResponseCapabilities_October2018_Final_508.pdf. [Accessed 1 July 2022].

[50] The World Bank. Agriculture and rural development. People, pathogens and our planet (volume 2): the economics of one health. Washington, DC; 2012.

[51] Marquez PV. How to prevent more pandemics? Focus on the relationship between humans, wildlife, and viruses. Blog; 2022. http://www.pvmarquez.com/howtopreventpandemics. [Accessed 1 July 2022].

[52] Nieuwkoop Van M, Eloit M. We must invest in pandemic prevention to build an effective global health architecture. Voices World Bank Blogs; 2021. https://blogs.worldbank.org/voices/we-must-invest-pandemic-prevention-build-effective-global-health-architecture. [Accessed 1 July 2022].

[53] Manuel D, Amadei CA, Campbell JR, Brault J-M, Veillard J. Strengthening public health surveillance through wastewater testing: an essential investment for the COVID-19 pandemic and future health threats. Washington: World Bank Group; 2022.

[54] Hocking L, George J, Broberg EK, Struelens MJ, Leitmeyer KC, Deshpande A, et al. Assessment of point-of-care testing devices for infectious disease surveillance, prevention and control—a mapping exercise. Stockholm: European Centre for Disease Prevention and Control; 2022. https://doi.org/10.2900/853921.

[55] Bremner B. Man versus microbe: what will it take to win?—Brian Bremner—Google books. London: World Scientific Publishing UK Limited; 2022.

[56] Bremner B. To save humanity, listen to the microbes | user walls. Politics; 2022. https://www.userwalls.news/n/save-humanity-listen-microbes-3258985/. [Accessed 1 July 2022].

[57] Wang H, Marquez PV, Figueras A. The value of a regional harmonized approach in monitoring the safety of vaccines and other medicines. Investing in Health World Bank Blog; 2022. https://blogs.worldbank.org/health/value-regional-harmonized-approach-monitoring-safety-vaccines-and-other-medicines. [Accessed 6 July 2022].

[58] World Health Organization. WHO Collaborating Centre for International Drug Monitoring. The importance of pharmacovigilance. Safety monitoring of medicinal products. Report. 2002. p. 48. https://apps.who.int/iris/bitstream/handle/10665/42493/a75646.pdf?sequence=1 &isAllowed=y. [Accessed 1 July 2022].

[59] Habarugira JMV, Härmark L, Figueras A. Pharmacovigilance data as a trigger to identify antimicrobial resistance and inappropriate use of antibiotics: a study using reports from the Netherlands pharmacovigilance centre. Antibiotics 2021;10:1512. https://doi.org/10.3390/antibiotics10121512.

[60] United Nations. Sustainable development goals; 2015. https://sdgs.un.org/goals. [Accessed 1 July 2022].

[61] Marquez PV, Farrington JL. The challenge of non-communicable diseases and road traffic injuries in sub-Saharan Africa: an overview. Washington, DC; 2013.

[62] Marquez PV, Saxena S. Making mental health a global priority. Cerebrum Dana Foundation; 2016. https://www.dana.org/article/making-mental-health-a-global-priority/. [Accessed 1 July 2022].

[63] Marquez PV, Dutta S. On world mental health day: a call to invest in interventions for young people. Investing in Health World Bank Blog; 2018. https://blogs.worldbank.org/health/world-mental-health-day-call-invest-interventions-young-people. [Accessed 1 July 2022].

[64] World Health Organization. Alcohol. Health topics; 2022. https://www.who.int/health-topics/alcohol#tab=tab_1. [Accessed 1 July 2022].

[65] Zhao F, Bali S, Kovacevic R, Weintraub J. A three-layer system to win the war against COVID-19 and invest in health systems of the future. BMJ Glob Heal 2021;6:e007365. https://doi.org/10.1136/bmjgh-2021-007365.

[66] Frenk J, Gómez-Dantés O. False dichotomies in global health: the need for integrative thinking. Lancet 2017;389:667−70. https://doi.org/10.1016/S0140-6736(16)30181-7.

[67] Frenk J. Bridging the divide: global lessons from evidence-based health policy in Mexico. Lancet 2006;368:954−61. https://doi.org/10.1016/S0140-6736(06)69376-8.

[68] Knaul FM, Bhadelia A, Atun R, Frenk J. Achieving effective universal health coverage and diagonal approaches to care for chronic illnesses. Heal Aff 2015;34:1514−22. https://doi.org/10.1377/hlthaff.2015.0514.

[69] Barış E, Silverman R, Wang H, Zhao F, Pate MA. Walking the talk: reimagining primary health care after COVID-19. Report. 2021. p. 225. https://doi.org/10.1596/978-1-4648-1768-7.

[70] Marquez P, Chalkidou K, Cutler D, Doyle N. Redesigning health care in ECA: some lessons from the UK why redesign health systems in ECA? Fixing health care: a worldwide dilemma. Eur Cent Asia Knowl Brief 2010:4. https://web.worldbank.org/archive/website01419/WEB/0__C-13. [Accessed 1 July 2022].

[71] Marquez PV. Health information systems in developing countries: star wars or reality? Investing in Health World Bank Blog; 2012. https://blogs.worldbank.org/health/health-information-systems-in-developing-countries-star-wars-or-reality. [Accessed 1 July 2022].

[72] OECD. Better ways to pay for health care. Paris: OECD Health Policy Studies, OECD Publishing; 2016. p. 170.

[73] Marquez PV. The imperative of integrated health care delivery systems. Investing in Health World Bank Blog; 2011. https://blogs.worldbank.org/health/the-imperative-of-integrated-health-care-delivery-systems. [Accessed 1 July 2022].

[74] Marquez LR. Improving health care in low- and middle-income countries: a case book. 1st ed. Switzerland: Springer Open; 2020. https://doi.org/10.1007/978-3-030-43112-9/ COVER.

[75] Marquez PV. Healthier workplaces = healthy profits. Investing in Health World Bank Blog; 2013. https://blogs.worldbank.org/health/healthier-workplaces-healthy-profits. [Accessed 1 July 2022].

[76] Detels R, Karim QA, Baum F, Li L, Leyland AH. Oxford textbook of global public health. 7th ed. Oxford University Press; 2021. https://doi.org/10.1093/MED/9780198816805. 001.0001.

[77] Das M, Desikachari GBR, Somanathan TV, Padmanaban P. How to improve public health systems lessons from Tamil Nadu. Policy Res Work Pap (WPS 5073) 2009:25. https:// openknowledge.worldbank.org/handle/10986/4265. [Accessed 1 July 2022].

[78] United Nations Website. The impact of digital technologies. UN75 2020 beyond shape our future together; n.d. p. 2. https://www.un.org/en/un75/impact-digital-technologies. [Accessed 1 July 2022].

[79] Frenk J, Chen L, Bhutta ZA, Cohen J, Crisp N, Evans T, et al. Health professionals for a new century: transforming education to strengthen health systems in an interdependent world. Lancet 2010;376:1923−58. https://doi.org/10.1016/S0140-6736(10)61854-5.

[80] Marquez PV. An imperative: reforming medical and public health education. Investing in Health World Bank Blog; 2011. https://blogs.worldbank.org/health/an-imperative-reforming-medical-and-public-health-education. [Accessed 1 July 2022].

[81] Zuckerman GA. Shot to save the world: the inside story of the life-or-death race for a COVID-19 vaccine. 2021. p. 355. https://www.amazon.co.uk/Shot-Save-World-Life-Death/dp/059342039X. [Accessed 1 July 2022].

[82] Mallapaty S, Callaway E, Kozlov M, Ledford H, Pickrell J, Van Noorden R. How COVID vaccines shaped 2021 in eight powerful charts. Nature 2021;600:580−3. https://doi.org/ 10.1038/d41586-021-03686-x.

[83] Marquez PV. To vaccinate or not to vaccinate? The COVID-19 vaccination dilemma. Blog; 2021. http://www.pvmarquez.com/vaccinehesitancy. [Accessed 1 July 2022].

[84] Doraiswamy PM, Fox C, Gordon J. How the pandemic is changing mental health. Opin Sci Am 2021. https://www.scientificamerican.com/article/how-the-pandemic-is-changing-mental-health/. [Accessed 1 July 2022].

[85] Staglin B, Herrman H. Inventive ways of delivering mental health care thrive during the pandemic. Opin Sci Am 2021. https://www.scientificamerican.com/article/inventive-ways-of-delivering-mental-health-care-thrive-during-the-pandemic/. [Accessed 1 July 2022].

[86] Case J. Making room for medical technologies—facilities management insights. Faciili-tiesnet Build Oper Manag 2004. https://www.facilitiesnet.com/healthcarefacilities/article/ Making-Room-for-Medical-Technologies−2132. [Accessed 1 July 2022].

[87] MedlinePlus. What is precision medicine?: MedlinePlus genetics. Natl Libr Med 2022. https://medlineplus.gov/genetics/understanding/precisionmedicine/definition/. [Accessed 1 July 2022].

[88] World Trade Organization. Waiver from certain provisions of the trips agreement for the prevention, containment and treatment of COVID-19. Comunication from India and South Africa. IP/C/W/669. 2020. p. 4. https://docs.wto.org/dol2fe/Pages/SS/directdoc.aspx? filename=q:/IP/C/W669.pdf&Open=True. [Accessed 1 July 2022].

[89] Marquez PV. Waiving COVID-19 vaccine patents is only the first step; need to develop production capacity and eliminate export controls on raw materials and equipment as well. Blog; 2021. http://www.pvmarquez.com/covid-19vaccinespatents. [Accessed 1 July 2022].

[90] Shafqat S, Kishwer S, Rasool RU, Qadir J, Amjad T, Ahmad HF. Big data analytics enhanced healthcare systems: a review. J Supercomput 2018;76:1754−99. https://doi.org/10.1007/S11227-017-2222-4.

[91] Gomez J. What is IoMT? (Internet of medical things). Koombea; 2021. https://www.koombea.com/blog/what-is-iomt/. [Accessed 1 July 2022].

[92] Hassanien AE, Dey N, Borra S. Medical big data and internet of medical things: advances, challenges and applications. 1st ed. London: Boca Raton and CRC Press, Taylor and Francis Group; 2018.

[93] Hoffman JD, Shayegani R, Spoutz PM, Hillman AD, Smith JP, Wells DL, et al. Virtual academic detailing (e-Detailing): a vital tool during the COVID-19 pandemic. J Am Pharm Assoc 2020;60:e95−9. https://doi.org/10.1016/j.japh.2020.06.028.

[94] Morrison A. Using data analytics for better health decisions. Blog; 2020. https://www.appnovation.com/blog/2020-02-using-data-analytics-better-health-decisions. [Accessed 1 July 2022].

[95] Martin G, Martin P, Hankin C, Darzi A, Kinross J. Cybersecurity and healthcare: how safe are we? BMJ 2017;358:j3179. https://doi.org/10.1136/bmj.j3179.

[96] Mayo Clinic. U.S. COVID-19 map: what do the trends mean for you?. 2020. https://www.mayoclinic.org/coronavirus-covid-19/map. [Accessed 1 July 2022].

[97] Bayuk J. CyberForensics: understanding information security investigations. 2010th ed. Totowa, NJ: Humana; 2010. https://doi.org/10.1007/978-1-60761-772-3.

[98] World Bank Group. High-performance health-financing for universal health coverage (vol. 2): driving sustainable, inclusive growth in the 21st century. 2019. p. 82. https://www.worldbank.org/en/topic/universalhealthcoverage/publication/high-performance-health-financing-for-uni versal-health-coverage-driving-sustainable-inclusive-growth-in-the-21st-century. [Accessed 1 July 2022].

[99] Junquera-Varela RF, Verhoeven M, Shukla GP, Haven B, Awasthi R, Moreno-Dodson B. Strengthening domestic resource mobilization: moving from theory to practice in low- and middle-income countries. 1st ed. Washington, DC: World Bank; 2017. https://doi.org/10.1596/978-1-4648-1073-2.

[100] Joseph Stiglitz BE, Tucker TN, Zucman January G. The starving state-why capitalism's salvation depends on taxation. 2020. p. 1. foreignaffairs.com/articles/united-states/2019-12-10/starving-state. [Accessed 1 July 2022].

[101] The Task Force on Fiscal Policy for Health. Health taxes to save lives: employing effective excise taxes on tobacco, alcohol, and sugary beverages. Report. 2019. p. 1−28. https://www.bbhub.io/dotorg/sites/2/2019/04/Health-Taxes-to-Save-Lives.pdf. [Accessed 1 July 2022].

[102] The White House. 2nd global COVID-19 summit commitments. Brief Room Statements Releases; 2022. https://www.whitehouse.gov/briefing-room/statements-releases/2022/05/12/2nd-global-covid-19-summit-commitments/. [Accessed 1 July 2022].

[103] Tangcharoensathien V, Srisookwatana O, Pinprateep P, Posayanonda T, Patcharanarumol W. Multisectoral actions for health: challenges and opportunities in complex policy environments. Int J Heal Policy Manag 2017;6:359−63. https://doi.org/10.15171/ijhpm.2017.61.

[104] Vora NM, Hannah L, Lieberman S, Vale MM, Plowright RK, Bernstein AS. Want to prevent pandemics? Stop spillovers. Nature 2022;605:419−22. https://doi.org/10.1038/d41586-022-01312-y.

[105] Loh EH, Zambrana-Torrelio C, Olival KJ, Bogich TL, Johnson CK, Mazet JAK, et al. Targeting transmission pathways for emerging zoonotic disease surveillance and control. Vector Borne Zoonotic Dis 2015;15:432−7. https://doi.org/10.1089/vbz.2013.1563.

[106] Weigang S, Fuchs J, Zimmer G, Schnepf D, Kern L, Beer J, et al. Within-host evolution of SARS-CoV-2 in an immunosuppressed COVID-19 patient as a source of immune escape variants. Nat Commun 2021;12:6405. https://doi.org/10.1038/s41467-021-26602-3.

[107] Jamison DT, Alwan A, Mock CN, Nugent R, Watkins DA, Adeyi O, et al. Universal health coverage and intersectoral action for health. In: Jamison DT, Gelband H, Horton S, Jha P, Laxminarayan R, Mock CN, et al., editors. Disease control priorities. Improving health and reducing poverty. 3rd ed., vol. 9. The International Bank for Reconstruction and Development/The World Bank; 2017. p. 1−21. https://doi.org/10.1596/978-1-4648-0527-1_CH1.

[108] Marquez PV, Nagpal S, Ndebele L. Taxes for better health: making the case at the joint learning network. Investing in Health World Bank Blog; 2018. https://blogs.worldbank.org/health/taxes-better-health-making-case-joint-learning-network. [Accessed 1 July 2022].

[109] Irwin A, Marquez PV, Jha PK, Peto R, Moreno-Dodson B, Goodchild M, et al. Tobacco tax reform at the crossroads of health and development: a multisectoral perspective. Washington, DC; 2017.

[110] World Bank Group. The high toll of traffic injuries: unacceptable and preventable. The macro-economic and welfare benefits of reducing road traffic injuries in low-& middle-income countries. Washington, DC; 2017.

[111] Marquez PV. Expanding the global tax base: taxing to promote public goods: tobacco taxes: summary report. Work Pap World Bank Gr 2016:48. https://documents.worldbank.org/en/publication/documents-reports/documentdetail/820951485943150390/summary-report. [Accessed 1 July 2022].

[112] Marquez PV. Running away from "tobacco road.". Investing in Health World Bank Blog; 2015. https://blogs.worldbank.org/health/running-away-tobacco-road. [Accessed 1 July 2022].

[113] Fuchs Tarlovsky A, Marquez PV, Dutta S, Icaza G, Fernanda M, Fuchs A, et al. Is tobacco taxation regressive? Evidence on public health, domestic resource mobilization, and equity improvements. Policy Notes 2019:72. https://doi.org/10.1596/31575.

[114] Marquez PV, Banjo GA, Chesheva EY, Muzira S. Confronting 'death on wheels': making roads safe in ECA. Eur Cent Asia Knowl Brief 2010;15:4. https://openknowledge.worldbank.org/handle/10986/10213. [Accessed 1 July 2022].

[115] Marquez PV. The case for physical and mental wellness programs in the workplace. Investing in Health World Bank Blog; 2017. https://blogs.worldbank.org/health/case-physical-and-mental-wellness-programs-workplace. [Accessed 1 July 2022].

[116] Kleinman A, Estrin GL, Usmani S, Chisholm D, Marquez PV, Evans TG, et al. Time for mental health to come out of the shadows. Lancet 2016;387:2274−5. https://doi.org/10.1016/S0140-6736(16)30655-9.

[117] OECD. Fit mind, fit job: from evidence to practice in mental health and work. Paris: OECD Publishing; 2015. https://doi.org/10.1787/9789264228283-en.

[118] OECD. Sick on the job?: myths and realities about mental health and work. Paris: OECD Publishing; 2012.

[119] Banerjee A, Duflo E. Poor economics: rethinking poverty and the ways to end it. Noida; London: Random House Publishers India Private Limited; 2011.

[120] Marquez PV. Time to put "health" into universal health coverage. Investing in Health World Bank Blog; 2016. https://blogs.worldbank.org/health/time-put-health-universal-health-coverage. [Accessed 1 July 2022].

[121] Marquez PV. Bikes, cities and health: a good combination. Transport for Development World Bank Blogs; 2013. https://blogs.worldbank.org/transport/bikes-cities-and-health-a-good-combination. [Accessed 1 July 2022].

[122] G7 Germany. Closing of G7 health ministers' meeting: far-reaching decisions adopted. News; 2022. https://www.bundesgesundheitsministerium.de/en/ministry/news/lauterbach-g7-health-ministers-to-discuss-pandemic-pact.html?s=08&cHash=27025db8253170789e043a6c3c8c3 8b0. [Accessed 1 July 2022].

[123] Kemm J. The limitations of "evidence-based" public health. J Eval Clin Pract 2006;12:319−24. https://doi.org/10.1111/j.1365-2753.2006.00600.x.

[124] Pullin AS, Knight TM. Effectiveness in conservation practice: pointers from medicine and public health. Conserv Biol 2001;15:50−4.

[125] Clift S. Creative arts as a public health resource: moving from practice-based research to evidence-based practice. Perspect Public Health 2012;132:120−7. https://doi.org/10.1177/1757913912442269.

[126] Lienhardt C, Cobelens FGJ. Operational research for improved tuberculosis control: the scope, the needs and the way forward. Int J Tuberc Lung Dis 2011;15:6−13.

[127] Saatkamp H, Coleman M. George Santayana (Stanford encyclopedia of philosophy). In: Zalta EN, editor. Stanford encyclopedia of philosophy; 2020.

[128] Sridhar DL. Preventable: how a pandemic changed the world & how to stop the next one. 1st ed. United Kingdom: Penguin; 2022.

[129] Marquez PV. How to accelerate universal health coverage in Latin America and the Caribbean? Blog; 2020. http://pvmarquez.com/health_reform. [Accessed 1 July 2022].

[130] Sigerist HE. The philosophy of hygiene. Bull Inst Hist Med 1933;1:323−31.

Index